Stability of Motion

Stability, Oscillations and Optimization of Systems
An International Series of Scientific Monographs, Textbooks, and Lecture Notes

Founder and Editor-in-Chief
A.A.Martynyuk
Institute of Mechanics NAS of Ukraine,
Kiev, Ukraine

Co-Editors
P.Borne
Ecole Centrale de Lille,
Villeneuve d'Ascq, France

C. Cruz-Hernandez
Telematics Direction, CICESE,
San Diego, CA, USA

Volume 1 Stability of Motion: The Role of Multicomponent Liapunov's Functions
A.A. Martynyuk

Additional Volumes in Preparation

Volume 2 Matrix Equations, Spectral Problems and Stability of Dynamic Systems
A.G. Mazko

Volume 3 Dynamics of Compressible Viscous Fluid
A.N. Guz

Stability of Motion

The Role of Multicomponent
Liapunov's Functions

A. A. MARTYNYUK
Institute of Mechanics
National Academy of Sciences of Ukraine
Kiev, Ukraine

C S P

© 2007 Cambridge Scientific Publishers

Printed and bound by TJ International Ltd, Padstow, Cornwall, UK

All rights reserved. No part of this book may be reprinted or reproduced or utilised in any form or by any electronic, mechanical, or other means, now known or hereafter invented, including photocopying and recording, or in any information storage or retrieval system, without prior permission in writing from the publisher.

British Library Cataloguing in Publication Data
A catalogue record for this book has been requested

Library of Congress Cataloging in Publication Data
A catalogue record has been requested

ISBN 978-1-904868-45-3

Cambridge Scientific Publishers Ltd
PO Box 806
Cottenham, Cambridge CB4 8RT
UK

www.cambridgescientificpublishers.com

CONTENTS

Introduction to the Series	ix
Preface	xiii
Notations	xvii
1 Stability Analysis of Continuous Systems	**1**
1.1 Introduction	1
1.2 The Direct Liapunov's Method via Multicomponent Auxiliary Functions	2
1.2.1 The general stability definitions	2
1.2.2 Classes of Liapunov's functions	6
1.2.3 Tests for stability	14
1.3 Stability Conditions of a Class of Large Scale Systems	42
1.4 Large-Scale Linear Systems	46
1.4.1 Non-autonomous linear systems	46
1.4.2 Time invariant linear systems	51
1.5 Polystability of Motion Analysis	54
1.5.1 The problem of exponential polystability	54
1.5.2 Approaches to the solution of the problem	56
1.5.3 Autonomous system	64
1.5.4 Polystability by the first order approximations	68
1.5.5 Polystability by pseudolinear approximations	73
1.6 Stability of Dynamical Systems in Metric Space	76
1.6.1 Basic concepts and definitions	76
1.6.2 Sufficient conditions for the stability of dynamical system	78
1.6.3 Stability analysis of hybrid system	82
1.6.4 Stability of the two-component systems analysis	85

1.7	Problems for Investigations	88
1.8	Comments and References	89
	References	92

2 Stability Analysis of Discrete-Time Systems — 99

2.1	Introduction	99
2.2	Large-Scale Discrete-Time Systems	100
2.3	Tests for Stability Analysis	101
	2.3.1 Liapunov's like theorems via matrix-valued functions	101
	2.3.2 A trend of application of vector Liapunov's functions	108
2.4	Stability Analysis of Linear Systems	116
	2.4.1 A new Liapunov's function for linear system	116
	2.4.2 Examples	120
2.5	Some Applications of Hierarchical Liapunov's Functions	124
	2.5.1 Hierarchical decomposition and stability conditions	124
	2.5.2 Hierarchical connective stability	128
	2.5.3 Application of hierarchical matrix-valued function	132
2.6	On Polydynamics of Systems on Time Scale	136
	2.6.1 Auxiliary results	136
	2.6.2 Problems of polydynamics	137
	2.6.3 Polydynamics analysis in general	138
2.7	Problems for Investigation	142
2.8	Comments and References	143
	References	144

3 Stability in Functional Differential Systems — 149

3.1	Introduction	149
3.2	Preliminaries	150
	3.2.1 Functional differential equations	150
	3.2.2 Auxiliary results	151
	3.2.3 Some estimates for the Liapunov matrix-valued functional	152
3.3	Theorems on Stability	154
3.4	Large-Scale System of Functional Differential Equations	161
	3.4.1 Approach A_1	161
	3.4.2 Approach A_2	164

	3.4.3	Approach B_1	165
	3.4.4	Approach B_2	167
	3.4.5	Some applications	168
3.5	An Approach to Construction of Liapunov-Krasovskii Functionals		173
3.6	Stability Analysis of Quasilinear Delay Systems		186
3.7	The Problems for Investigation		192
3.8	Comments and References		193
References			195

4 Stability Analysis of Impulsive Systems — 199

4.1	Introduction		199
4.2	What is a model of impulsive systems?		200
	4.2.1	Systems with impulses at fixed times	201
	4.2.2	Systems with impulses at variable times	202
	4.2.3	Autonomous systems with impulses	202
4.3	Existence of solutions and the problem of "beating"		203
4.4	Stability in the Sense of Liapunov		207
	4.4.1	Stability analysis in the first approximation	208
4.5	Matrix Liapunov Functions Method		213
	4.5.1	A scalar approach	213
	4.5.2	Vector approach	220
4.6	Stability Analysis of Large Scale Impulsive Systems		224
	4.6.1	Auxiliary estimations	225
	4.6.2	Tests for stability	232
	4.6.3	The conditions of instability	235
4.7	Novel Approach to Stability Analysis of Uncertain Impulsive Systems		237
	4.7.1	Statement of the problem	237
	4.7.2	Comparison principle with the block-diagonal matrix function	240
	4.7.3	Test for strict stability	242
	4.7.4	Vector approach	245
	4.7.5	Uncertain linear impulsive systems	247
4.8	Problems for Investigation		254

	4.9 Comments and References	255
	References	256
5	**Applications**	**261**
	5.1 Introduction	261
	5.2 Stabilization of motion of a rigid body	262
	5.2.1 Statement of the problem	262
	5.2.2 Motion equations of mechanical system	263
	5.2.3 Orientation stabilization of mechanical system in a prescribed direction	265
	5.2.4 Stabilization of dynamically symmetric rigid body in prescribed direction	275
	5.2.5 Some comments on Theorem 5.2.4	276
	5.3 Stability of Uncertain Discrete-Time Neural Networks	280
	5.3.1 Stability of uncertain discrete-time systems	280
	5.3.2 Exponential stability of a discrete-time neural system	285
	5.3.3 Neural system with perturbed equilibrium	290
	5.3.4 Numerical results	292
	5.4 Stability in the Extended Dynamical Systems	295
	5.4.1 Preliminary results	295
	5.4.2 Application of Liapunov's matrix function to extended systems	297
	5.4.3 Stability of linear large scale systems	298
	5.4.4 Numerical example	303
	5.5 Interval Stability of Mechanical Systems	305
	5.5.1 Auxiliary results	305
	5.5.2 Interval stability of linear system	306
	5.5.3 Interval stability conditions for a mechanical system	307
	5.6 Analysis of Population Growth Models of Kolmogorov Type	310
	5.6.1 Generalized Kolmogorov model	310
	5.6.2 Boundedness with respect to two measures	310
	5.6.3 Stability with respect to two measures	315
	5.6.4 Application to a generalized Lotka-Volterra system	317
	5.7 Comments and References	319
	References	320
Author Index		**323**
Subject Index		**327**

INTRODUCTION TO THE SERIES

Modern stability theory, oscillations and optimization of nonlinear systems have developed in response to the practical problems of celestial mechanics and applied engineering has become an integral part of human activity and development at the end of the 20th century.

For a process or a phenomenon, such as atom oscillations or a supernova explosion, if a mathematical model is constructed in the form of a system of differential equations, then the investigation is possible either by a direct (numerical as a rule) integration of the equations or by analysis by qualitative methods.

In the 20th century, the fundamental works by Euler (1707–1783), Lagrange (1736–1813), Poincaré (1854–1912), Liapunov (1857–1918) and others have been thoroughly developed and applied in the investigation of stability and oscillations of natural phenomena and the solution of many problems of technology.

In particular, the problems of piloted space flights and those of astrodynamics were solved due to the modern achievements of stability theory and motion control. The Poincaré and Liapunov methods of qualitative investigation of solutions to nonlinear systems of differential equations in macroworld study have been refined to a great extent though not completed. Also modelling and establishing stability conditions for microprocesses are still at the stage of accumulating ideas and facts and forming the principles; for examples, the stability problem of thermonuclear synthesis.

Obviously, this is one of the areas for the application of stability and control theory in the 21th century. The development of efficient methods and algorithms in this area requires the interaction and publication of ideas and results of various mathematical theories as well as the co-operation of scientists specializing in different areas of mathematics and engineering.

The mathematical theory of optimal control (of moving objects, water resources, global processes in world economy, etc.) is being developed in terms of basic ideas and results obtained in 1956–1961 and formulated in Pontryagin's principle of optimality and Bellman's principle of dynamical programming. The efforts of many scholars and engineers in the framework of these ideas resulted in the efficient methods of control for many concrete systems and technological processes.

Thus, the development of classical ideas and results of stability and control theory remains the principle direction for scholars and experts modern stage of the mathematical theories. The aim of the International book series; **Stability, Oscillations and Optimization of Systems** is to provide a medium for the rapid publication of high quality original monographs in the following areas:

Development of the theory and methods of stability analysis:
a. Nonlinear Systems (ordinary differential equations, partial differential equations, stochastic differential equations, functional differential equations, integral equations, difference equations, etc.)
b. Nonlinear operators (bifurcations and singularity, critical point theory, polystability, etc.)

Development of up-to-date methods of the theory of nonlinear oscillations:
a. Analytical methods.
b. Qualitative methods.
c. Topological methods.
d. Numerical and computational methods, etc.

Development of the theory and up-to-date methods of optimization of systems:
a. Optimal control of systems involving ODE, PDE, integral equations, equations with retarded argument, etc.
b. Nonsmooth analysis.
c. Necessary and sufficient conditions for optimality.
d. Hamilton-Jacobi theories.
e. Methods of successive approximations, etc.

Applications:
a. Physical sciences (classical mechanics, including fluid and solid mechanics, quantum and statistical mechanics, plasma physics, astrophysics, etc.).

b. Engineering (mechanical engineering, aeronautical engineering, electrical engineering, chemical engineering).
c. Mathematical biology and life sciences (molecular biology, population dynamics, theoretical ecology).
d. Social sciences (economics, philosophy, sociology).

In the forthcoming publications of the series the readers will find fundamental results and survey papers by international experts presenting the results of investigations in many directions of stability and control theory including uncertain systems and systems with chaotic behaviour of trajectories.

It is in this spirit that we see the importance of the "Stability, Oscillations and Optimization of Systems" series, and we are would like to thank Cambridge Scientific Publishers, Ltd. for their interest and cooperation in publishing this series.

PREFACE

Motion stability theory plays an important part in many branches of natural sciences and engineering because more and more experts need modern tools of analysis and synthesis of real physical and other natural systems. There are several remarkable monographs presenting the results of development of motion stability theory since publication of Liapunov's dissertation paper in 1892. This very period is characterized by an intensive development of motion stability theory provoked by the engineering problems of aviation, rocket building, submarine dynamics, economics, traffic, construction, etc. Mathematical analysis of stability includes geometric methods such as phase planes or vectors, analytical methods such as iterations, perturbations and transformations, and topological methods.

As is known, one of the main methods of motion stability analysis is the direct Liapunov method (method of auxiliary functions). This method was developed by A.M. Liapunov in 1892 and based first on the auxiliary scalar function which can be the total energy of the system.

In the late 1960s the theory of direct Liapunov method was extended by the ideas of multicomponent Liapunov functions (vector-function) and comparison technique. Further stability theory was developed in several directions, among which we note stability investigation of solutions for systems of differential equations with quasimonotone right-side parts and large scale systems of various nature.

Supersufficiency of stability conditions established in terms of the method of vector Liapunov functions was an impetus to search for new approaches in motion stability theory. One of the approaches was developed in the late 1970ties in terms of the two-index system of functions (matrix-valued function) being an appropriate medium for constructing scalar or vector Liapunov function. Since the 1980s, this idea has been intensively worked out at the Stability of Processes Department of S.P.Timoshenko

Institute of Mechanics, National Academy of Sciences of Ukraine, Kiev and other world scientific centers.

Many results have now been obtained in this new direction of direct Liapunov method development based on multicomponent auxiliary functions.

The monograph presented to the readers consists of five chapters.

It is the prime purpose of this book to present the results of the direct Liapunov method development based on multicomponent auxiliary functions for the following classes of equations: continuous, discrete-time systems, functional differential and uncertain systems with impulsive perturbations.

A secondary purpose of this book is to collect together and unify recent advances in stability theory via multicomponent auxiliary functions and to highlight those results which are directly applicable to practical dynamic systems.

The book is self-contained.

In Chapter 1 we consider the direct Liapunov's method via multicomponent auxiliary functions.

In Chapter 2 the stability and boundedness of solutions of discrete-time systems and systems on time scales are discussed.

Chapter 3 shows how Liapunov's direct method may be generalized to functional-differential equations via multicomponent auxiliary functions.

Chapter 4 provides the results obtained in a new direction of the theory of impulsive equations. Namely, the impulsive system is considered as a hybrid and moreover, continuous and discrete components have an equal effect on the dynamical properties of the whole system. The efficiency of the approach is shown by the example of linear uncertain impulsive system.

Lastly in Chapter 5 we discuss some applications of general results to particular problems of mechanics and neurodynamics.

Some important features of this text are as follows:

* it presents at an introductory level the basic principles of the method of Liapunov's matrix functions (functionals) of various types of nonlinear systems, e.g. continuous, discrete-time, functional differential systems, impulsive systems and their applications.

* it discusses the new developments in methods of constructing the elements of matrix-valued function (functional). The approaches proposed in this problem are illustrated by examples and some applications of independent importance.

* it demonstrates the manifestations of the general Liapunov method via matrix-valued functions and shows how this efficient technique can be adopted to study qualitative and quantitative properties of motions of various types of systems.

* it emphasises the analysis of stability in various types of nonlinear large scale systems via Liapunov matrix-valued functions and illustrates the practical use of general theoretical results. This is the first text that explains systematically the use of Liapunov's matrix functions method in the study of stability for functional differential equations and uncertain impulsive systems.

* it illustrates the practical use of Liapunov's matrix functions method through the study of motion stabilization in a prescribed direction of a solid bearing moving material points, exponential stability of discrete-time neural systems, interval stability of linear mechanical systems, and the mathematical Kolmogorov model of population dynamics.

In our opinion, the book should serve as a suitable supplementary text for courses and seminars in stability theory by Liapunov's direct method at the advanced undergraduate and the beginning of graduate level, in mathematics, physics and engineering.

Acknowledgements

I am indebted to many people in connection with the publication of this text. I wish, first of all, to thank Professor Yu.A. Mitropolskii for his interest and kind attention to these investigations. I wish to thank the publishers of the journals *Doklady Mathematics* (Mrs. I.V. Isavnina), *International Applied Mechanics* (Professor A.N. Guz), and *Nonlinear Analysis* (Professor V. Lakshmikantham) for their kind attention and consideration of some of the results.

My thanks go to L.N. Chernetskaya, A.N. Chernienko, and S.N. Rasshivalova who skillfully typed the entire manuscript with its many many revised versions in translation from Russian. And finally, it is a pleasure to thank Mrs. Janie Wardle from Cambridge Scientific Publishers for her enthusiastic support of this project.

A.A.Martynyuk

NOTATION

R — the set of all real numbers
$R_+ = [0, +\infty) \subset R$ — the set of all nonnegative numbers
R^k — k-th dimensional real vector space
$R \times R^n$ — the Cartesian product of R and R^n
$G_1 \times G_1$ — topological product
(a, b) — open interval $a < t < b$
$[a, b]$ — closed interval $a \leq t \leq b$
$A \cup B$ — union of sets A and B
$A \cap B$ — intersection of sets A and B
$\overline{D}$ — closure of set D
∂D — boundary of set D
$N_\tau^+ \triangleq \{\tau_0, \ldots, \tau_0 + k, \ldots\}, \quad \tau_0 \geq 0, \quad k = 1, 2, \ldots$
$\{x \colon \Phi(x)\}$ — set of x's for which the proposition Φ is true
$\mathcal{T} = [-\infty, +\infty] = \{t \colon -\infty \leq t \leq +\infty\}$ — the largest time interval
$\mathcal{T}_\tau = [\tau, +\infty) = \{t \colon \tau \leq t < +\infty\}$ — the right semi-open unbounded interval associated with τ
$\mathcal{T}_i \subseteq R$ — a time interval of all initial moments to under consideration (or, all admissible t_0)
$\mathcal{T}_0 = [t_0, +\infty) = \{t \colon t_0 \leq t < +\infty\}$ — the right semi-open unbounded interval associated with t_0
$\|x\|$ — the Euclidean norm of vector x in R^n
$\chi(t; t_0, x_0)$ — a motion of a system at $t \in R$ iff $x(t_0) = x_0$, $\chi(t_0; t_0, x_0) \equiv x_0$
$B_\varepsilon = \{x \in R^n \colon \|x\| < \varepsilon\}$ — open ball with center at the origin and radius $\varepsilon > 0$
$\delta_M(t_0, \varepsilon) = \max\{\delta \colon \delta = \delta(t_0, \varepsilon) \ni x_0 \in B_\delta(t_0, \varepsilon) \Rightarrow \chi(t; t_0, x_0) \in B_\varepsilon, \forall t \in \mathcal{T}_0\}$ — the maximal δ obeying the definition of stability
$\Delta_M(t_0) = \max\{\Delta \colon \Delta = \Delta(t_0), \forall \rho > 0, \forall x_0 \in B_\Delta, \exists \tau(t_0, x_0, \rho) \in (0, +\infty) \ni \chi(t; t_0, x_0) \in B_\rho, \forall t \in \mathcal{T}_\tau\}$ — the maximal Δ obeying the definition of attractivity

$\tau_m(t_0, x_0, \rho) = \min\{\tau \colon \tau = \tau(t_0, x_0, \rho) \ni \chi(t; t_0, x_0) \in B_\rho, \ \forall t \in \mathcal{T}_\tau\}$ —
the minimal τ satisfying the definition of attractivity

$\mathcal{N}$ — a time-invariant neighborhood of original of R^n

$f \colon R \times \mathcal{N} \to R^n$ — a vector function mapping $R \times \mathcal{N}$ into R^n

$C(\mathcal{T}_\tau \times \mathcal{N})$ — the family of all functions continuous on $\mathcal{T}_\tau \times \mathcal{N}$

$C^{(i,j)}(\mathcal{T}_\tau \times \mathcal{N})$ — the family of all functions i-times differentiable on $\mathcal{T}_\tau$ and j-times differentiable on $\mathcal{N}$

$\mathcal{C} = C([-\tau, 0], R^n)$ — the space of continuous functions which map $[-\tau, 0]$ into R^n

$U(t, x)$, $U \colon \mathcal{T}_\tau \times R^n \to R^{s \times s}$ — matrix-valued Liapunov function, $s = 2, 3, \ldots, m$

$V(t, x)$, $V \colon \mathcal{T}_\tau \times R^n \to R^s$ — vector Liapunov function

$v(t, x)$, $v \colon \mathcal{T}_\tau \times R^n \to R_+$ — scalar Liapunov function

$D^+v(t, x)$ $(D^-v(t, x))$ — the upper right (left) Dini derivative of v along $\chi(t; t_0, x_0)$ at (t, x)

$D_+v(t, x)$ $(D_-v(t, x))$ — the lower right (left) Dini derivative of v along $\chi(t; t_0, x_0)$ at (t, x)

$D^*v(t, x)$ — denotes that both $D^+v(t, x)$ and $D_+v(t, x)$ can be used

$Dv(t, x)$ — the Eulerian derivative of v along $\chi(t; t_0, x_0)$ at (t, x)

$\lambda_i(\cdot)$ — the i-th eigenvalue of a matrix $(\cdot)$

$\lambda_M(\cdot)$ — the maximal eigenvalue of a matrix $(\cdot)$

$\lambda_m(\cdot)$ — the minimal eigenvalue of a matrix $(\cdot)$

1

STABILITY ANALYSIS OF CONTINUOUS SYSTEMS

1.1 Introduction

This Chapter contains the main results of qualitative analysis of motions for large scale dynamical systems described by ordinary differential equations in terms of multicomponent Liapunov's functions.

The Chapter is arranged as follows.

Section 1.2 deals with stability problems for continuous dynamical systems. The definitions and sufficient conditions for various types of motion stability of nonautonomous and nonlinear systems are presented via matrix-valued and vector auxiliary functions. The main theorems of the Section are supplied with corollaries which illustrate the generality of the results obtained and indicate the sources for the assertions.

In Section 1.3 some general theorems of Section 1.2 are suplied with a constructive algorithm of constructing the Liapunov functions in terms of matrix-valued auxiliary function. The conditions for various types of stability of zero solution for a class of large-scale systems are formulated in terms of the property of having a fixed sign of special matrices.

Section 1.4 sets out conditions of stability analysis of linear systems. These conditions are established in terms of a matrix-valued function constructed by the approach proposed in Section 1.3.

Section 1.5 contains the results of polistability analysis via Liapunov scalar and matrix-valued functions for linear (pseudolinear) nonautonomous and autonomous systems of the first approximation.

Section 1.6 presents an approach to stability analysis of dynamical systems in metric spaces. The main idea in this Section is using multicomponent mapping together with principle of comparison. As an example of application we consider some hybrid system via matrix-valued auxiliary functions. For an example of a two-component hybrid system we demonstrate the efficiency of the proposed approach of the Liapunov matrix-valued functions.

In the final Section 1.7 some unsolved problems of the method of matrix-valued Liapunov functions are presented.

Thus, this chapter provides a development of the direct Liapunov method consisting in both the establishment of general theorems and the proposition of a new approach of constructing the appropriate Liapunov functions for some classes of linear and nonlinear systems.

1.2 The Direct Liapunov's Method via Multicomponent Auxiliary Functions

In this Section the notions of motion stability corresponding to the motion properties of nonautonomous systems are presented being necessary in subsequent presentation. Basic notions of the method of multicomponent Liapunov functions are discussed and general theorems and some corollaries are set out.

Throughout this Section, real systems of ordinary differential equations will be considered.

1.2.1 The general stability definitions
We consider systems which can appropriately be described by ordinary differential equations of the form

$$(1.2.1) \qquad \frac{dy_i}{dt} = Y_i(t, y_1, \ldots, y_n), \quad i = 1, 2, \ldots, n,$$

or in the equivalent vector form

$$(1.2.2) \qquad \frac{dy}{dt} = Y(t, y),$$

where $x \in R^n$, $Y(t,y) = (Y_1(t,y), \ldots, Y_n(t,y))^{\mathrm{T}}$, $Y \colon \mathcal{T} \times R^n \to R^n$. In the present section we will assume that the right-hand part of (1.2.2) satisfies the solution existence and uniqueness conditions of the Cauchy problem

$$(1.2.3) \qquad \frac{dy}{dt} = Y(t, y), \quad y(t_0) = y_0,$$

for any $(t_0, y_0) \in \mathcal{T} \times \Omega$, where $\Omega \subset R^n$, $0 \in \Omega$ and Ω is an open connected subset of R.

It is clear that the solution of problem (1.2.3) may not exist on R (on R_+), even if the right-hand part $Y(t,y)$ of system (1.2.3) is definite and continuous for all $(t,y) \in \mathcal{T} \times R^n$.

For example, the Cauchy problem for the equation

$$\frac{dy}{dt} = 1 + y^2, \quad y(0) = 0, \tag{1.2.4}$$

where y is a scalar, has a unique solution $y(t) = \operatorname{tg} t$, existing on the interval $(-\frac{\pi}{2}, \frac{\pi}{2})$, only, while the right-hand part of equation (1.2.4) is definite on the whole plane (t, y).

Let $y(t) = \psi(t; t_0, y_0)$ be the solution of system (1.2.2), definite on the interval $[t_0, \tau)$ and noncontinuable behind the point τ, i.e. $y(t)$ is not definite for $t = \tau$. Then

$$\overline{\lim} \, \|y(t)\| = +\infty \quad \text{as} \quad t \to \tau - 0. \tag{1.2.5}$$

Using solution $y(t)$ and the right-hand part of system (1.2.2) we construct the vector-function

$$f(t, x) = Y(t, x + \psi(t)) - Y(t, \psi(t)) \tag{1.2.6}$$

and consider the system

$$\frac{dx}{dt} = f(t, x). \tag{1.2.7}$$

It is easy to verify that the solutions of systems (1.2.2) and (1.2.7) are correlated as

$$x(t) = y(t) - \psi(t)$$

on the general interval of existence of solutions $y(t)$ and $\psi(t)$. It is clear that system (1.2.7) has a trivial solution $x(t) \equiv 0$. This solution corresponds to the solution $y(t) = \psi(t)$ of system (1.2.2). Obviously, the reduction of system (1.2.2) to system (1.2.7) is possible only when the solution $y(t) = \psi(t)$ is known.

Qualitative investigation of solutions of system (1.2.2) with respect to solution $\psi(t)$ is reduced to the investigation of behavior of solution $x(t)$ to system (1.2.7) which differs "little" from the trivial one for $t = t_0$.

In the case when stability of unperturbed motion is discussed with respect to some continuously differentiable functions $Q_s(t, \psi_1, \ldots, \psi_n)$ the perturbed motion equations are found by the system of equations

$$\frac{dx_i}{dt} = \frac{\partial Q_i}{\partial t} + \sum_{s=1}^{n} \frac{\partial Q_i}{\partial \psi_s} Y_s(t, \psi_1, \ldots, \psi_n) - \dot{F}(t),$$

where $x_i = Q_i(t, \psi_1(t), \ldots, \psi_n(t)) - F_s(t)$, and $F_i(t) = Q_i(t, \psi_1(t), \ldots, \psi_n(t))$ are some known time functions. The system of equations of (1.2.7) type obtained here satisfies the condition $f(t, 0) = 0$ for all $t \in \mathcal{T}$ and therefore this system has a trivial solution in this case as well.

In motion stability theory system (1.2.7) is called the system of perturbed motion equations.

Since equations (1.2.7) can generally not be solved analytically in closed form, the qualitative properties of the equilbrium state are of great practical interest. We begin with a series of definitions.

A very large number of definitions of stability exist for the system (1.2.7). Of course, the various definitions of stability can be broadly classified as those which deal with the trajectory, or motion, or the equilibrium of the null solution of free or unforced systems, and those which consider the dynamic response of systems subject to various classes of forcing functions or inputs. In the following, the equilibrium state of (1.2.7) can always be set equal to zero by a linear state transformation, so that the equilibrium state and the null solution to (1.2.7) are considered throughout as equivalent.

Definition 1.2.1 The equilibrium state $x = 0$ of the system (1.2.7) is:

(i) *stable* iff for every $t_0 \in \mathcal{T}_i$ and every $\varepsilon > 0$ there exists $\delta(t_0, \varepsilon) > 0$, such that $\|x_0\| < \delta(t_0, \varepsilon)$ implies

$$\|x(t; t_0, x_0)\| < \varepsilon \quad \text{for all} \quad t \in \mathcal{T}_0;$$

(ii) *uniformly stable* iff both (i) holds and for every $\varepsilon > 0$ the corresponding maximal δ_M obeying (i) satisfies

$$\inf \left[\delta_M(t, \varepsilon) \colon t \in \mathcal{T}_i\right] > 0;$$

(iii) *stable in the whole* iff both (i) holds and

$$\delta_M(t, \varepsilon) \to +\infty \quad \text{as} \quad \varepsilon \to +\infty \quad \text{for all} \quad t \in R;$$

(iv) *uniformly stable in the whole* iff both (ii) and (iii) hold;

(v) *unstable* iff there are $t_0 \in \mathcal{T}_i$, $\varepsilon \in (0, +\infty)$ and $\tau \in \mathcal{T}_0$, $\tau > t_0$, such that for every $\delta \in (0, +\infty)$ there is x_0, $\|x_0\| < \delta$, for which

$$\|x(\tau; t_0, x_0)\| \geq \varepsilon.$$

Definition 1.2.2 The equilibrium state $x = 0$ of the system (1.2.7) is:

(i) *attractive* iff for every $t_0 \in \mathcal{T}_i$ there exists $\Delta(t_0) > 0$ and for every $\zeta > 0$ there exists $\tau(t_0; x_0, \zeta) \in [0, +\infty)$ such that $\|x_0\| < \Delta(t_0)$ implies $\|x(t; t_0, x_0)\| < \zeta$ for all $t \in (t_0 + \tau(t_0; x_0, \zeta), +\infty)$;

(ii) x_0-*uniformly attractive* iff both (i) is true and for every $t_0 \in R$ there exists $\Delta(t_0) > 0$, and for every $\zeta \in (0, +\infty)$ there exists $\tau_u[t_0, \Delta(t_0), \zeta] \in [0, +\infty)$ such that
$$\sup[\tau_m(t_0; x_0, \zeta) \colon x_0 \in B_\Delta(t_0)] = \tau_u(t_0, x_0, \zeta);$$

(iii) t_0-*uniformly attractive* iff both (i) is true, there is $\Delta > 0$, and for every $(x_0, \zeta) \in B_\Delta \times (0, +\infty)$ there exists $\tau_u(R, x_0, \zeta) \in [0, +\infty)$ such that
$$\sup[\tau_m(t_0); x_0, \zeta) \colon t_0 \in \mathcal{T}_i] = \tau_u(\mathcal{T}_i, x_0, \zeta);$$

(iv) *uniformly attractive* iff both (ii) and (iii) hold, that is, that (i) is true, there exists $\Delta > 0$ and for every $\zeta \in (0, +\infty)$ there is $\tau_u(R, \Delta, \zeta) \in [0, +\infty)$ such that
$$\sup[\tau_m(t_0; x_0, \zeta) \colon (t_0, x_0) \in \mathcal{T}_i \times B_\Delta] = \tau(\mathcal{T}_i, \Delta, \zeta);$$

The properties (i) – (iv) hold "*in the whole*" iff (i) is true for every $\Delta(t_0) \in (0, +\infty)$ and every $t_0 \in \mathcal{T}_i$.

Definition 1.2.3 The equilibrium state $x = 0$ of the system (1.2.7) is:

(i) *asymptotically stable* iff it is both stable and attractive;
(ii) *equi-asymptotically stable* iff it is both stable and x_0-uniformly attractive;
(iii) *quasi-uniformly asymptotically stable* iff it is both uniformly stable and t_0-uniformly attractive;
(iv) *uniformly asymptotically stable* iff it is both uniformly stable and uniformly attractive;
(v) the properties (i) – (iv) hold "*in the whole*" iff both the corresponding stability of $x = 0$ and the corresponding attraction of $x = 0$ hold in the whole;
(vi) *exponentially stable* iff there are $\Delta > 0$ and real numbers $\alpha \geq 1$ and $\beta > 0$ such that $\|x_0\| < \Delta$ implies

$$\|x(t; t_0, x_0)\| \leq \alpha \|x_0\| \exp[-\beta(t - t_0)], \quad \text{for all} \quad t \in \mathcal{T}_0, \quad \text{for all} \quad t_0 \in \mathcal{T}_i.$$

This holds *in the whole* iff it is true for $\Delta = +\infty$.

Let $g: R^n \to R^n$ define the time invariant system

$$\frac{dx}{dt} = g(x), \tag{1.2.8}$$

where $g(0) = 0$ and the components of g are smooth functions of the components of x for x near zero. Every stability property of $x = 0$ of (1.2.8) is uniform in $t_0 \in R$.

Note that the nonperturbed motion equations of the time invariant system can be reduced to the time invariant system (1.2.8) iff the solution $\psi(t) = \text{const}$. Otherwise, i.e. if $\psi(t) \neq \text{const}$, equations (1.2.8) can be nonstationary.

In the investigation of both system (1.2.2) and (1.2.8) the solution $x(t)$ is assumed to be definite for all $t \in \mathcal{T}$ (for all $t \in \mathcal{T}_0$).

1.2.2 Classes of Liapunov's functions Presently the Liapunov direct method (see Liapunov [1]) in terms of three classes of auxiliary functions: scalar, vector and matrix ones are intensively applied in qualitative theory. In this point we shall present the description of the above mentioned classes of functions.

1.2.2.1 Matrix-valued Liapunov function For the system (1.2.7) we shall consider a continuous matrix-valued function

$$U(t,x) = [v_{ij}(t,x)], \quad i,j = 1,2,\ldots,m, \tag{1.2.9}$$

where $v_{ij} \in C(\mathcal{T}_\tau \times R^n, R)$ for all $i,j = 1,2,\ldots,m$.

We assume that the following conditions are fulfilled
  (i) $v_{ij}(t,x)$, $i,j = 1,2,\ldots,m$, are locally Lipschitzian in x;
  (ii) $v_{ij}(t,0) = 0$ for all $t \in R_+$ $(t \in \mathcal{T}_\tau)$, $i,j = 1,2,\ldots,m$;
  (iii) $v_{ij}(t,x) = v_{ji}(t,x)$ in any open connected neighbourhood $\mathcal{N}$ of point $x = 0$ for all $t \in R_+$ $(t \in \mathcal{T}_\tau)$.

Definition 1.2.4 All functions of the type

$$v(t,x,\alpha) = \alpha^T U(t,x)\alpha, \quad \alpha \in R^m, \tag{1.2.10}$$

where $U \in C(\mathcal{T}_\tau \times \mathcal{N}, R^{m\times m})$, are attributed to the *class SL*.

Here the vector α can be specified as follows:
  (i) $\alpha = y \in R^m$, $y \neq 0$;
  (ii) $\alpha = \xi \in C(R^n, R^m_+)$, $\xi(0) = 0$;
  (iii) $\alpha = \psi \in C(\mathcal{T}_\tau \times R^n, R^m_+)$, $\psi(t,0) = 0$;
  (iv) $\alpha = \eta \in R^m_+$, $\eta > 0$.

Note that the choice of vector a can influence the property of having a fixed sign of function (1.2.13) and its total derivative along solutions of system (1.2.7).

1.2.2.2 Comparison functions Comparison functions are used as upper or lower estimates of the function v and its total time derivative. They are usually denoted by φ, $\varphi\colon R_+ \to R_+$. The main contributor to the investigation of properties of and use of the comparison functions is Hahn [1]. What follows is mainly based on his definitions and results.

Definition 1.2.5 *A function* φ, $\varphi\colon R_+ \to R_+$, *belongs to*
  (i) *the class* $K_{[0,\alpha)}$, $0 < \alpha \leq +\infty$, iff both it is defined, continuous and strictly increasing on $[0,\alpha)$ and $\varphi(0) = 0$;
  (ii) *the class* K iff (i) holds for $\alpha = +\infty$, $K = K_{[0,+\infty)}$;
  (iii) *the class* KR iff both it belongs to the class K and $\varphi(\zeta) \to +\infty$ as $\zeta \to +\infty$;
  (iv) *the class* $L_{[0,\alpha)}$ iff both it is defined, continuous and strictly decreasing on $[0,\alpha)$ and $\lim[\varphi(\zeta)\colon \zeta \to +\infty] = 0$;
  (v) *the class* L iff (iv) holds for $\alpha = +\infty$, $L = L_{[0,+\infty)}$.

Let φ^{-1} denote the *inverse function* of φ, $\varphi^{-1}[\varphi(\zeta)] \equiv \zeta$. The next result was established by Hahn [1].

Proposition 1.2.1
  (1) If $\varphi \in K$ and $\psi \in K$ then $\varphi(\psi) \in K$;
  (2) If $\varphi \in K$ and $\sigma \in L$ then $\varphi(\sigma) \in L$;
  (3) If $\varphi \in K_{[0,\alpha)}$ and $\varphi(\alpha) = \xi$ then $\varphi^{-1} \in K_{[0,\xi)}$;
  (4) If $\varphi \in K$ and $\lim[\varphi(\zeta)\colon \zeta \to +\infty] = \xi$ then φ^{-1} is not defined on $(\xi, +\infty]$;
  (5) If $\varphi \in K_{[0,\alpha)}$, $\psi \in K_{[0,\alpha)}$ and $\varphi(\zeta) > \psi(\zeta)$ on $[0,\alpha)$ then $\varphi^{-1}(\zeta) < \psi^{-1}(\zeta)$ on $[0,\beta]$, where $\beta = \psi(\alpha)$.

Definition 1.2.6 *A function* φ, $\varphi\colon R_+ \times R_+ \to R_+$, *belongs to:*
  (i) *the class* $KK_{[0;\alpha,\beta)}$ iff both $\varphi(0,\zeta) \in K_{[0,\alpha)}$ for every $\zeta \in [0,\beta)$ and $\varphi(\zeta,0) \in K_{[0,\beta)}$ for every $\zeta \in [0,\alpha)$;
  (ii) *the class* KK iff (i) holds for $\alpha = \beta = +\infty$;
  (iii) *the class* $KL_{[0;\alpha,\beta)}$ iff both $\varphi(0,\zeta) \in K_{[0,\alpha)}$ for every $\zeta \in [0,\beta)$ and $\varphi(\zeta,0) \in L_{[0,\beta)}$ for every $\zeta \in [0,\alpha)$;
  (iv) *the class* KL iff (iii) holds for $\alpha = \beta = +\infty$;
  (v) *the class* CK iff $\varphi(t,0) = 0$, $\varphi(t,u) \in K$ for every $t \in R_+$;

(vi) the *class* $\mathcal{M}$ iff $\varphi \in C(R_+ \times R^n, R_+)$, $\inf \varphi(t,x) = 0$, $(t,x) \in R_+ \times R^n$;

(vii) the *class* $\mathcal{M}_0$ iff $\varphi \in C(R_+ \times R^n, R_+)$, $\inf_x \varphi(t,x) = 0$ for each $t \in R_+$;

(viii) the *class* Φ iff $\varphi \in C(K, R_+)$: $\varphi(0) = 0$, and $\varphi(w)$ is increasing with respect to cone K.

Definition 1.2.7 Two *functions* $\varphi_1, \varphi_2 \in K$ (or $\varphi_1, \varphi_2 \in KR$) are said to be *of the same order of magnitude* if there exist positive constants α, β, such that

$$\alpha\varphi_1(\zeta) \leq \varphi_2(\zeta) \leq \beta\varphi_1(\zeta) \quad \text{for all} \quad \zeta \in [0, \zeta_1] \quad (\text{or for all} \quad \zeta \in [0, \infty)).$$

1.2.2.3 Properties of matrix-valued functions For the functions of the class SL we shall cite some definitions which are applied in the investigation of dynamics of system (1.2.7).

Definition 1.2.8 The *matrix-valued function* $U: \mathcal{T}_\tau \times R^n \to R^{m \times m}$ is:

(i) *positive semi-definite on* $\mathcal{T}_\tau = [\tau, +\infty)$, $\tau \in R$, iff there are time-invariant connected neighbourhood $\mathcal{N}$ of $x = 0$, $\mathcal{N} \subseteq R^n$, and vector $y \in R^m$, $y \neq 0$, such that
  (a) $v(t,x,y)$ is continuous in $(t,x) \in \mathcal{T}_\tau \times \mathcal{N} \times R^m$;
  (b) $v(t,x,y)$ is non-negative on $\mathcal{N}$, $v(t,x,y) \geq 0$ for all $(t,x,y \neq 0) \in \mathcal{T}_\tau \times \mathcal{N} \times R^m$, and
  (c) vanishes at the origin: $v(t,0,y) = 0$ for all $t \in \mathcal{T}_\tau \times R^m$;
  (d) iff the conditions (a)–(c) hold and for every $t \in \mathcal{T}_\tau$, there is $w \in \mathcal{N}$ such that $v(t,w,y) > 0$, then v is strictly positive semi-definite on $\mathcal{T}_\tau$;

(ii) *positive semi-definite on* $\mathcal{T}_\tau \times \mathcal{G}$ iff (i) holds for $\mathcal{N} = \mathcal{G}$;

(iii) *positive semi-definite in the whole on* $\mathcal{T}_\tau$ iff (i) holds for $\mathcal{N} = R^n$;

(iv) *negative semi-definite (in the whole) on* $\mathcal{T}_\tau$ (on $\mathcal{T}_\tau \times \mathcal{N}$) iff $(-v)$ is positive semi-definite (in the whole) on $\mathcal{T}_\tau$ (on $\mathcal{T}_\tau \times \mathcal{N}$) respectively.

The expression "on $\mathcal{T}_\tau$" is omitted iff all corresponding requirements hold for every $\tau \in R$.

Definition 1.2.9 The *matrix-valued function* $U: \mathcal{T}_\tau \times R^n \to R^{m \times m}$ is:

(i) *positive definite on* $\mathcal{T}_\tau$, $\tau \in R$, iff there are a time-invariant connected neighbourhood $\mathcal{N}$ of $x = 0$, $\mathcal{N} \subseteq R^n$ and a vector $y \in R^m$,

$y \neq 0$, such that both it is positive semi-definite on $\mathcal{T}_\tau \times \mathcal{N}$ and there exists a positive definite function w on $\mathcal{N}$, $w \colon R^n \to R_+$, obeying $w(x) \leq v(t, x, y)$ for all $(t, x, y) \in \mathcal{T}_\tau \times \mathcal{N} \times R^m$;

(ii) *positive definite on* $\mathcal{T}_\tau \times \mathcal{G}$ iff (i) holds for $\mathcal{N} = \mathcal{G}$;

(iii) *positive definite in the whole on* $\mathcal{T}_\tau$ iff (i) holds for $\mathcal{N} = R^n$;

(iv) *negative definite (in the whole) on* $\mathcal{T}_\tau$ (on $\mathcal{T}_\tau \times \mathcal{N} \times R^m$) iff $(-v)$ is positive definite (in the whole) on $\mathcal{T}_\tau$ (on $\mathcal{T}_\tau \times \mathcal{N} \times R^m$) respectively;

(v) *weakly decrescent* if there exists a $\Delta_1 > 0$ and a function $a \in CK$ such that $v(t, x, y) \leq a(t, \|x\|)$ as soon as $\|x\| < \Delta_1$;

(vi) *asymptotically decrescent* if there exists a $\Delta_2 > 0$ and a function $b \in KL$ such that $v(t, x, y) \leq b(t, \|x\|)$ as soon as $\|x\| < \Delta_2$.

The expression "*on* $\mathcal{T}_\tau$" is omitted iff all corresponding requirements hold for every $\tau \in R$.

Proposition 1.2.2 The matrix-valued function $U \colon R \times R^n \to R^{m \times m}$ is positive definite on $\mathcal{T}_\tau$, $\tau \in R$, iff it can be written as

$$y^\mathrm{T} U(t, x) y = y^\mathrm{T} U_+(t, x) y + a(\|x\|),$$

where $U_+(t, x)$ is a positive semi-definite matrix-valued function and $a \in K$.

Definition 1.2.10 (cf. Grujić, et al. [1]) Set $v_\zeta(t)$ is *the largest connected neighborhood of* $x = 0$ *at* $t \in R$ which can be associated with a function $U \colon R \times R^n \to R^{m \times m}$ so that $x \in v_\zeta(t)$ implies $v(t, x, y) < \zeta$, $y \in R^m$.

Definition 1.2.11 The matrix-valued function $U \colon R \times R^n \to R^{s \times s}$ is:

(i) *decreasing on* $\mathcal{T}_\tau$, $\tau \in R$, iff there is a time-invariant neighborhood $\mathcal{N}$ of $x = 0$ and a positive definite function w on $\mathcal{N}$, $w \colon R^n \to R_+$, such that $y^\mathrm{T} U(t, x) y \leq w(x)$ for all $(t, x) \in \mathcal{T}_\tau \times \mathcal{N}$;

(ii) *decreasing on* $\mathcal{T}_\tau \times \mathcal{G}$ iff (i) holds for $\mathcal{N} = \mathcal{G}$;

(iii) *decreasing in the whole on* $\mathcal{T}_\tau$ iff (i) holds for $\mathcal{N} = R^n$.

The expression "*on* $\mathcal{T}_\tau$" is omitted iff all corresponding conditions still hold for every $\tau \in R$.

Proposition 1.2.3 The matrix-valued function $U \colon R \times R^n \to R^{m \times m}$ is decreasing on $\mathcal{T}_\tau$, $\tau \in R$, iff it can be written as

$$y^\mathrm{T} U(t, x) y = y^\mathrm{T} U_-(t, x) y + b(\|x\|), \quad (y \neq 0) \in R^m,$$

where $U_-(t, x)$ is a negative semi-definite matrix-valued function and $b \in K$.

Definition 1.2.12 The matrix-valued function $U\colon R \times R^n \to R^{m\times m}$ is:

(i) *radially unbounded on* $\mathcal{T}_\tau$, $\tau \in R$, iff $\|x\| \to \infty$ implies $y^{\mathrm{T}}U(t,x)y \to +\infty$ for all $t \in \mathcal{T}_\tau$, $y \in R^m$, $y \neq 0$;

(ii) *radially unbounded*, iff $\|x\| \to \infty$ implies $y^{\mathrm{T}}U(t,x)y \to +\infty$ for all $t \in \mathcal{T}_\tau$ for all $\tau \in R$, $y \in R^m$, $y \neq 0$.

Proposition 1.2.4 The matrix-valued function $U\colon \mathcal{T}_\tau \times R^n \to R^{m\times m}$ is radially unbounded in the whole (on $\mathcal{T}_\tau$) iff it can be written as

$$y^{\mathrm{T}}U(t,x)y = y^{\mathrm{T}}U_+(t,x)y + a(\|x\|) \quad \text{for all} \quad x \in R^n,$$

where $U_+(t,x)$ is a positive semi-definite matrix-valued function in the whole (on $\mathcal{T}_\tau$) and $a \in KR$.

According to Liapunov function (1.2.17) is applied in motion investigation of system (1.2.7) together with its total derivative along solutions $x(t) = x(t;t_0,x_0)$ of system (1.2.7). Assume that each element $v_{ij}(t,x)$ of the matrix-valued function (1.2.16) is definite on the open set $\mathcal{T}_\tau \times \mathcal{N}$, $\mathcal{N} \subset R^n$, i.e. $v_{ij}(t,x) \in C(\mathcal{T}_\tau \times \mathcal{N}, R)$.

If $\gamma(t;t_0,x_0)$ is a solution of system (1.2.7) with the initial conditions $x(t_0) = x_0$, i.e. $\gamma(t_0;t_0,x_0) = x_0$, the right-hand upper derivative of function (1.2.17) for $\alpha = y$, $y \in R^m$, with respect to t along the solution of (1.2.7) is determined by the formula

$$(1.2.11) \qquad D^+v(t,x,y) = y^{\mathrm{T}}D^+U(t,x)y,$$

where $D^+U(t,x) = [D^+v_{ij}(t,x)]$, $i,j = 1,2,\ldots,m$, and

$$(1.2.12) \quad D^+v_{ij}(t,x) = \limsup\Big\{\sup_{\gamma(t,t,x)=x}[v_{ij}(t+\sigma, \gamma(t+\sigma, t, x)) \\ - v_{ij}(t,x)]\sigma^{-1}\colon \sigma \to 0^+\Big\}, \quad i,j=1,2,\ldots,m.$$

In the case when system (1.2.7) has a unique solution for every initial value of $x(t_0) = x_0$ $((t_0,x_0) \in \mathcal{T}_\tau \times \mathcal{N})$, the expression (1.2.12) is equivalent to

$$(1.2.13) \quad D^+v_{ij}(t,x) = \limsup\{[v_{ij}(t+\sigma, \gamma(t+\sigma, t, x)) \\ - v_{ij}(t,x)]\sigma^{-1}\colon \sigma \to 0^+\}, \quad i,j=1,2,\ldots,m.$$

Further we assume that for all $i, j = 1, 2, \ldots, m$ the functions $v_{ij}(t, x)$ are continuous and locally Lipschitzian in x, i.e. for every point in $\mathcal{N}$ there exists a neighbourhood Δ and a positive number $L = L(\Delta)$ such that

$$|v_{ij}(t, x) - v_{ij}(t, y)| \leq L\|x - y\|, \quad i, j = 1, 2, \ldots, m,$$

for any $(t, x) \in \mathcal{T}_\tau \times \Delta$, $(t, y) \in \mathcal{T}_\tau \times \Delta$. Besides, the expression (1.2.12) is equivalent to

$$\begin{aligned}(1.2.14) \quad D^+ v_{ij}(t, x) = \limsup \{[v_{ij}(t + \sigma, x + \sigma f(t, x)) \\ - v_{ij}(t, x)] \sigma^{-1} \colon \sigma \to 0^+\}, \quad i, j = 1, 2, \ldots, m.\end{aligned}$$

If the matrix-valued function $U(t, x) \in C^1(\mathcal{T}_\tau \times \mathcal{N}, R^{m \times m})$, i.e. all its elements $v_{ij}(t, x)$ are functions continuously differentiable in t and x, then the expression (1.2.14) is equivalent to

$$(1.2.15) \quad Dv_{ij}(t, x) = \frac{\partial v_{ij}}{\partial t}(t, x) + \sum_{s=1}^{n} \frac{\partial v_{ij}}{\partial x_s}(t, x) f_s(t, x),$$

where $f_s(t, x)$ are components of the vector-function $f(t, x) = (f_1(t, x), \ldots, f_n(t, x))^T$.

Function (1.2.15) has the Euler derivative (1.2.10) at point (t, x) along solution $x(t; t_0, x_0)$ of system (1.2.7) iff

$$(1.2.16) \quad \begin{aligned} D^+ v(t, x, y) = D_+ v(t, x, y) = D^- v(t, x, y) \\ = D_- v(t, x, y) = Dv(t, x, y). \end{aligned}$$

Note that the application of any of the expressions (1.2.12), (1.2.13) or (1.2.15) in (1.2.11) is admissible.

1.2.2.4 Vector Liapunov functions A vector-valued Liapunov function

$$(1.2.17) \quad V(t, x) = (v_1(t, x), v_2(t, x), \ldots, v_m(t, x))^T$$

can be obtained via matrix-valued function (1.2.9) in several ways.

Definition 1.2.13 All vector functions of the type

$$(1.2.18) \quad L(t, x, b) = AU(t, x)b,$$

where $U \in C(\mathcal{T}_\tau \times R^n, R^{s\times s})$, A is a constant matrix $s \times s$, and vector b is defined according to (i)–(iv) similarly to the definition of the vector a, are attributed to the *class VL*.

If in two-index system of functions (1.2.9) for all $i \neq j$ the elements $v_{ij}(t,x) = 0$, then

$$v(t,x) = \mathrm{diag}\,(v_{11}(t,x), \ldots, v_{mm}(t,x))^\mathrm{T},$$

where $v_{ii} \in C(\mathcal{T}_\tau \times R^n, R)$, $i = 1, 2, \ldots, m$, , is a vector-valued function. Besides, the function (1.2.18) has the components

$$L_k(t,x,b) = \sum_{i=1}^{m} a_{ki} b_i v_{ii}(t,x), \quad k = 1, 2, \ldots, m.$$

The methods of application of Liapunov's vector functions in motion stability theory are presented in a number of monographs some of which are mentioned at the end of this Chapter.

1.2.2.5 Scalar Liapunov function The simplest type of auxiliary function for system (1.2.7) is the function

(1.2.19) $$v(t,x) \in C(\mathcal{T}_0 \times R^n, R_+),$$

for which

(a) $v(t,0) = 0$ for all $t \in \mathcal{T}_\tau$;
(b) $v(t,x)$ is locally Lipschitzian in x.

In stability theory both sign-definite in the sense of Liapunov and semi-definite functions are applied. We shall set out some examples.

Example 1.2.1 (i) The function

$$v(t,x) = (1 + \sin^2 t)x_1^2 + (1 + \cos^2 t)x_2^2$$

is positive definite and decreasing, while the function

$$v(t,x) = (x_1^2 + x_2^2)\sin^2 t$$

is decreasing and positive semi-definite.

(ii) The function
$$v(t,x) = x_1^2 + (1+t)x_2^2$$
is positive definite but not non-decreasing, while the function
$$v(t,x) = x_1^2 + \frac{x_2^2}{1+t}$$
is decreasing but not positive definite.

(iii) The function
$$v(t,x) = (1+t)(x_1 - x_2)^2$$
is positive semi-definite and non-decreasing.

Among the variety of the Liapunov functions the quadratic forms

(1.2.20) $\qquad v(t,x) = x^{\mathrm{T}} P(t) x, \quad P^{\mathrm{T}}(t) = P(t),$

are of special importance, where $P(t)$ is $n \times n$-matrix with continuous and bounded elements for all $t \in \mathcal{T}_\tau$.

Proposition 1.2.5 For the quadratic form (1.2.20) to be positive definite it is necessary and sufficient that

(1.2.21) $\qquad \begin{vmatrix} p_{11}(t) & \cdots & p_{1s}(t) \\ \cdots & \cdots & \cdots \\ p_{s1}(t) & \cdots & p_{ss}(t) \end{vmatrix} > k > 0, \quad s = 1, 2, \ldots, n,$

for all $t \in \mathcal{T}_\tau$.

Note that if conditions (1.2.21) are satisfied, the "quasi-quadratic" form

(1.2.22) $\qquad v(t,x) = x^{\mathrm{T}} P(t) x + \psi(t,x), \quad \psi(t,0) = 0,$

is positive definite, provided that some constants a, b ($a > 0$, $b \geq 2$) exist, such that $|\psi(t,x)| \leq a r^b$, where $r = (x^{\mathrm{T}} x)^{1/2}$.

To calculate total derivative of function (1.2.22) along solutions of system (1.2.7) either (1.2.12)–(1.2.15) is applied for $i = j = 1$, depending on the assumptions on system (1.2.7) and function (1.2.22).

It is well known (see Yoshizawa [1]) that if $D^+ v(t,x) \leq 0$ and consequently $D^+ v(t,x(t)) \leq 0$, then the function $v(t,x)$ is a nonincreasing function of $t \in \mathcal{T}_\tau$. Further, if $D^+ v(t,x) \geq 0$, then $v(t,x(t))$ is nondecreasing along any solution of (1.2.7) and vice versa.

We shall formulate these observations as follows.

Proposition 1.2.6 Suppose $m(t) = v(t, x(t))$ is continuous on (a, b). Then $m(t)$ is nondecreasing (nonincreasing) on (a, b) if and only if $D^+ m(t) \geq 0$ (≤ 0) for every $t \in (a, b)$, where

$$D^+ m(t) = \lim \sup \{[m(t + \sigma) - m(t)] \sigma^{-1} \colon \sigma \to 0^+\}.$$

Further all auxiliary functions allowing the solution of the problem on stability (instability) of the equilibrium state $x = 0$ of system (1.2.7) are called the Liapunov functions. The construction of the Liapunov functions still remains one of the central problems of stability theory.

1.2.3 Tests for stability

1.2.3.1 Matrix Liapunov functions applications Functions (1.2.10), (1.2.18) and (1.2.20) together with their total derivatives (1.2.11) along solutions of system (1.2.7) allow us to establish the existence conditions for the motion properties of system (1.2.7) of various types such as stability, instability, boundedness, etc. Below we shall set out some results in these areas.

Theorem 1.2.1 *Let the vector-function f in system (1.2.7) be continuous on $R \times \mathcal{N}$ (on $\mathcal{T}_\tau \times \mathcal{N}$). If there exist*

(1) *an open connected time-invariant neighborhood $\mathcal{G} \subset \mathcal{N}$ of the point $x = 0$;*

(2) *a matrix-valued function $U \in C(R \times \mathcal{N}, R^{m \times m})$ and a vector $y \in R^m$ such that the function $v(t, x, y) = y^T U(t, x) y$ is locally Lipschitzian in x for all $t \in R$ ($t \in \mathcal{T}_\tau$);*

(3) *functions $\psi_{i1}, \psi_{i2}, \psi_{i3} \in K$, $\widetilde{\psi}_{i2} \in CK$, $i = 1, 2, \ldots, m$;*

(4) *$m \times m$ matrices $A_j(y)$, $j = 1, 2, 3$, $\widetilde{A}_2(y)$ such that*

(a) $\psi_1^T(\|x\|) A_1(y) \psi_1(\|x\|) \leq v(t, x, y) \leq \widetilde{\psi}_2^T(t, \|x\|) \widetilde{A}_2(y) \widetilde{\psi}_2(t, \|x\|)$
for all $(t, x, y) \in R \times \mathcal{G} \times R^m$ (for all $(t, x, y) \in \mathcal{T}_\tau \times \mathcal{G} \times R^m$);

(b) $\psi_1^T(\|x\|) A_1(y) \psi_1(\|x\|) \leq v(t, x, y) \leq \psi_2^T(\|x\|) A_2(y) \psi_2(\|x\|)$
for all $(t, x, y) \in R \times \mathcal{G} \times R^m$ (for all $(t, x, y) \in \mathcal{T}_\tau \times \mathcal{G} \times R^m$);

(c) $D^+ v(t, x, y) \leq \psi_3^T(\|x\|) A_3(y) \psi_3(\|x\|)$
for all $(t, x, y) \in R \times \mathcal{G} \times R^m$ (for all $(t, x, y) \in \mathcal{T}_\tau \times \mathcal{G} \times R^m$).

Then, if the matrices $A_1(y)$, $A_2(y)$, $\widetilde{A}_2(y)$, $(y \neq 0) \in R^m$ are positive definite and $A_3(y)$ is negative semi-definite, then

(a) *the state $x = 0$ of system (1.2.7) is stable (on $\mathcal{T}_\tau$), provided condition (4)(a) is satisfied;*

(b) the state $x = 0$ of system (1.2.7) is uniformly stable (on $\mathcal{T}_\tau$), provided condition (4)(b) is satisfied.

Corollary 1.2.1 Let

(1) condition (1) of Theorem 1.2.1 be satisfied;
(2) there exist at least one couple of indices $(p, q) \in [1, m]$ for which $(v_{pq}(t, x) \neq 0) \in U(t, x)$ and function $v(t, x, e) = e^{\mathrm{T}} U(t, x) e = v(t, x)$ for all $(t, x) \in R \times \mathcal{G}$ (for all $(t, x) \in \mathcal{T}_\tau \times \mathcal{G}$) satisfy the conditions

 (a) $\psi_1(\|x\|) \leq v(t, x)$;
 (b) $v(t, x) \leq \psi_2(\|x\|)$;
 (c) $D^+ v(t, x)|_{(1.2.7)} \leq 0$,

where ψ_1, ψ_2 are some functions of the class K.

Then, the state $x = 0$ of system (1.2.7) is stable (on $\mathcal{T}_\tau$) under conditions (a) and (c), and uniformly stable (on $\mathcal{T}_\tau$) under conditions (a) – (c).

Theorem 1.2.2 Let the vector-function f in system (1.2.7) be continuous on $R \times R^n$ (on $\mathcal{T}_\tau \times R^n$). If there exist

(1) a matrix-valued function $U \in C(R \times R^n, R^{m \times m})$ $(U \in C(\mathcal{T}_\tau \times R^n, R^{m \times m}))$ and a vector $y \in R^m$ such that the function $v(t, x, y) = y^{\mathrm{T}} U(t, x) y$ is locally Lipschitzian in x for all $t \in R$ $(t \in \mathcal{T}_\tau)$;
(2) functions $\varphi_{1i}, \varphi_{2i}, \varphi_{3i} \in KR$, $\widetilde{\varphi}_{2i} \in CKR$, $i = 1, 2, \ldots, m$;
(3) $m \times m$ matrices $B_j(y)$, $j = 1, 2, 3$, $\widetilde{B}_2(y)$ such that

 (a) $\varphi_1^{\mathrm{T}}(\|x\|) B_1(y) \varphi_1(\|x\|) \leq v(t, x, y) \leq \widetilde{\varphi}_2^{\mathrm{T}}(t, \|x\|) \widetilde{B}_2(y) \widetilde{\varphi}_2(t, \|x\|)$
 for all $(t, x, y) \in R \times R^n \times R^m$ (for all $(t, x, y) \in \mathcal{T}_\tau \times R^n \times R^m$);
 (b) $\varphi_1^{\mathrm{T}}(\|x\|) B_1(y) \varphi_1(\|x\|) \leq v(t, x, y) \leq \varphi_2^{\mathrm{T}}(\|x\|) B_2(y) \varphi_2(\|x\|)$
 for all $(t, x, y) \in R \times R^n \times R^m$ (for all $(t, x, y) \in \mathcal{T}_\tau \times R^n \times R^m$);
 (c) $D^+ v(t, x, y) \leq \varphi_3^{\mathrm{T}}(\|x\|) B_3(y) \varphi_3(\|x\|)$ for all $(t, x, y) \in R \times R^n \times R^m$ (for all $(t, x, y) \in \mathcal{T}_\tau \times R^n \times R^m$).

Then, provided that matrices $B_1(y)$, $B_2(y)$ and $\widetilde{B}_2(y)$ for all $(y \neq 0) \in R^m$ are positive definite and matrix $B_3(y)$ is negative semi-definite,

 (a) under condition (3)(a) the state $x = 0$ of system (1.2.7) is stable in the whole (on $\mathcal{T}_\tau$);
 (b) under condition (3)(b) the state $x = 0$ of system (1.2.7) is uniformly stable in the whole (on $\mathcal{T}_\tau$).

Corollary 1.2.2 Let for function $v(t,x,e) = v(t,x)$, mentioned in condition (2) of Corollary 1.2.1 for all $(t,x) \in R \times R^n$ (for all $(t,x) \in \mathcal{T}_\tau \times R^n$) the following conditions hold

(a) $\varphi_1(\|x\|) \leq v(t,x)$;
(b) $v(t,x) \leq \varphi_2(\|x\|)$, for some function φ_2;
(c) $D^+ v(t,x)|_{(1.2.7)} \leq 0$,

where φ_1, φ_2 are of class KR.

Then the state $x = 0$ of system (1.2.7) is stable in the whole (on $\mathcal{T}_\tau$) under conditions (a) and (c) and uniformly stable in the whole (on $\mathcal{T}_\tau$) under conditions (a) – (c).

Theorem 1.2.3 Let the vector-function f in system (1.2.7) be continuous on $R \times \mathcal{N}$ (on $\mathcal{T}_\tau \times \mathcal{N}$). If there exist

(1) an open connected time-invariant neighborhood $\mathcal{G} \subset \mathcal{N}$ of the point $x = 0$;
(2) a matrix-valued function $U \in C(R \times \mathcal{N}, R^{m \times m})$ ($U \in C(\mathcal{T}_\tau \times \mathcal{N}, R^{m \times m})$) and a vector $y \in R^m$ such that the function $v(t,x,y) = y^\mathrm{T} U(t,x) y$ is locally Lipschitzian in x for all $t \in R$ ($t \in \mathcal{T}_\tau$);
(3) functions $\eta_{1i}, \eta_{2i}, \eta_{3i} \in K$, $\tilde{\eta}_{2i} \in CK$, $i = 1, 2, \ldots, m$;
(4) $m \times m$ matrices $C_j(y)$, $j = 1, 2, 3$, $\tilde{C}_2(y)$ such that

   (a) $\eta_1^\mathrm{T}(\|x\|) C_1(y) \eta_1(\|x\|) \leq v(t,x,y) \leq \tilde{\eta}_2^\mathrm{T}(t,\|x\|) \tilde{C}_2(y) \tilde{\eta}_2(t,\|x\|)$
   for all $(t,x,y) \in R \times \mathcal{G} \times R^m$ (for all $(t,x,y) \in \mathcal{T}_\tau \times \mathcal{G} \times R^m$);
   (b) $\eta_1^\mathrm{T}(\|x\|) C_1(y) \eta_1(\|x\|) \leq v(t,x,y) \leq \eta_2^\mathrm{T}(\|x\|) C_2(y) \eta_2(\|x\|)$
   for all $(t,x,y) \in R \times \mathcal{G} \times R^m$ (for all $(t,x,y) \in \mathcal{T}_\tau \times \mathcal{G} \times R^m$);
   (c) $D^* v(t,x,y) \leq \eta_3^\mathrm{T}(\|x\|) C_3(y) \eta_3(\|x\|) + m(t, \eta_3(\|x\|))$
   for all $(t,x,y) \in R \times \mathcal{G} \times R^m$ (for all $(t,x,y) \in \mathcal{T}_\tau \times \mathcal{G} \times R^m$),

   where function $m(t,\cdot)$ satisfies the condition

   $$\lim \frac{|m(t, \eta_3(\|x\|))|}{\|\eta_3\|} = 0 \quad \text{as} \quad \|\eta_3\| \to 0$$

   uniformly in $t \in R$ ($t \in \mathcal{T}_\tau$).

Then, provided the matrices $C_1(y)$, $C_2(y)$, $\tilde{C}_2(y)$ are positive definite and matrix $C_3(y)$ ($y \neq 0$) $\in R^m$ is negative definite, then

(a) under condition (4)(a) the state $x = 0$ of the system (1.2.7) is asymptotically stable (on $\mathcal{T}_\tau$);
(b) under condition (4)(b) the state $x = 0$ of the system (1.2.7) is uniformly asymptotically stable (on $\mathcal{T}_\tau$).

Corollary 1.2.3 Let

(1) condition (1) of Theorem 1.2.2 be satisfied;
(2) for function $v(t,x,e) = v(t,x)$, mentioned in condition (2) of Corollary 1.2.1 for all $(t,x) \in R \times \mathcal{G}$ (for all $(t,x) \in \mathcal{T}_\tau \times \mathcal{G}$))
   (a) $\psi_1(\|x\|) \leq v(t,x) \leq \psi_2(\|x\|)$;
   (b) $D^+v(t,x)|_{(1.2.7)} \leq -\psi_3(\|x\|)$,

where ψ_1, ψ_2, ψ_3 are of class K.

Then the state $x = 0$ of system (1.2.7) is uniformly asymptotically stable (on $\mathcal{T}_\tau$).

Theorem 1.2.4 Let the vector-function f in system (1.2.7) be continuous on $R \times R^n$ (on $\mathcal{T}_\tau \times R^n$) and conditions (1)–(3) of Theorem 1.2.2 be satisfied.

Then, provided that matrices $B_1(y)$, $B_2(y)$ and $\widetilde{B}_2(y)$ are positive definite and matrix $B_3(y)$ for all $(y \neq 0) \in R^m$ is negative definite,

(a) under condition (3)(a) of Theorem 1.2.2 the state $x = 0$ of system (1.2.7) is asymptotically stable in the whole (on $\mathcal{T}_\tau$);
(b) under condition (3)(b) of Theorem 1.2.2 the state $x = 0$ of system (1.2.7) is uniformly asymptotically stable in the whole (on $\mathcal{T}_\tau$).

Corollary 1.2.4 For function $v(t,x,e) = v(t,x)$, mentioned in condition (2) of Corollary 1.2.1 for all $(t,x) \in R \times R^n$ (for all $(t,x) \in \mathcal{T}_\tau \times R^n$) let

(a) $\varphi_1(\|x\|) \leq v(t,x) \leq \varphi_2(\|x\|)$;
(b) $D^+v(t,x)|_{(1.2.7)} \leq -\psi_3(\|x\|)$,

where φ_1, φ_2 are of class KR and ψ_3 is of class K.

Then the state $x = 0$ of system (1.2.7) is uniformly stable in the whole (on $\mathcal{T}_\tau$).

Theorem 1.2.5 Let the vector-function f in system (1.2.7) be continuous on $R \times \mathcal{N}$ (on $\mathcal{T}_\tau \times \mathcal{N}$). If there exist

(1) an open connected time-invariant neighborhood $\mathcal{G} \subset \mathcal{N}$ of the point $x = 0$;
(2) a matrix-valued function $U \in C(R \times \mathcal{N}, R^{m \times m})$ and a vector $y \in R^m$ such that the function $v(t,x,y) = y^T U(t,x)y$ is locally Lipschitzian in x for all $t \in R$ ($t \in \mathcal{T}_\tau$);
(3) functions $\sigma_{2i}, \sigma_{3i} \in K$, $i = 1, 2, \ldots, m$, a positive real number Δ_1 and positive integer p, $m \times m$ matrices $F_2(y)$, $F_3(y)$ such that

(a) $\Delta_1\|x\|^p \leq v(t,x,y) \leq \sigma_2^T(\|x\|)F_2(y)\sigma_2(\|x\|)$ for all $(t,x,y \neq 0) \in R \times \mathcal{G} \times R^m$ (for all $(t,x,y \neq 0) \in \mathcal{T}_\tau \times \mathcal{G} \times R^m$);

(b) $D^+v(t,x,y) \leq \sigma_3^T(\|x\|)F_3(y)\sigma_3(\|x\|)$ for all $(t,x,y \neq 0) \in R \times \mathcal{G} \times R^m$ (for all $(t,x,y \neq 0) \in \mathcal{T}_\tau \times \mathcal{G} \times R^m$).

Then, provided that the matrices $F_2(y)$, $(y \neq 0) \in R^m$ are positive definite, the matrix $F_3(y)$, $(y \neq 0) \in R^m$ is negative definite and functions σ_{2i}, σ_{3i} are of the same magnitude, then the state $x = 0$ of system (1.2.7) is exponentially stable (on $\mathcal{T}_\tau$).

Corollary 1.2.5 Let

(1) condition (1) of Theorem 1.2.1 be satisfied;
(2) for function $v(t,x,e) = v(t,x)$, mentioned in condition (2) of Corollary 1.2.1 for all $(t,x) \in R \times \mathcal{G}$ (for all $(t,x) \in \mathcal{T}_\tau \times \mathcal{G}$)

(a) $c_1\|x\|^p \leq v(t,x) \leq \varphi_1(\|x\|)$,

(b) $D^+v(t,x)|_{(1.2.7)} \leq -\varphi_2(\|x\|)$.

Then, if the functions φ_1, φ_2 are of class K and of the same magnitude, the state $x = 0$ of system (1.2.7) is exponentially stable (on $\mathcal{T}_\tau$).

Theorem 1.2.6 Let the vector-function f in system (1.2.7) be continuous on $R \times R^n$ (on $\mathcal{T}_\tau \times R^n$). If there exist

(1) a matrix-valued function $U \in C(R \times R^n, R^{m \times m})$ ($U \in C(\mathcal{T}_\tau \times R^n, R^{m \times m})$) and a vector $y \in R^m$ such that the function $v(t,x,y) = y^T U(t,x) y$ is locally Lipschitzian in x for all $t \in R$ (for all $t \in \mathcal{T}_\tau$);

(2) functions $\nu_{2i}, \nu_{3i} \in KR$, $i = 1,2,\ldots,m$, a positive real number $\Delta_2 > 0$ and a positive integer q;

(3) $m \times m$ matrices H_2, H_3 such that

(a) $\Delta_2\|x\|^q \leq v(t,x,y) \leq \nu_2^T(\|x\|)H_2(y)\nu_2(\|x\|)$ for all $(t,x,y \neq 0) \in R \times R^n \times R^m$ (for all $(t,x,y) \in \mathcal{T}_\tau \times R^n \times R^m$);

(b) $D^+v(t,x,y) \leq \nu_3^T(\|x\|)H_3(y)\nu_3(\|x\|)$ for all $(t,x,y \neq 0) \in R \times R^n \times R^m$ (for all $(t,x,y \neq 0) \in \mathcal{T}_\tau \times R^n \times R^m$).

Then, if the matrix $H_2(y)$ for all $(y \neq 0) \in R^m$ is positive definite, the matrix $H_3(y)$ for all $(y \neq 0) \in R^m$ is negative definite and functions ν_{2i}, ν_{3i} are of the same magnitude, the state $x = 0$ of system (1.2.7) is exponentially stable in the whole (on $\mathcal{T}_\tau$).

Corollary 1.2.6 For function $v(t,x,e) = v(t,x)$, mentioned in condition (2) of Corollary 1.2.1 for all $(t,x) \in R \times R^n$ (for all $(t,x) \in R^n \times \mathcal{G}$) let

(a) $c_2\|x\|^q \leq v(t,x) \leq \psi_1(\|x\|)$,

(b) $D^+v(t,x)|_{(1.2.7)} \leq -\psi_2(\|x\|)$,

where $\psi_1, \psi_2 \in KR$–class and are of the same magnitude.

Then the state $x = 0$ of system (1.2.7) is exponentially stable in the whole (on $\mathcal{T}_\tau$).

Theorem 1.2.7 Let the vector-function f in system (1.2.7) be continuous on $R \times \mathcal{N}$ (on $\mathcal{T}_\tau \times \mathcal{N}$). If there exist

(1) an open connected time-invariant neighborhood $\mathcal{G} \subset \mathcal{N}$ of the point $x = 0$;

(2) a matrix-valued function $U \in C^1(R \times \mathcal{N}, R^{m \times m})$ ($U \in C^1(\mathcal{T}_\tau \times \mathcal{N}, R^{m \times m})$) and a vector $y \in R^m$;

(3) functions $\psi_{1i}, \psi_{2i}, \psi_{3i} \in K$, $i = 1, 2, \ldots, m$, $m \times m$ matrices $A_1(y)$, $A_2(y)$, $G(y)$ and a constant $\Delta > 0$ such that

(a) $\psi_1^T(\|x\|)A_1(y)\psi_1(\|x\|) \leq v(t,x,y) \leq \psi_2^T(\|x\|)A_2(y)\psi_2(\|x\|)$
for all $(t,x,y) \in R \times \mathcal{G} \times R^m$ (for all $(t,x,y) \in \mathcal{T}_\tau \times \mathcal{G} \times R^m$);

(b) $D^+v(t,x,y) \geq \psi_3^T(\|x\|)G(y)\psi_3(\|x\|)$ for all $(t,x,y) \in R \times \mathcal{G} \times R^m$ (for all $(t,x,y) \in \mathcal{T}_\tau \times \mathcal{G} \times R^m$);

(4) point $x = 0$ belongs to $\partial\mathcal{G}$;

(5) $v(t,x,y) = 0$ on $\mathcal{T}_0 \times (\partial\mathcal{G} \cap B_\Delta)$, where $B_\Delta = \{x \colon \|x\| < \Delta\}$.

Then, if matrices $A_1(y)$, $A_2(y)$ and $G(y)$ for all $(y \neq 0) \in R^m$ are positive definite, the state $x = 0$ of system (1.2.7) is unstable (on $\mathcal{T}_\tau$).

Corollary 1.2.7 Let

(1) condition (1) of Theorem 1.2.7 be satisfied;

(2) there exist at least one couple of indices $(p,q) \in [1,m]$ such that $(v_{pq}(t,x) \neq 0) \in U(t,x)$ and a function $v(t,x,e) = v(t,x) \in C^1(R \times B_\Delta, R_+)$, $\overline{B}_\Delta \subset \mathcal{G}$, such that on $\mathcal{T}_0 \times \mathcal{G}$

(a) $0 < v(t,x) \leq a < +\infty$, for some $a > 0$;

(b) $D^+v(t,x)|_{(1.2.7)} \geq \varphi(v(t,x))$ for some function φ of class K;

(c) point $x = 0$ belongs to $\partial\mathcal{G}$;

(d) $v(t,x) = 0$ on $\mathcal{T}_0 \times (\partial\mathcal{G} \cap B_\Delta)$.

Then the state $x = 0$ of the system (1.2.7) is unstable.

We shall pay our attention to some specific features of the functions applied in Corollary 1.2.7.

Function $v(t,x)$ specifies the domain $v(t,x) > 0$, which is changing for $t \in \mathcal{T}_\tau$. Clearly this domain may cease its existence before the instability of motion is discovered.

If the function $v(t,x)$ is positive definite (strictly positive semi-definite), then the domain $v(t,x) > 0$ exists for all $t \in \mathcal{T}_\tau$.

If the function $v(t,x)$ is constant negative, the domain $v(t,x) > 0$ does not exist.

Example 1.2.2 (i) Function
$$v(t,x) = \sin t \, x_1 x_2$$
is of variable sign and domain $v(t,x) > 0$ exists but not for all $t \in \mathcal{T}_\tau$.

(ii) For the function
$$v(t,x) = (\cos t - 2)\, x_1^2 x_2$$
the domain $v(t,x) > 0$ exists for all $t \in \mathcal{T}_\tau$.

(iii) For the function
$$v(t,x) = \left(\frac{1}{t} - a\right) x_1 x_2 - x_2^2, \quad a > 0,$$
the domain $v(t,x) > 0$ exists for all $t \geq t_0$, and for $t_0 > 1/a$.

Corollary 1.2.8 Let condition (1) of Theorem 1.2.7 be satisfied. If there exist $t_0 \in \mathcal{T}_0$, $\Delta > 0$, $(\overline{B}_\Delta \subset \mathcal{N})$ and an open set $\mathcal{G} \subset B_\Delta$ and the function $v(t,x,e) = v(t,x) \in C^1(\mathcal{T}_0 \times B_\Delta, R)$, mentioned in Corollary 1.2.7 such that on $\mathcal{T}_0 \times \mathcal{G}$

(a) $0 < v(t,x) \leq \varphi_1(\|x\|)$;

(b) $D^+ v(t,x)|_{(1.2.7)} \geq \varphi_2(\|x\|)$ for some φ_1, φ_2 of class K;

(c) point $x = 0$ belongs to $\partial \mathcal{G}$;

(d) $v(t,x) = 0$ on $\mathcal{T}_0 \times (\partial \mathcal{G} \cap B_\Delta)$.

Then the state $x = 0$ of (1.2.7) is unstable.

Corollary 1.2.9 If in Corollary 1.2.8 condition (b) is replaced by

(b') $D^+ v(t,x)|_{(1.2.7)} \geq k v(t,x) + w(t,x)$

on $\mathcal{T}_0 \times \mathcal{G}$, where $k > 0$ and function $w(t, x) \geq 0$ is continuous on $\mathcal{T}_0 \times \mathcal{G}$, then the state $x = 0$ of system (1.2.7) is unstable.

1.2.3.2 Nonautonomous problem of Lefschetz We consider a problem on stability in a product space for a system of differential equations of the perturbed motion (cf. Lefschetz [1])

(1.2.23)
$$\frac{dy}{dt} = g(t, y) + G(t, y, z),$$
$$\frac{dz}{dt} = h(t, z) + H(t, y, z).$$

Here $y \in R^p$, $z \in R^q$, $g \colon R_+ \times R^p \to R^p$, $G \colon R_+ \times R^p \times R^q \to R^p$, $h \colon R_+ \times R^q \to R^q$, $H \colon R_+ \times R^p \times R^q \to R^q$. In addition, function g, G; h, H are continuous on $R_+ \times R^p$, $R_+ \times R^q$, $R_+ \times R^p \times R^q$ and they vanish for $y = z = 0$.

The problem itself is to point out the connection between the stability properties of equilibrium state $y = z = 0$ with respect to system (1.2.23) on $R^p \times R^q$ and its nonlinear approximation

(1.2.24)
$$\frac{dy}{dt} = g(t, y),$$
$$\frac{dz}{dt} = h(t, z).$$

Assumption 1.2.1 Let there exist the time-invariant neighborhood $\mathcal{N}_y \subseteq R^p$ and $\mathcal{N}_z \subseteq R^q$ of the equilibrium state $y = 0$ and $z = 0$, respectively and let there exist a matrix-valued function

(1.2.25)
$$U(t, y, z) = \begin{pmatrix} v_{11}(t, y) & v_{12}(t, y, z) \\ v_{21}(t, y, z) & v_{22}(t, z) \end{pmatrix}$$

the element v_{ij} of which satisfy the estimations (cf. Krasovskii [1], Djordjević [1])
(1.2.26)
$$\underline{c}_{11}\|y\|^2 \leq v_{11}(t, y) \leq \bar{c}_{11}\|y\|^2 \quad \text{for all} \quad (t, y \neq 0) \in R_+ \times \mathcal{N}_y;$$
$$\underline{c}_{22}\|z\|^2 \leq v_{22}(t, z) \leq \bar{c}_{22}\|z\|^2 \quad \text{for all} \quad (t, z \neq 0) \in R_+ \times \mathcal{N}_z;$$
$$\underline{c}_{12}\|y\|\|z\| \leq v_{12}(t, y, z) \leq \bar{c}_{12}\|y\|\|z\|$$
$$\text{for all} \quad (t,, y \neq 0, z \neq 0) \in R_+ \times \mathcal{N}_y \times \mathcal{N}_z;$$
$$v_{12}(t, y, z) = v_{21}(t, y, z) \quad \text{for all} \quad (t,, y \neq 0, z \neq 0) \in R_+ \times \mathcal{N}_y \times \mathcal{N}_z;$$
$$\underline{c}_{ii} > 0, \quad \bar{c}_{ii} > 0, \quad \underline{c}_{ij}, \bar{c}_{ij} = \text{const} \in R, \quad i \neq j.$$

Assumption 1.2.2 Let there exist functions $\alpha_{ij}(t)$, $i = 1, 2$; $j = 1, 2, \ldots, 8$ which are bounded on any finite interval on R_+ such that

(1.2.27)
$$\frac{\partial v_{11}}{\partial t} + \left(\frac{\partial v_{11}}{\partial y}\right)^{\mathrm{T}} g \leq \alpha_{11}\|y\|^2;$$

$$\frac{\partial v_{22}}{\partial t} + \left(\frac{\partial v_{11}}{\partial y}\right)^{\mathrm{T}} G \leq \alpha_{12}\|y\|^2 + \alpha_{13}\|y\|\|z\|;$$

$$\left(\frac{\partial v_{22}}{\partial z}\right)^{\mathrm{T}} h \leq \alpha_{21}\|z\|^2;$$

$$\left(\frac{\partial v_{22}}{\partial z}\right)^{\mathrm{T}} H \leq \alpha_{22}\|z\|^2 + \alpha_{23}\|y\|\|z\|;$$

$$\frac{\partial v_{12}}{\partial t} + \left(\frac{\partial v_{12}}{\partial y}\right)^{\mathrm{T}} g \leq \alpha_{14}\|y\|^2 + \alpha_{15}\|y\|\|z\|;$$

$$\left(\frac{\partial v_{12}}{\partial y}\right)^{\mathrm{T}} G \leq \alpha_{16}\|y\|^2 + \alpha_{17}\|y\|\|z\| + \alpha_{18}\|z\|^2;$$

$$\left(\frac{\partial v_{12}}{\partial z}\right)^{\mathrm{T}} h \leq \alpha_{24}\|z\|^2 + \alpha_{25}\|y\|\|z\|;$$

$$\left(\frac{\partial v_{12}}{\partial z}\right)^{\mathrm{T}} H \leq \alpha_{26}\|y\|^2 + \alpha_{27}\|y\|\|z\| + \alpha_{28}\|z\|^2.$$

For system (1.2.23) we have the following result.

Theorem 1.2.8 *Suppose that*
(1) *all conditions of Assumptions 1.2.1, 1.2.2 are fulfilled;*
(2) *the matrices*
$$\underline{C} = \begin{pmatrix} \underline{c}_{11} & \underline{c}_{12} \\ \underline{c}_{21} & \underline{c}_{22} \end{pmatrix}, \quad \underline{c}_{12} = \underline{c}_{21}, \quad \overline{C} = \begin{pmatrix} \overline{c}_{11} & \overline{c}_{12} \\ \overline{c}_{21} & \overline{c}_{22} \end{pmatrix}, \quad \overline{c}_{12} = \overline{c}_{21}$$
are positive definite;
(3) *the matrix* $M(t) = \dfrac{1}{2}(S^{\mathrm{T}}(t) + S(t))$ *where*
$$S(t) = \begin{pmatrix} \sigma_{11} & \sigma_{12} \\ \sigma_{21} & \sigma_{22} \end{pmatrix}, \quad \sigma_{12} = \sigma_{21}$$

and

$$\sigma_{11} = \eta_1^2(\alpha_{11} + \alpha_{12}) + 2\eta_1\eta_2(\alpha_{14} + \alpha_{16} + \alpha_{26});$$
$$\sigma_{22} = \eta_2^2(\alpha_{21} + \alpha_{22}) + 2\eta_1\eta_2(\alpha_{18} + \alpha_{24} + \alpha_{28});$$
$$\sigma_{12} = \frac{1}{2}\left(\eta_1^2\alpha_{13} + \alpha_{23}\eta_2^2\right) + \eta_1\eta_2(\alpha_{15} + \alpha_{25} + \alpha_{17} + \alpha_{27}),$$

η_1, η_2 being positive numbers, have the characteristic roots $\beta_1(t)$, $\beta_2(t)$ and let $\operatorname{Re}\beta_i(t) \leq \delta$, for all $t \geq t_0$.

Then the state of equilibrium $y = z = 0$ of the system (1.2.23) is uniformly stable if the number δ is equal to zero and exponentially stable if $\delta < 0$.

If conditions of Assumptions 1.2.1, 1.2.2 are fulfilled for $\mathcal{N}_y = R^p$, $\mathcal{N}_y = R^q$ and conditions (2), (3) of the theorem hold, then the equilibrium state $y = z = 0$ of the system (1.2.23) is uniformly stable in the whole if $\delta = 0$ and exponentially stable in the whole if $\delta < 0$.

Proof The proof uses the direct method. On the basis of estimations (1.2.26), it is not difficult to show that the function $v(t,y,z) = \eta^T U(t,y,z)\eta$ satisfies the estimates

$$(1.2.28) \qquad u^T \Phi^T \underline{C} \Phi u \leq v(t,y,z) \leq u^T \Phi^T \overline{C} \Phi u,$$

where $u^T = (\|y\|, \|z\|)$, $\Phi = \operatorname{diag}[\eta_1, \eta_2]$.

Also, in view of Assumption 1.2.1 and the estimates (1.2.27), the derivative $Dv(t,y,z)$ defined by $Dv(t,y,z) = \eta^T DU(t,y,z)\eta$ satisfies

$$(1.2.29) \qquad Dv(t,y,z) \leq u^T S(t) u = u^T M(t) u \leq \delta u^T u$$

By virtue of (2) and (3) and the inequalities (1.2.28), (1.2.29), we see that all conditions of Theorems 1.2.1, 1.2.5 are verified for the function $v(t,y,z)$ and its derivative. Hence the proof is complete.

If in estimate (1.2.27) we change the sign of inequality for the opposite one, then by means of the method similar to the given one we can obtain the estimate

$$Dv(t,y,z) \geq u^T \widetilde{S}(t) u$$

which allows us to formulate instability conditions for the equilibrium state $y = z = 0$ of system (1.2.23) on the basis of Theorem 1.2.7.

The statement of Theorem 1.2.8 shows that uniform stability or exponential stability of the equilibrium state $y = z = 0$ of system (1.2.23) can hold even if the equilibrium state $y = z = 0$ of system (1.2.24) has no properties of asymptotic quasi-stability (cf. Lefschetz [1]).

1.2.3.3 Vector Liapunov functions applications The idea of vector Liapunov function was developed in several directions in many papers and books. The papers by Corduneanu [1, 2] and Conti [1] where the scalar

Liapunov function was applied together with differential inequalities have become a background for a series of important results in motion stability theory obtained via the principle of comparison with vector Liapunov function. In view of the existence of several excellent books on Liapunov's vector functions method in this section we review basic ideas and results developed lately while working out the method of vector Liapunov functions since it helps to compare the advantages achieved by using matrix-valued and/or hierarchical auxiliary functions.

1.2.3.3.1 Scalar approach We return back to the system (1.2.7) and consider also a *vector function*

$$V(t,x) = (v_1(t,x), v_2(t,x), \ldots, v_m(t,x))^T, \quad 1.2.30$$

where $v_s \in C(\mathcal{T}_0 \times R^n, R)$, $s = 1, 2, \ldots, m$, and its total derivative along solutions of the system (1.2.7)

(1.2.31) $D^+V(t,x) = \lim \sup \{[V(t+\theta, x+\theta f(t,x)) - V(t,x)]\theta^{-1} : \theta \to \theta^+\}$

for $(t,x) \in \mathcal{T}_0 \times R^n$.

The notion of the property of having a fixed sign of function (1.2.30) is included as a special case of Definitions 1.2.8, 1.2.9, 1.2.11, and 1.2.12 by means of the measures such as

(a) $v_0(t,x) = e^T V(t,x)$, $e = (1, 1, \ldots, 1) \in R_+^m$;
(b) $v_0(t,x) = \alpha^T V(t,x)$, $(t,x) \in \mathcal{T}_0 \times R^n$, $\alpha \in R^m$;
(c) $v_0(t,x) = \max_{1 \le i \le m} v_i(t,x)$;
(d) $v_0(t,x) = Q(V(t,x))$ where $Q \in C(R_+^m, R_+)$, $Q(0) = 0$ and $Q(u)$ is nondecreasing in u.

The state vector x of system (1.2.7) is divided into m subvectors, i.e. $x = (x_1^T, \ldots, x_m^T)^T$, where $x_s \in R^{n_s}$ and $n_1 + n_2 + \cdots + n_m = n$.

Assume that

(1.2.32) $\quad a_{i1}\psi_{i1}^{1/2}(\|x_i\|) \le v_i(t,x) \le a_{i2}\psi_{i2}^{1/2}(\|x_i\|), \quad i = 1, 2, \ldots, m,$

where a_{i1} and a_{i2} are some positive constants and ψ_{i1} and ψ_{i2} are of class K (KR).

Actually the condition (1.2.32) means that the components $v_i(t,x)$ of the vector function (1.2.30) are positive definite and decreasing with respect to a part of variables.

Let us introduce designations

(1.2.33)
$$A_1 = \mathrm{diag}\,[a_{11}, a_{12}, \ldots, a_{1m}],$$
$$A_2 = \mathrm{diag}\,[a_{21}, a_{22}, \ldots, a_{2m}].$$

Proposition 1.2.7 For the vector function (1.2.30) to be positive definite and decreasing, it is sufficient that matrices (1.2.33) in the bilateral inequalities

(1.2.34) $$u_1^T A_1 u_1 \leq v(t, x, \alpha) \leq u_2^T A_2 u_2$$

be positive definite, where

$$u_1 = \left(\psi_{11}^{1/2}(\|x_1\|), \ldots, \psi_{1m}^{1/2}(\|x_m\|)\right)^T,$$

$$u_2 = \left(\psi_{21}^{1/2}(\|x_1\|), \ldots, \psi_{2m}^{1/2}(\|x_m\|)\right)^T.$$

If $\psi_{i1} = \psi_{i2} = \|x_i\|$, then the estimates (1.2.34) are known (see Krasovskii [1]) as the estimates characteristics of the quadratic forms.

Taking into account (1.2.31) we get for the function (1.2.32)

(1.2.35) $$D^+ v(t, x, \alpha) = \alpha^T D^+ V(t, x).$$

Let for $(t, x) \in \mathcal{T}_0 \times R^n$ there exist an $m \times m$ matrix $S(t, x)$, for which

(1.2.36) $$D^+ v(t, x, \alpha) \leq \psi_3^T S(t, x) \psi_3,$$

where $\psi_3 = \left(\psi_{13}^{1/2}(\|x_1\|), \psi_{23}^{1/2}(\|x_2\|), \ldots, \psi_{m3}^{1/2}(\|x_m\|)\right)^T.$

Estimates (1.2.34)–(1.2.36) allows us to establish stability conditions for the state $x = 0$ of system (1.2.7) as follows.

Theorem 1.2.9 Let the vector function f in system (1.2.7) be continuous on $R \times \mathcal{N}$ (on $\mathcal{T}_\tau \times \mathcal{N}$). If there exist

(1) an open connected time-invariant neighborhood $\mathcal{G}$ of point $x = 0$;
(2) the decreasing positive definite vector function V on $\mathcal{G}$ (on $\mathcal{T}_\tau \times \mathcal{G}$);
(3) the $m \times m$-matrix $S(t, x)$ on $\mathcal{G}$ (on $\mathcal{T}_\tau \times \mathcal{G}$) such that inequality (1.2.36) is satisfied.

Then

(a) the state $x = 0$ of system (1.2.7) is uniformly stable if the matrix $M(t, x) = \frac{1}{2}(S T(t, x) + S(t, x))$ is negative semidefinite on $\mathcal{G}$ (on $\mathcal{T}_\tau \times \mathcal{G}$);

(b) the state $x = 0$ of system (1.2.7) is uniformly asymptotically stable (on $\mathcal{T}_\tau$) providing the matrix $M(t, x) + \varepsilon E$, $\varepsilon > 0$, E is $m \times m$ identity matrix, is negative definite on $\mathcal{G}$ (on $\mathcal{T}_\tau \times \mathcal{G}$);

(c) the state $x = 0$ of system (1.2.7) is exponentially stable if there exists a constant $a > 0$ and $b > 0$ such that $a\|x\|^b \leq u_1^T A_1 u_1$, function $v(t, x, \alpha)$ is decrescent, and matrix $M(t, x) + \varepsilon E$ is negative definite.

Proof Formula (1.2.32) and estimates (1.2.34) and (1.2.36) allow us to repeat all points of the proof of Theorems 2.3.1, 2.3.3 and 2.3.5 by Martynyuk [11] (cf. Yoshizawa [1] and Michel and Miller [1]). The theorem is proved.

Remark 1.2.1 New points of Theorem 1.2.9 resulting from the application of vector function (1.2.30) are

(a) a possibility to apply the components $v_i(t,x)$, $i = 1, 2, \ldots, m$, being of a fixed sign with respect to a part of variables;

(b) a possibility to check the property of having a fixed sign of the matrix $M(t,x)$ via the algebraic method.

The Theorem 1.2.9 proves to be quite universal in the framework of the scalar approach of the vector Liapunov function application (cf. Grujic *et al.* [1], Michel and Miller [1], Šiljak [1], etc.)

Also, within the scalar approach the application of the vector Liapunov function together with the comparison equation are developed.

In order to present some results in this direction let us first introduce some notions from the theory of differential inequalities (see Olech and Opial [1], Walter [1], Schroder [1], etc.)

We consider the scalar differential equation

$$(1.2.37) \qquad \frac{du}{dt} = g(t,u), \quad u(t_0) = u_0 \geq 0,$$

where $t \in R_+$, $u \in R$, $g \colon R_+ \times B(r) \to R$ for some $r > 0$. Assume that $g(t,u)$ is continuous on $R_+ \times B(r)$ and $g(t,0) = 0$ for all $t \geq t_0$. Our assumptions on $g(t,u)$ imply that the solution $u(t) = u(t;t_0,u_0)$ of (1.2.37) is nonnegative for all $t \geq t_0$.

Definition 1.2.14 Let $w_M(t)$ be a solution of (1.2.37) in the interval $[t_0, a)$. Then $w_M(t)$ is called a *maximal solution* of (1.2.37) if for any other solution $u(t)$ existing on $[t_0, a)$ and passing through the point (t_0, u_0) we have $u(t) \leq w_M(t)$ for all $t \in [t_0, a)$.

Definition 1.2.15 Let $w_m(t)$ be a solution of (1.2.37) in the interval $[t_0, a)$. Then $w_m(t)$ is called a *minimal solution* of (1.2.37) if for any other solution $u(t)$ existing on $[t_0, a)$ and passing through the point (t_0, u_0) we have $u(t) \geq w_m(t)$ for all $t \in [t_0, a)$.

It is well known that if $g(t,u)$ is continuous on $R_+ \times B(r)$ and if $u_0 \in B(r)$, then equation (1.2.37) has both a maximal solution $w_M(t)$ and a minimal solution $w_m(t)$ for any $w_M(t_0) = w_m(t_0) = u_0$. Each of these solutions either exists for all $t \in [t_0, \infty)$ or else must leave the domain of definition of at some time $t_1 > t_0$.

The following result involves estimating a function satisfying a differential inequality by the maximum solution of the equation corresponding to the the inequality.

Theorem 1.2.10 *Assume that $g(t, u)$ is continuous on $R_+ \times B(r)$ and $w_M(t)$ is the maximal solution of (1.2.37) on interval $[t_0, a)$ with $w_M(t_0) = u_0$. Also, let $m(t)$ be a continuous function on $[t_0, a)$ such that $m(t_0) \leq u_0$, and $(t, m(t)) \in R_+ \times B(r)$ satisfying the differential inequality*

$$(1.2.38) \qquad Dm(t) \leq g(t, m(t)), \quad t \in [t_0, a).$$

Then, on the common interval of existence $m(t)$ and $w_M(t)$ the inequality

$$(1.2.39) \qquad m(t) \leq w_M(t)$$

holds.

Remark 1.2.2 If (1.2.38) is reversed and $m(t_0) \geq u_0$, then the conclusion (1.2.39) must be replaced by $m(t) \geq w_m(t)$ for all $[t_0, a)$, where $w_m(t)$ is the minimal solution of (1.2.37).

The Theorem 1.2.10 and Remark 1.2.2 are very important tools for applications, because they can be used to reduce the problem of determining the behavior of solution of (1.2.7) to the solution of a scalar equation (1.2.37) and the properties of the auxiliary Liapunov function.

We can easily see that the following results are true.

Theorem 1.2.11 *Let the vector function f in system (1.2.7) be continuous on $R \times \mathcal{N}$ (on $\mathcal{T}_\tau \times \mathcal{N}$) and scalar function g in system (1.2.37) be continuous on $R \times B(r)$. If there exist*

(1) *an open connected time-invariant neighborhood $\mathcal{G}$ of point $x = 0$;*
(2) *the vector function $V(t, x)$ and a vector $\alpha \in R^m$ for which inequalities (1.2.34) are satisfied;*
(3) *the function $g \in C(\mathcal{T}_\tau \times R_+, R)$, $g(t, 0) = 0$ such that*

$$D^+ v(t, x, \alpha) \leq g(t, v(t, x, \alpha)) \quad \text{for all} \quad (t, x) \in \mathcal{T}_\tau \times \mathcal{N};$$

(4) *the solution $u = 0$ of the comparison equation (1.2.37) existing for $t \geq t_0$.*

Then, if the zero solution of (1.2.37)

(a) *is stable and $v(t, x, \alpha)$ is positive definite, state $x = 0$ of (1.2.7) is stable;*

(b) *is uniformly stable and $v(t, x, \alpha)$ is positive definite and decrescent the state $x = 0$ of (1.2.7) is uniformly stable;*

(c) *is asymptotically stable and $v(t, x, \alpha)$ is positive definite, the state $x = 0$ of (1.2.7) is asymptotically stable;*

(d) *is uniformly asymptotically stable and $v(t, x, \alpha)$ is positive definite and decrescent, the state $x = 0$ of (1.2.7) is uniformly asymptotically stable;*

(e) *is exponentially stable and there are constants $a > 0$ and $b > 0$ such that $a\|x\|^b \leq u_1^T A_1 u_1$, $v(t, x, \alpha)$ is decrescent, the state $x = 0$ of (1.2.7) is exponentially stable;*

(f) *is bounded (uniformly bounded, uniformly ultimately bounded) and $f\colon R_+ \times R^n \to R^n$; $g\colon R_+ \times R \to R$, function $v(t, x, \alpha)$ is positive definite and decrescent and (3) holds for all $(t, x) \in R_+ \times R^n$, then the solutions of (1.2.7) are bounded (uniformly bounded, uniformly ultimately bounded).*

The proof of the Theorem 1.2.11 is similar to the proof of Theorem 5.3.7 from the book of Rama Mohana Rao [1] (cf. also Corduneanu [1]).

Applying Remark 1.2.2 to the function $v(t, x, \alpha)$ we can easily obtain the next result.

Theorem 1.2.12 *Let the vector function f in the system (1.2.7) and g in the system (1.2.37) be continuous on their respective domains of definition. Assume that the function $v(t, x, \alpha)$ be continuous on $\mathcal{T}_\tau \times \mathcal{N}$ and such that*

$$D^+ v(t, x, \alpha) \geq g(t, v(t, x, \alpha)) \quad \text{for all} \quad (t, x) \in \mathcal{T}_\tau \times \mathcal{N}.$$

Then, if the zero solution of (1.2.37) is unstable and $v(t, x, \alpha)$ is positive definite, the state $x = 0$ of (1.2.7) is unstable.

One of the Theorem 1.2.11 generalizations is based on the application of a majorizing function $g \in C(\mathcal{T}_\tau \times R^n \times R_+, R)$, $g(t, r, x) = 0$ when $r = 0$ and $x = 0$.

Besides an extended system

(1.2.40)
$$\frac{dx}{dt} = f(t, x), \qquad x(t_0) = x_0,$$
$$\frac{dr}{dt} = g(t, r, x), \qquad r(t_0) = r_0 \geq 0$$

is treated for which certain type of r-stability of the zero solution $(x^T, r)^T = 0$ yields an appropriate type of stability of the state $x = 0$ of (1.2.7).

The idea has been developed and applied for the cases when the function $w(t, r) = w(r)$, i.e. it is independent of $t \in \mathcal{T}_\tau$. These and other results obtained in this direction are set out by Hatvani [1], Martynyuk [3], etc.

Example 1.2.3 (cf. Michel and Miller [1]) We will show now how the matrix $S(t, x)$ in the inequality (1.2.36) could constructing for the system (1.2.7) admit a decomposition to the form

$$(1.2.41) \qquad \frac{dx_i}{dt} = f_i(t, x_i) + h_i(t, x_1, \ldots, x_m), \quad i = 1, 2, \ldots, m,$$

where $x_i \in R^{n_i}$, $f_i \in C(R_+ \times R^{n_i}, R^{n_i})$, $h_i \in C(R_+ \times R^{n_1} \times \ldots \times R^{n_m}, R^{n_i})$, $\sum_{i=1}^{m} n_i = n$.

The functions $f_i(t, x_i)$ in (1.2.40) represent isolated decoupled subsystems

$$(1.2.42) \qquad \frac{dx_i}{dt} = f_i(t, x_i), \quad x_i(t_0) = x_{i0}, \quad i = 1, 2, \ldots, m,$$

and $h_i(t, x)$ are interconnections among the subsystems (1.2.41) which have the form $h_i(t, x, E) = h_i(t, e_{i1}x_1, e_{i2}x_2, \ldots, e_{im}x_m)$, with $e_{ij} \in C(R_+ \times R^n, [0, 1])$ as the elements of interconnection matrix E (see Šiljak [1]).

If in the system (1.2.7) all possible interconnections are present and its structure is described by a matrix $E = (e_{ij})$, then another choice of interconnection matrix $E_0 = (e_{ij}^0)$, represents a structural perturbation of the system (1.2.7).

We assume that there exist continuous functions $v_i \colon R_+ \times R^n \to R$, comparison functions $\psi_{i1}, \psi_{i2} \in KR$ and $\psi_{i3} \in K$, constants $\sigma_i \in R$ and a_{ij} for all $i, j = 1, 2, \ldots, m$, such that for all $(t, x) \in R_+ \times R^n$ and $x_i \in R^{n_i}$, the inequalities

(a) $\psi_{i1}(\|x_i\|) \leq v_i(t, x) \leq \psi_{i2}(\|x_i\|)$;

(b) $D^+ v_i(t, x)|_{(1.2.41)} \leq \sigma_i \psi_{i3}(\|x_i\|)$;

(c) for any interconnections $e_{ij} \in E$, where $e_{i,j}$ are the elements of fundamental matrix of connections

$$D^+ v_i(t, x)\big|_{(1.2.41)} - D^+ v_i(t, x)\big|_{(1.2.42)} \leq \psi_{i3}^{1/2}(\|x_i\|) \sum_{j=1}^{m} a_{ij} \psi_{j3}^{1/2}(\|x_j\|)$$

hold for all $i = 1, 2, \ldots, m$.

In view of assumptions (a)–(c) we arrive at

$$D^+v(t, x, \alpha) \leq \psi_3^T(\|x\|)\widetilde{S}\psi_3(\|x\|),$$

where $\psi_3 = \left(\psi_{13}^{1/2}(\|x_1\|), \ldots, \psi_{m3}^{1/2}(\|x_m\|)\right)^T$, and $\widetilde{S} = [s_{ij}]$ with

$$s_{ij} = \begin{cases} \alpha_i(\sigma_i + a_{ii}) & \text{for } i = j, \\ \dfrac{1}{2}(\alpha_i a_{ij} + \alpha_j a_{ji}) & \text{for } i \neq j. \end{cases}$$

Clearly if $\widetilde{M} = \dfrac{1}{2}(\widetilde{S}^T + \widetilde{S})$ is negative semidefinite (negative definite) matrix, then we derive the conditions of connective stability of (1.2.7) from Theorem 1.2.11.

1.2.3.3.2 Vector approach The combination of vector function (1.3.3) with the comparison system

(1.2.43) $$\dfrac{du}{dt} = G(t, u), \quad u(t_0) = u_0 \geq 0,$$

where $u \in R_+^m$, $G \in C(\mathcal{T}_\tau \times R_+^m, R^m)$, $G(t, 0) = 0$ for all $t \in \mathcal{T}_\tau$, leads to the general result of the method of vector Liapunov function. We note that under the above mentioned assumptions equation (1.2.42) possesses solution $u(t)$ for every $u_0 = u(t_0) \in B(r)$ which is not necessarily unique. These solutions either exist for all $t \in [t_0, \infty)$ or else must leave the domain of definition of $G(t, u)$ at some finite time $t_1 > t_0$. We assume that $u = 0$ is an isolated equilibrium of (1.2.42).

In order to extend Definitions 1.2.14, 1.2.15 to system (1.2.42) we require the following additional concept.

Definition 1.2.16 The function $G(t, u) = (G_1(t, u), \ldots, G_m(t, u))^T$ is said to be *quasi-monotone nondecreasing* if for each component G_j, $j = 1, 2, \ldots, m$, the inequality $G_j(t, u) \leq G_j(t, z)$ is true whenever $u_i \leq z_i$ for all $i \neq j$ and $u_j = z_j$ for each $t \in \mathcal{T}_\tau$.

To appreciate this definition, note that if $G(t, u) = Au$, where A is an $m \times m$ matrix, $G(t, u)$ is quasi-monotone, implies that $a_{ij} \leq 0$, $i, j = 1, 2, \ldots, m$, $i \neq j$.

We recall the following results whose proof may be found in Walter [1], Szarski [1], etc.

Proposition 1.2.8 Let $G(t, u)$ be continuous and quasi-monotone nondecreasing in u for each t. Then, if $(t_0, u_0) \in \mathcal{T}_\tau \times B(r)$, the system (1.2.42) has a maximal and minimal solution which can be extended to boundary of $B(r)$. Each of these solutions must either be defined for all $t \in \mathcal{T}_\tau$ or else leave the domain of definition of $G(t, u)$ at some finite time $t_1 > t_0$.

Now we have the following generalization of Theorem 1.2.10.

Theorem 1.2.13 Assume that the vector function $G(t, u)$ is continuous on $\mathcal{T}_\tau \times B(r,)$ $G(t, u)$ is quasi-monotone nondecreasing in u for each $t \in \mathcal{T}_\tau$ and $r_M(t)$ is the maximal solution of (1.2.42) existing on $[t_0, a)$ with $r_M(t_0) = u_0$. Let $m(t)$ be a continuous function such that $m(t_0) \leq u_0$, and

(1.2.44) $$Dm(t) \leq G(t, m(t)), \quad t \in \mathcal{T}_\tau,$$

holds for a fixed Dini derivative. Then $m(t_0) \leq u_0$ implies

$$m(t) \leq r_M(t)$$

for all $t \in [t_0, a)$.

We note that the inequalities in Theorem 1.2.12 are componentwise.

Remark 1.2.3 If (1.2.43) is reversed and $m(t_0) \geq u_0$, the conclusion (1.2.41) must be replaced by $m(t) \geq r_m(t)$ for all $t \in [t_0, a)$, where $R_m(t)$ is a minimal solution of (1.2.42).

Applying Theorem 1.2.13 to vector-valued function $V(t, x)$ the following results can be easily established.

Theorem 1.2.14 Let the vector function f in system (1.2.7) be continuous on $R \times \mathcal{N}$ (on $\mathcal{T}_\tau \times \mathcal{N}$) and vector function G in system (1.2.42) be continuous on $R \times B(r)$ (on $\mathcal{T}_\tau \times B(r)$). If there exist

(1) an open connected time-invariant neighborhood $\mathcal{G}$ of point $x = 0$;
(2) the vector function $V \in C(\mathcal{T}_\tau \times \mathcal{N}, R_+^m)$, $V(t, x)$ is locally Lipschitzian in x and a real vector $\alpha \in R^m$ such that function $v(t, x, \alpha)$ satisfies bilateral inequality (1.2.34);
(3) the function $G(t, x) \in C(R \times B(r), R^m)$ $(G(t, x) \in C(\mathcal{T}_\tau \times B(r), R^m))$, $G(t, 0) = 0$ for $t \in R$ $(t \in \mathcal{T}_\tau)$ and $G(t, u)$ is quasi-monotone nondecreasing in u for all $t \in \mathcal{T}_\tau$, so that

$$D^+ V(t, x) \leq G(t, V(t, x)), \quad (t, x) \in \mathcal{T}_\tau \times \mathcal{N}.$$

Then, if the zero solution of (1.2.42)

(a) *is stable and $v(t,x,\alpha)$ is positive definite, the state $x = 0$ of (1.2.7) is stable;*
(b) *is uniformly stable and $V(t,x)$ is positive definite and decrescent the state $x = 0$ of (1.2.7) is uniformly stable;*
(c) *is asymptotically stable and $V(t,x)$ is positive definite, the state $x = 0$ of (1.2.7) is asymptotically stable;*
(d) *is uniformly asymptotically stable and $V(t,x)$ is positive definite and decrescent, the state $x = 0$ of (1.2.7) is uniformly asymptotically stable;*
(e) *is exponentially stable and there are constants $a > 0$ and $b > 0$ such that $a\|x\|^b < u_1^T A_1 u_1$, $V(t,x)$ is decrescent, the state $x = 0$ of (1.2.7) is exponentially stable;*
(f) *is bounded (uniformly bounded, uniformly ultimately bounded) and $f: R_+ \times R^n \to R^n$; $g: R_+ \times R^m \to R^m$; function $V(t,x)$ is positive definite and decrescent and (3) holds for all $(t,x) \in R_+ \times R^n$, then the solutions of (1.2.7) are bounded (uniformly bounded, uniformly ultimately bounded).*

Proof We shall cite first an assertion that establishes a relationship between the vector function variation and maximal solution to comparison system (1.3.13).

Proposition 1.2.9 Let $V \in C(\mathcal{T}_\tau \times \mathcal{N}, R_+^m,)$ and $V(t,x)$ be locally Lipschitzian in x. Let the vector function $D^+V(t,x)$ specified by (1.3.4) satisfy the inequality

$$D^+V(t,x) \leq G(t, V(t,x)), \quad \text{for all} \quad (t,x) \in \mathcal{T}_\tau \times \mathcal{N},$$

where $G \in C(\mathcal{T}_\tau \times R_+^m, R^m)$ and the function $\Omega(t,u)$ be quasi-monotone nondecreasing in u. Assume that the maximal solution $u_M(t; t_0, r_0)$ of the comparison system

$$\frac{du}{dt} = G(t,u), u(t_0) = u_0,$$

exists on the interval $\mathcal{T}_\tau$ and passes through the point $(t_0, r_0) \in \mathcal{T}_\tau \times R_+^m$. If $x(t; t_0, x_0)$ is any solution to system (1.2.7) defined on $[t_0, t_0 + \delta)$, $t_0 \in \mathcal{T}_\tau$ and passing through the point $(t_0, x_0) \in \mathcal{T}_\tau \times \mathcal{N}$, then the condition

(1.2.45) $$V(t_0, x_0) \leq r_0$$

yields the estimate

(1.2.46) $\quad V(t, x(t; t_0, x_0)) \leq u_M(t; t_0, r_0) \quad$ for all $\quad t \in [t_0, t_0 + \delta)$.

Further the fact that function $v(t, x, \alpha)$ satisfies bilateral inequality (1.2.34) with positive definite matrices A_1 and A_2 implies that the vector function $V(t, x)$ is positive definite and decreasing.

Estimate (1.2.36) and the fact that the solution $u = 0$ of the system (1.2.42) possesses a certain type of stability lead to the conclusion that the state $x = 0$ of the system (1.2.7) has a corresponding type of stability.

Applying the Remark 1.2.3 to vector-valued function $V(t, x)$ we can prove the following instability result.

Theorem 1.2.15 *Let the vector functions f in system (1.2.7) and G in system (1.2.40) be continuous on their respective domains of definition. Assume that the function $V(t, x)$ be continuous on $\mathcal{T}_\tau \times \mathcal{N}$ and such that*

$$D^+ V(t, x) \geq G(t, V(t, x)) \quad \text{for all} \quad (t, x) \in \mathcal{T}_\tau \times \mathcal{N}.$$

Then, if the zero solution of (1.2.7) is unstable and $v(t, x, \alpha)$ is positive definite, the state $x = 0$ of (1.2.7) is unstable.

Let us now see how these results apply to the stability analysis of large scale systems.

Example 1.2.4 (cf. Šiljak [1]) We assume that for the system (1.2.40) there exist functions $v_i \colon R_+ \times R^n \to R_+$ which is continuous on the domain $R_+ \times R^n$, as well as the function $v(t, x, \alpha) = \alpha^T V(t, x)$, $V(t, x) = (v_1(t, x), \ldots, v_m(t, x))^T$, is positive definite and decrescent, functions $\phi_{i1}, \phi_{i2} \in K(KR)$ and $\phi_{i3} \in K$, and functions $\xi_{ij}(t, x) \colon R_+ \times R^n \to R$, with $\sup_{R_+ \times R^n} |\xi_{ij}(t, x)| = \nu_{ij} < +\infty$ for all $i, j = 1, 2, \ldots, m$ such that for all $(t, x) \in R_+ \times R^n$ and $x_i \in R^{n_i}$ the following inequalities

(a) $c_{i1}\phi_{i1}(\|x_i\|) \leq v_i(t, x) \leq c_{i2}\phi_{i2}(\|x_i\|)$, $c_{i1}, c_{i2} > 0$, $i = 1, 2, \ldots, m$;

(b) $D^+ v_i(t, x)|_{(1.2.41)} \leq -\phi_{i3}(\|x_i\|)$;

(c) for any interconnections $e_{ij} \in E$

$$D^+ v_i(t, x)\big|_{(1.2.41)} - D^+ v_i(t, x)\big|_{(1.2.42)} \leq \sum_{j=1}^{m} \bar{e}_{ij} \xi_{ij}(t, x) \phi_{j3}(\|x_j\|)$$

hold.

In view of assumptions (a)–(c) we get

$$D^+v_i(t,x) \leq -\phi_{i3}(\|x_i\|) + \sum_{j=1}^{m} \bar{e}_{ij}\xi_{ij}(t,x)\phi_{j3}(\|x_j\|), \quad i = 1, 2, \ldots, m,$$

and then

$$D^+V(t,x)|_{(1.2.7)} \leq WQ(V(t,x))$$

where

$$Q(V) = \left(\phi_{13}(c_{11}^{-1}\phi_{11}^{-1}(v_1(t,x))),\ \phi_{23}(c_{21}^{-1}\phi_{21}^{-1}(v_2(t,x))),\ \ldots,\right.$$
$$\left.\phi_{m3}(c_{m1}^{-1}\phi_{m1}^{-1}(v_m(t,x)))\right)^{\mathrm{T}}$$

and $W = [w_{ij}]$ with

$$w_{ij} = -\delta_{ij} + \delta_{ij} \sup_{R_+ \times R^n} \xi_{ij}(t,x)$$
$$+ (1 - \delta_{ij}) \max\left(0,\ \sup_{R_+ \times R^n} \xi_{ij}(t,x)\right).$$

Then, the stability properties of zero solution of

(1.2.47) $$\frac{du}{dt} = WQ(u), \quad u(t_0) = u_0 \geq 0,$$

imply the corresponding stability properties of state $x = 0$ of (1.2.7) by Theorem 1.2.13

Further we assume that the vector function $G(t, u)$ has bounded partial derivatives in u.

Designate

$$\left.\frac{\partial G}{\partial u}\right|_{u=0} = P(t), \quad \Phi(t, u) = G(t, u) - P(t)u.$$

Consider a system of comparison equations

(1.2.48) $$\frac{du}{dt} = P(t)u + \Phi(t, u), \quad u(t_0) = u_0 \geq 0,$$

and its linear approximation

(1.2.49) $$\frac{d\xi}{dt} = P(t)\xi, \quad \xi(t_0) = \xi_0 \geq 0.$$

Definition 1.2.17 An $m \times m$ matrix $P(t)$ is a nonautonomous Metzler matrix (M-matrix) if

$$p_{ij}(t) \begin{cases} < 0 & \text{for all } t \in \mathcal{T}_0, \ i = j; \\ \geq 0 & \text{for all } t \in \mathcal{T}_0, \ i \neq j, \ i, j = 1, 2, \ldots, m. \end{cases}$$

Definition 1.2.18 Nonautonomous linear system (1.2.49) is called a *reducible comparison system*, provided that there exists a Liapunov transformation $\xi = Q(t)y$ by means of which it can be reduced to the system

$$\frac{dy}{dt} = By,$$

with a constant M-matrix B. Moreover

$$B = Q^{-1}\left(PQ - \frac{dQ}{dt}\right).$$

Recall that for the *Liapunov transformation*

(1.2.50) $$\xi = Q(t)y$$

there exists $Q^{-1}(t)$ and $Q \in C^1(\mathcal{T}_0, R^{m \times m})$.

Besides, the values

$$k = \sup_{t \geq 0} \|Q(t)\| \quad \text{and} \quad l = \sup_{t \geq 0} \|Q^{-1}(t)\|$$

are finite.

Theorem 1.2.16 *Let for the system (1.2.7) the following conditions hold true*

(1) *there exists a positive definite decreascent vector function $V(t, x)$ such that*

(1.2.51) $$D^+V(t, x) \leq P(t)V(t, x) + \Phi(t, V(t, x)),$$

where $P(t)$ is a nonautonomous M-matrix and $\Phi(t, u)$ is quasi-monotone in u and

$$\lim_{\|u\| \to 0} \frac{\|\Phi(t, u)\|}{\|u\|} = 0 \quad \text{uniformly in } t \geq t_0;$$

(2) *a matrix $P(t)$ reducible in the sense of Liapunov.*

Then the following assertions are valid

(a) *if the matrix B in the system*

(1.2.52) $$\frac{dy}{dt} = By + Q^{-1}\Phi(t, Qy)$$

has all eigenvalues with negative real parts, then the zero solution of comparison system (1.2.48) is uniformly asymptotically stable;

(b) *if the matrix B in the system (1.2 52) has all eigenvalues with negative real parts and in addition $v(t, x, \alpha) \geq a\,\|x\|^2$ for some $a > 0$, then the zero solution of comparison equation (1.2.48) is exponentially stable;*

(c) *if the inequality (1.2.51) holds with a reversed sign and the matrix B in system (1.2.52) has at least one eigenvalue with positive real part, then the zero solution of comparison system (1.2.48) is unstable.*

For the proof of Theorem 1.2.16 see Martynyuk [11].

If in Theorem 1.2.16 inequality (1.2.48) is satisfied with a constant matrix P being an M-matrix, then all assertions of Theorem 1.2.16 remain valid without the Liapunov transformation of system (1.2.48) to (1.3.21) (see Michel and Miller [1]).

1.2.3.3.3 An application of the nonlinear comparison systems This supplement of the direct Liapunov method theory is based on the application of vector function

(1.2.53) $$V(t, x, \eta) = U(t, x)\eta, \quad \eta \in R_+^m,$$

constructed in terms of matrix-valued function (1.2.9) and the comparison principle.

In the space R^m, where $m < n$, a cone $K = \{u \in R^m\colon u_i \geq 0,\ i = 1, 2, \ldots m\}$ with the interior K^0 is considered.

Function $G \in C(R_+^m, R^m)$ is quasi-monotone nondecreasing with respect to a cone K, if $x \overset{K}{\leq} y$ and $\varphi(x - y) = 0$ for some $\varphi \in K_0^*$ imply $\varphi(G(x) - G(y)) \leq 0$, where K_0^* is a cone conjugated with K^0 and $\varphi(x) \geq 0$ for all $x \in K$.

Definition 1.2.19 The autonomous system

(1.2.54) $$\frac{du}{dt} = G(u_1, \ldots, u_m), \quad u(t_0) = u_0 \geq 0$$

is called the *comparison system* for system (1.2.7) iff its maximal solution $u_M(t, u_0)$ and minimal solution $u_m(t, u_0)$ are connected via function (1.2.53) and (1.2.10) with the solutions of initial system (1.2.7) by the inequalities

$$(1.2.55) \qquad u_m(t, u_0) \leq V(t, x(t; t_0, x_0), \eta) \leq u_M(t, u_0),$$

$$(1.2.56) \qquad e^{\mathrm{T}} u_m(t, u_0) \leq v(t, x(t; t_0, x_0), \eta) \leq e^{\mathrm{T}} u_M(t, u_0)$$

for all $t \in R_+$, $e = (1, \ldots, 1)^{\mathrm{T}} \in R_+^m$. Inequalities (1.2.55) are fulfilled component-wise.

The vector function G in system (1.2.54) satisfies the conditions below.

Assumption 1.2.3 Comparison system (1.2.54) is such that:
 (1) the vector-function $G \in C(R_+^m, R^m)$ is quasi-monotone nondecreasing with respect to a cone K (satisfies W^0-conditions);
 (2) the solution of Cauchy problem (1.2.54) is locally unique;
 (3) there exists a neighborhood D of point $u = 0$ such that for all $u \in \overline{D}$, $u \neq 0$, $G(u) \neq 0$ and $G(0) = 0$, where $\overline{D}$ is a closure of domain D.

Assumption 1.2.4 Diagonal elements $v_{ii}(t, x)$ and off-diagonal elements $v_{ij}(t, x)$ of matrix-valued function (1.2.9) satisfy the estimates

$$(1.2.57) \qquad \underline{\gamma}_{ii} \psi_{1i}^2(\|x_i\|) \leq v_{ii}(t, x) \leq \overline{\gamma}_{ii} \psi_{2i}^2(\|x_i\|),$$

$$(1.2.58) \quad \underline{\gamma}_{ij} \psi_{1i}(\|x_i\|) \psi_{1j}(\|x_j\|) \leq v_{ij}(t, x) \leq \overline{\gamma}_{ij} \psi_{2i}(\|x_i\|) \psi_{2j}(\|x_j\|)$$

for all $(t, x_i, x_j) \in R_+ \times R^{n_i} \times R^{n_j}$, $x_i \in R^{n_i}$, $x_j \in R^{n_j}$, $\sum_{i=1}^{m} n_i = n$, $\underline{\gamma}_{ii}, \overline{\gamma}_{ii} > 0$ for all $i = 1, 2, \ldots m$, $\underline{\gamma}_{ij}, \overline{\gamma}_{ij}$ are arbitrary constants and $(\psi_{1i}, \psi_{2i}) \in K(KR)$-class.

Proposition 1.2.10 Under all conditions of Assumption 1.2.4 for the function

$$(1.2.59) \qquad v(t, x, \eta) = \eta^{\mathrm{T}} U(t, x) \eta$$

the bilateral estimate

$$(1.2.60) \qquad \psi_1^{\mathrm{T}}(\|x\|) A \psi_1(\|x\|) \leq v(t, x, \eta) \leq \psi_2^{\mathrm{T}}(\|x\|) B \psi_2(\|x\|)$$

holds true for all $(t, x) \in R_+ \times R^n$, where

$$\psi_1(\|x\|) = (\psi_{11}(\|x_1\|), \ldots, \psi_{1m}(\|x_m\|))^{\mathrm{T}},$$
$$\psi_2(\|x\|) = (\psi_{21}(\|x_1\|), \ldots, \psi_{2m}(\|x_m\|))^{\mathrm{T}},$$
$$A = Y^{\mathrm{T}} \underline{G} Y, \quad B = Y^{\mathrm{T}} \overline{G} Y, \quad Y = \mathrm{diag}\,[\eta_1, \ldots, \eta_m],$$
$$\underline{G} = [\underline{\gamma}_{ij}], \quad \overline{G} = [\overline{\gamma}_{ij}], \quad i, j = 1, 2, \ldots m.$$

Proof This assertion is proved by direct calculation of upper and lower estimates for function (1.2.59) in view of inequalities (1.2.57) and (1.2.58).

Theorem 1.2.17 *Assume that the vector-function f in system (1.2.7) is continuous on $R_+ \times R^n$, the vector-function G in system (1.2.54) satisfies conditions of Assumption 1.2.3 and, moreover,*

(1) *there exist matrix-valued function (1.2.9) with elements (1.2.57), (1.2.58) and vector $\eta \in R_+^m$ such that*

$$D^+ V(t, x, \eta)|_{(1.2.7)} \leq G(V_1(t, x, \eta), \ldots, V_m(t, x, \eta))$$

for all $(t, x) \in R_+ \times R^n$;

(2) *in estimate (1.2.60) the matrices A and B are positive definite;*

(3) *for any $\delta > 0$ the system of inequalities*

$$G_j(\theta_1, \ldots, \theta_m) < 0, \quad j = 1, 2, \ldots m,$$

possesses solution $\theta_1, \ldots, \theta_m$ such that $0 < \theta_j < \delta$ for all $j = 1, 2, \ldots m$.

Then the zero solution of system (1.2.7) is uniformly asymptotically stable.

Proof Under condition (1) of Theorem 1.2.17 it is easy to obtain comparison system (1.2.54). Condition (3) of Theorem 1.2.17 is necessary and sufficient for uniform asymptotic stability of the isolated zero solution of system (1.2.54) (see Martynyuk and Obolenskii [1], and Martynyuk [19]). Then we shall apply the comparison principle and show the validity of assertion of Theorem 1.2.17.

Uniform asymptotic stability of the isolated zero solution of system (1.2.54) implies that for any δ_1: $0 < \delta_1 < r < +\infty$ and $\varepsilon > 0$ there exists $T = T(\varepsilon) > 0$ such that for any $t_1 \geq t_0$ and initial values $e^{\mathrm{T}} u(t_1, u_0) < \delta_1$, $e = (1, \ldots, 1) \in R_+^m$ the estimate

(1.2.61) $$e^{\mathrm{T}} u(t, u_0) \leq \lambda_m(A) a(\varepsilon)$$

holds for all $t \geq t_1 + T(\varepsilon)$, where $0 < \lambda_m(A)$ is a minimal eigenvalue of matrix A from estimate (1.2.60) and function $a(\|x\|) \leq \psi_1^T(\|x\|)\psi_1(\|x\|)$ for all $x \in N$ is a function of class K.

We note that estimate (1.2.60) implies a sequence of inequalities

$$v(t,x,\eta) \leq \psi_2^T(\|x\|)B\psi_2(\|x\|) \leq \lambda_M(B)\psi_2^T(\|x\|)\psi_2(\|x\|) \leq \lambda_M(B)b(\|x\|)$$

for all $(t,x) \in R_+ \times N$, and $b(\|x\|) \geq \psi_2^T(\|x\|)\psi_2(\|x\|)$ for $x \in N$, $b(\|x\|) \in K$-class. We take δ from condition $\lambda_M(B)b(\delta) \leq \delta_1$. If $t_1 \geq t_0$ and $\|x(t_1)\| \leq \delta$, then $u(t_1) = v(t_1,x,\eta) \leq \lambda_M(B)b(\|x(t_1)\|)$, where $\lambda_M(B) > 0$ by condition (2) of Theorem 1.2.17.

When inequality (1.2.61) is fulfilled for all $t \geq t_1 + T(\varepsilon)$ by comparison principle we have

$$v(t,x,\eta) = \eta^T V(t,x,\eta) \leq e^T u(t,u).$$

Therefore $\lambda_m(A)a(\|x(t)\|) < \lambda_m(A)a(\varepsilon)$ for all $t \geq t_1 + T(\varepsilon)$ and consequently $\|x(t)\| < \varepsilon$ for all $t \geq t_1 + T(\varepsilon)$, if $t_1 \geq t_0$ than $\|x(t_1)\| \leq \delta$. This proves Theorem 1.2.17.

Corollary 1.2.10 Assume that in system (1.2.7) vector function $f \in C(R_+ \times R^n, R^n)$, $f(t,0) = 0$, and moreover

(1) there exist matrix-valued function (1.2.9) with elements (1.2.57) and (1.2.58), vector $\eta \in R_+^m$ and constant $m \times m$ matrix P with nonnegative off-diagonal elements such that

$$D^+V(t,x,\eta)|_{(1.2.7)} \leq PV(t,x,\eta)$$

for all $(t,x) \in R_+ \times R^n$;

(2) in estimate (1.2.60) the matrices A and B are positive definite;

(3) for any $\delta > 0$ the system of inequalities

$$\sum_{j=1}^{m} p_{ij}\theta_j < 0, \quad i = 1, 2, \ldots m,$$

possesses solution $\theta_1, \ldots, \theta_m$ such that $0 < \theta_j < \delta$ for all $j = 1, 2, \ldots m$.

Then the zero solution of system (1.2.7) is uniformly asymptotically stable.

Corollary 1.2.11 Let system (1.2.54) describe perturbed motion of a real system with a finite number of degrees of freedom, vector function G satisfy conditions of Assumption 1.2.3 and condition (3) of Theorem 1.2.17 be satisfied. Then and only then the isolated zero solution of system (1.2.54) is uniformly asymptotically stable.

Example 1.2.5 Let us consider the longitudinal motion of an aircraft which can be represented by the system of differential equations

(1.2.62)
$$\frac{dx_i}{dt} = -\rho_i x_i + \sigma, \quad i = 1, 2, \ldots, n,$$
$$\frac{d\sigma}{dt} = \sum_{i=1}^{n} a_i x_i - p\sigma - f(\sigma),$$

where $\rho_i > 0$, $p > 0$, $\sigma f(\sigma) > 0$ for $\sigma \neq 0$, $f(0) = 0$.

Let us illustrate estimations of the domain of the parameters values for which the state $(x = 0, \sigma = 0)$ of the system (1.2.62) is asymptotically stable.

After some transformations of system (1.2.62) we get the estimation

(1.2.63)
$$\sum_{i=1}^{n} \frac{|a_i|}{\rho_i} \leq p$$

which determines restrictions on parameters for which the state $(y = 0, z = 0)$ is asymptotically stable.

Remark 1.2.4 For the system of equations (1.2.62) for $n = 4$ Piontkovskii and Rutkovskaya [1] determined the following estimation of the domain of parameters

(1.2.64)
$$\frac{1}{\left(\min\limits_{i}(\rho_i)\right)^2} \sum_{i=1}^{4} |a_i| < \left(\frac{p}{4}\right)^2.$$

It is obvious that both for $n = 4$ and $n > 4$ the estimation (1.2.63) defines a larger domain as compared with that ensured by the estimation (1.2.64).

Remark 1.2.5 Though the condition (3) of Theorem 1.2.17 is necessary and sufficient condition of uniform asymptotic stability of the isolated zero solution of comparison system (1.2.54) the accuracy of estimates of system (1.2.7) domain of parameters ensuring uniform asymptotic stability of its zero solution depends on the accuracy of component majorants of total derivative of vector function (1.2.53) along solutions of system (1.2.7). Below we present an example illustrating our discussion.

Example 1.2.6 Consider the system

(1.2.65)
$$\dot{x}_1 = -x_1(x_1^2 + x_2^2) - x_1^2 x_2,$$
$$\dot{x}_2 = 100\, x_1^3 - 100\, x_2(x_1^2 + x_2^2) + a\, x_3^9,$$
$$\dot{x}_3 = -x_3^9 + b\|\hat{x}\|^3,$$

where a and b are some constant values and $\hat{x} = (x_1, x_2)^{\mathrm{T}}$. Let the components of vector function V be of the form

(1.2.66)
$$v_1 = 100\, \frac{x_1^2}{2} + \frac{x_2^2}{2}, \quad v_2 = x_3^\gamma,$$

where γ is a rational number with odd denominator and even numerator. By means of function (1.2.66) it is easy to obtain the comparison system

(1.2.67)
$$\frac{du_1}{dt} = -0{,}04\, u_1^2 + |a|(2u_1)^{1/2} u_2^{9/8},$$
$$\frac{du_2}{dt} = -\gamma u_2^{1+\frac{8}{\gamma}} + |b|\gamma (2u_1)^{3/2} u_2^{1-\frac{1}{\gamma}}.$$

Condition (3) of Theorem 1.2.17 is reduced to the following. The system of inequalities

(1.2.68)
$$-0{,}04\theta_1^2 + |a|(2\theta_1)^{1/2} \theta_2^{9/\gamma} < 0,$$
$$-\gamma \theta_2^{1+\frac{8}{\gamma}} + |b|\gamma(2\theta_1)^{3/2} \theta_2^{1-\frac{1}{\gamma}} < 0$$

possesses positive solution in some neighbourhood of point $(u = 0) \in R_+^2$. It the parameters a and b are such that $|ab| < 0{,}01$, then this condition holds true and the zero solution of system (1.2.65) is uniformly asymptotically stable.

If for functions (1.2.66) we set $\gamma = 4$, then

(1.2.69)
$$\dot{v}_1(x_1, x_2) = -100\|\hat{x}\|^4 + |a|\, \|\hat{x}\|\, |x_3|^3,$$
$$\dot{v}_2(x_3) = -4|x_3|^{12} + 4|b|\, \|\hat{x}\|^3 |x|^3$$

and for uniform asymptotic stability of its zero solution it is necessary and sufficient that the system of inequalities

(1.2.70)
$$-100\,\theta_1^4 + |a|\,\theta_1 \theta_2^9 < 0,$$
$$-4\,\theta_2^{12} + 4|b|\,\theta_1^3 \theta_2^3 < 0$$

has a positive solution. We find from system (1.2.70) that this condition is satisfied for $|ab| < 100$.

Thus, by choosing an appropriate vector-function (1.2.66) and majorant one can essentially enlarge the domain of parameter values ensuring uniform asymptotic stability of zero solution of the initial system.

1.3 Stability Conditions of a Class of Large Scale Systems

We consider a system with a finite number of degrees of freedom whose motion is described by the equation

$$(1.3.1) \qquad \frac{dx_i}{dt} = f_i(x_i) + g_i(t, x_1, \ldots, x_m), \qquad i = 1, 2, \ldots, m$$

where $x_i \in R^{n_i}$, $t \in \mathcal{T}_\tau$, $\mathcal{T}_\tau = [\tau, +\infty)$, $f_i \in C(R^{n_i}, R^{n_i})$, $g_i \in C(\mathcal{T}_\tau \times R^{n_1} \times \cdots \times R^{n_m}, R^{n_i})$.

Introduce the designation

$$(1.3.2) \qquad G_i(t, x) = g_i(t, x_1, \ldots, x_m) - \sum_{j=1, j \neq i}^{m} g_{ij}(t, x_i, x_j),$$

where $g_{ij}(t, x_i, x_j) = g_i(t, 0, \ldots, x_i, \ldots, x_j, \ldots, 0)$ for all $i \neq j$; $i, j = 1, 2, \ldots, m$. Taking into consideration (1.3.2) system (1.3.1) is rewritten as

$$(1.3.3) \qquad \frac{dx_i}{dt} = f_i(x_i) + \sum_{j=1, j \neq i}^{m} g_{ij}(t, x_i, x_j) + G_i(t, x), \qquad i = 1, 2, \ldots, m.$$

Actually equations (1.3.3) describe the class of large-scale nonlinear nonautonomously connected systems. It is of interest to extend the method of matrix Liapunov functions to this class of equations in view of the new method of construction of nondiagonal elements of matrix-valued functions.

Assumption 1.3.1 There exist open connected neighborhoods $\mathcal{N}_i \subseteq R^{n_i}$ of the equilibriums state $x_i = 0$, functions $v_{ii} \in C^1(R^{n_i}, R_+)$, the comparison functions φ_{i1}, φ_{i2} and ψ_i of class $K(KR)$ and real numbers $\underline{c}_{ii} > 0$, $\bar{c}_{ii} > 0$ and γ_{ii} such that

(1) $v_{ii}(x_i) = 0$ for all $(x_i = 0) \in \mathcal{N}_i$;
(2) $\underline{c}_{ii} \varphi_{i1}^2(\|x_i\|) \leq v_{ii}(x_i) \leq \bar{c}_{ii} \varphi_{i2}^2(\|x_i\|)$;
(3) $(D_{x_i} v_{ii}(x_i))^T f_i(x_i) \leq \gamma_{ii} \psi_i^2(\|x_i\|)$ for all $x_i \in \mathcal{N}_i$, $i = 1, 2, \ldots, m$.

It is clear that under conditions of Assumption 1.3.1 the equilibrium states $x_i = 0$ of nonlinear isolated subsystems

$$\frac{dx_i}{dt} = f_i(x_i), \quad i = 1, 2, \ldots, m \tag{1.3.4}$$

are

(a) *uniformly asymptotically stable in the whole*, if $\gamma_{ii} < 0$ and $(\varphi_{i1}, \varphi_{i2}, \psi_i) \in KR$-class;
(b) *stable*, if $\gamma_{ii} = 0$ and $(\varphi_{i1}, \varphi_{i2}) \in K$-class;
(c) *unstable*, if $\gamma_{ii} > 0$ and $(\varphi_{i1}, \varphi_{i2}, \psi_i) \in K$-class.

The approach proposed in this section takes large scale systems (1.3.3) into consideration, subsystems (1.3.4) having various dynamical properties specified by conditions of Assumption 1.3.1.

Assumption 1.3.2 There exist open connected neighborhoods $\mathcal{N}_i \subseteq R^{n_i}$ of the equilibrium states $x_i = 0$, functions $v_{ij} \in C^{1,1,1}(\mathcal{T}_\tau \times R^{n_i} \times R^{n_j}, R)$, comparison functions $\varphi_{i1}, \varphi_{i2} \in K(KR)$, positive constants $(\eta_1, \ldots, \eta_m)^T \in R^m$, $\eta_i > 0$ and arbitrary constants $\underline{c}_{ij}, \bar{c}_{ij}$, $i, j = 1, 2, \ldots, m$, $i \neq j$ such that

(1) $v_{ij}(t, x_i, x_j) = 0$ for all $(x_i, x_j) = 0 \in \mathcal{N}_i \times \mathcal{N}_j$, $t \in \mathcal{T}_\tau$, $i, j = 1, 2, \ldots, m$, $(i \neq j)$;

(2) $\underline{c}_{ij}\varphi_{i1}(\|x_i\|)\varphi_{j1}(\|x_j\|) \leq v_{ij}(t, x_i, x_j) \leq \bar{c}_{ij}\varphi_{i2}(\|x_i\|)\varphi_{j2}(\|x_j\|)$ for all $(t, x_i, x_j) \in \mathcal{T}_\tau \times \mathcal{N}_i \times \mathcal{N}_j$, $i \neq j$,

where

$$\begin{aligned}
& D_t v_{ij}(t, x_i, x_j) + (D_{x_i} v_{ij}(t, x_i, x_j))^T f_i(x_i) \\
& + (D_{x_j} v_{ij}(t, x_i, x_j))^T f_j(x_j) + \frac{\eta_i}{2\eta_j}(D_{x_i} v_{ii}(x_i))^T g_{ij}(t, x_i, x_j) \\
& + \frac{\eta_j}{2\eta_i}(D_{x_j} v_{jj}(x_j))^T g_{ji}(t, x_i, x_j) = 0.
\end{aligned} \tag{1.3.5}$$

In particular cases equations (1.3.5) are transformed into the systems of ordinary differential equations or algebraic equations whose solutions can be constructed analytically.

Assumption 1.3.3 There exist open connected neighborhoods $\mathcal{N}_i \subseteq R^{n_i}$ of the equilibrium states $x_i = 0$, comparison functions $\psi \in K(KR)$,

$i = 1, 2, \ldots, m$, real numbers α_{ij}^1, α_{ij}^2, α_{ij}^3, ν_{ki}^1, ν_{kij}^1, μ_{kij}^1 and μ_{kij}^2, $i, j, k = 1, 2, \ldots, m$, such that

(1) $(D_{x_i} v_{ii}(x_i))^{\mathrm{T}} G_i(t, x) \leq \psi_i(\|x_i\|) \sum_{k=1}^{m} \nu_{ki}^1 \psi_k(\|x_k\|) + R_1(t, \psi)$
for all $(t, x_i, x_j) \in \mathcal{T}_\tau \times \mathcal{N}_i \times \mathcal{N}_j$;

(2) $(D_{x_i} v_{ij}(t, \cdot))^{\mathrm{T}} g_{ij}(t, x_i, x_j) \leq \alpha_{ij}^1 \psi_i^2(\|x_i\|) + \alpha_{ij}^2 \psi_i(\|x_i\|) \psi_j(\|x_j\|) + \alpha_{ij}^3 \psi_j^2(\|x_j\|) + R_2(t, \psi)$ for all $(t, x_i, x_j) \in \mathcal{T}_\tau \times \mathcal{N}_i \times \mathcal{N}_j$;

(3) $(D_{x_i} v_{ij}(t, \cdot))^{\mathrm{T}} G_i(t, x) \leq \psi_j(\|x_j\|) \sum_{k=1}^{m} \nu_{ijk}^2 \psi_k(\|x_k\|) + R_3(t, \psi)$
for all $(t, x_i, x_j) \in \mathcal{T}_\tau \times \mathcal{N}_i \times \mathcal{N}_j$;

(4) $(D_{x_i} v_{ij}(t, \cdot))^{\mathrm{T}} g_{ik}(t, x_i, x_k) \leq \psi_j(\|x_j\|)(\mu_{ijk}^1 \psi_k(\|x_k\|) + \mu_{ijk}^2 \psi_i(\|x_i\|)) + R_4(t, \psi)$ for all $(t, x_i, x_j) \in \mathcal{T}_\tau \times \mathcal{N}_i \times \mathcal{N}_j$.

Here $R_s(t, \psi)$ are polynomials in $\psi = (\psi_1(\|x_1\|), \ldots, \psi_m(\|x_m\|))$ in a power higher than three, $R_s(t, 0) = 0$, $s = 1, \ldots, 4$.

Under conditions (2) of Assumptions 1.3.1 and 1.3.2 it is easy to establish for function

$$(1.3.6) \qquad v(t, x, \eta) = \eta^{\mathrm{T}} U(t, x) \eta = \sum_{i,j=1}^{m} v_{ij}(t, \cdot) \eta_i \eta_j$$

the bilateral estimate

$$(1.3.7) \qquad u_1^{\mathrm{T}} H^{\mathrm{T}} \underline{C} H u_1 \leq v(t, x, \eta) \leq u_2^{\mathrm{T}} H^{\mathrm{T}} \overline{C} H u_2,$$

where

$$u_1 = (\varphi_{11}(\|x_1\|), \ldots, \varphi_{m1}(\|x_m\|))^{\mathrm{T}},$$
$$u_2 = (\varphi_{12}(\|x_1\|), \ldots, \varphi_{m2}(\|x_m\|))^{\mathrm{T}}$$

which holds true for all $(t, x) \in \mathcal{T}_\tau \times \mathcal{N}$, $\mathcal{N} = \mathcal{N}_1 \times \cdots \times \mathcal{N}_m$.

Based on conditions (3) of Assumptions 1.3.1, 1.3.2 and conditions (1)–(4) of Assumption 1.3.3 it is easy to establish the inequality estimating the auxiliary function variation along solutions of system (1.3.3). This estimate reads

$$(1.3.8) \qquad Dv(t, x, \eta)\big|_{(1.3.3)} \leq u_3^{\mathrm{T}} M u_3 + \widetilde{R}(t, \psi, \eta),$$

where $u_3 = (\psi_1(\|x_1\|), \ldots, \psi_m(\|x_m\|))$, $\widetilde{R}(t, \psi, \eta)$ is a polynomial in ψ in a power higher than three, $\widetilde{R}(t, 0, \eta) = 0$, and holds for all $(t, x) \in \mathcal{T}_\tau \times \mathcal{N}$.

Elements σ_{ij} of matrix M in the inequality (1.3.8) have the following structure

$$\sigma_{ii} = \eta_i^2 \gamma_{ii} + \eta_i^2 \nu_{ii} + \sum_{k=1, k\neq i}^{m} (\eta_k \eta_i \nu_{kii}^2 + \eta_i^2 \nu_{kii}^2) + 2 \sum_{j=1, j\neq i}^{m} \eta_i \eta_j (\alpha_{ij}^1 + \alpha_{ji}^3);$$

$$\sigma_{ij} = \frac{1}{2}(\eta_i^2 \nu_{ji}^1 + \eta_j^2 \nu_{ij}^1) + \sum_{k=1, k\neq j}^{m} \eta_k \eta_j \nu_{kij}^2 + \sum_{k=1, k\neq i}^{m} \eta_i \eta_j \nu_{kij}^2$$

$$+ \eta_i \eta_j (\alpha_{ij}^2 + \alpha_{ji}^2) + \sum_{\substack{k=1, k\neq i, \\ k\neq j}}^{m} (\eta_k \eta_j \mu_{kji}^1 + \eta_i \eta_j \mu_{ijk}^2 + \eta_i \eta_k \mu_{kij}^1 + \eta_i \eta_j \mu_{jik}^2),$$

$$i = 1, 2, \ldots, m, \quad i \neq j.$$

Sufficient criteria of various types of stability of the equilibrium state $x = 0$ of system (1.3.3) are formulated in terms of the sign definiteness of matrices $\underline{C}$, $\overline{C}$ and M from estimates (1.3.7), (1.3.8). We shall show that the following assertion is valid.

Theorem 1.3.1 *Assume that the perturbed motion equations are such that all conditions of Assumptions 1.3.1 – 1.3.3 are fulfilled and moreover*

(1) *matrices $\underline{C}$ and $\overline{C}$ in estimate (1.3.7) are positive definite;*
(2) *matrix M in inequality (1.3.8) is negative semi-definite and for all $(t, x) \in \mathcal{T}_\tau \times \mathcal{N}_\delta$, $\mathcal{N}_\delta = \{x \in R^n : \|x\| < \delta, \ \delta > 0\}$, the estimate*

$$|\widetilde{R}(t, \psi, \eta)| \leq |u_3^T M u_3|$$

is fulfilled;
(3) *matrix M in inequality (1.3.8) is negative definite.*

Then the equilibrium state $x = 0$ of system (1.3.3) is

(a) *uniformly stable for conditions (1) – (2) of the Theorem 1.3.1 and*
(b) *uniformly asymptotically stable for conditions (1), (3) of the Theorem 1.3.1.*

If, additionally, in conditions of Assumptions 1.3.1 – 1.3.3 all estimates are satisfied for $\mathcal{N}_i = R^{n_i}$, $R_k(t, \psi) = 0$, $k = 1, \ldots, 4$, and comparison functions $(\varphi_{i1}, \varphi_{i2}) \in KR$-class, then the equilibrium state of system

1.4 Large-Scale Linear Systems

Linear systems of perturbed motion equations are of essential interest in the description of various phenomena in physical and technical systems. General theory of such systems is developed well because in some cases such systems can be integrated precisely. On the other hand systems of this type are the first approximation of quasilinear equations in the investigation of which the information on the properties of the first approximation system is incorporated. For this class of systems of equations the construction of the Liapunov functions remains the focus of attention of many researchers.

1.4.1 Non-autonomous linear systems Consider a large-scale system whose motion is described by the equations

$$(1.4.1) \qquad \frac{dx_i}{dt} = A_{ii}x_i + \sum_{j=1, j \neq i}^{m} A_{ij}(t)x_j, \quad i = 1, 2, \ldots, m.$$

Here the state vectors $x_i \in R^{n_i}$ and $A_{ii} \in R^{n_i \times n_i}$ are constant matrices for all $i = 1, 2, \ldots, m$; $A_{ij}(t) \in C(R, R^{n_i \times n_j})$, $i, j = 1, 2, \ldots, m$, $i \neq j$, $n = \sum_{i=1}^{m} n_i$.

For the independent subsystems

$$(1.4.2) \qquad \frac{dx_i}{dt} = A_{ii}x_i, \quad i = 1, 2, \ldots, m,$$

the auxiliary functions $v_{ii}(x_i)$ are constructed as the quadratic forms

$$(1.4.3) \qquad v_{ii}(x_i) = x_i^T P_{ii} x_i, \quad i = 1, 2, 3,$$

whose constant matrices P_{ii} are determined by the algebraic Liapunov equations

$$(1.4.4) \qquad A_{ii}^T P_{ii} + P_{ii} A_{ii} = -G_{ii}, \quad i = 1, 2, \ldots, m,$$

where G_{ii} are pre-assigned matrices of constant sign. For the construction of non-diagonal elements $v_{ij}(t, x_i, x_j)$ of the matrix-valued function $U(t, x)$ we apply equation (1.3.5). Note that for the system (1.4.1)

$$f_i(x_i) = A_{ii}x_i, \quad f_j(x_j) = A_{jj}x_j,$$
$$g_{ij}(t, x_i, x_j) = A_{ij}(t)x_j, \quad g_{ji}(t, x_i, x_j) = A_{ji}(t)x_j, \quad G_i(t, x) = 0.$$

Suppose that at least one of the matrices A_{ij} or A_{ji} is not equal to constant. Then we determine function $v_{ij}(t, x_i, x_j)$ as

$$(1.4.5) \qquad v_{ij}(t, x_i, x_j) = v_{ji}(t, x_j, x_i) = x_i^T P_{ij}(t) x_j,$$

where $P_{ij} \in C^1(R, R^{n_i \times n_j})$.

Since for the bilinear forms (1.4.5)

$$D_t v_{ij}(t, x_i, x_j) = x_i^T \frac{dP_{ij}}{dt} x_j, \quad D_{x_i} v_{ij}(t, x_i, x_j) = x_j^T P_{ij}(t)^T,$$
$$D_{x_j} v_{ij}(t, x_i, x_j) = x_i^T P_{ij}(t),$$

the equation (1.3.5) becomes

$$x_i^T \left(\frac{dP_{ij}}{dt} + A_{ii}^T P_{ij} + P_{ij} A_{jj} + \frac{\eta_i}{\eta_j} P_{ii} A_{ij}(t) + \frac{\eta_j}{\eta_i} A_{ji}^T(t) P_{jj} \right) x_j = 0.$$

For determination of matrices P_{ij} this correlation yields a system of matrix differential equations

$$(1.4.6) \qquad \frac{dP_{ij}}{dt} + A_{ii}^T P_{ij} + P_{ij} A_{jj} = -\frac{\eta_i}{\eta_j} P_{ii} A_{ij}(t) - \frac{\eta_j}{\eta_i} A_{ji}^T(t) P_{jj}$$
$$i, j = 1, 2, \ldots, m, \quad i \neq j.$$

Note that equations (1.4.6) can be solved in the explicit form. To this end we consider a linear operator (general information on linear operators can be found, for example, in Daletskii and Krene [1])

$$F_{ij} \colon R^{n_i \times n_j} \to R^{n_i \times n_j}, \quad F_{ij} X = A_{ii}^T X + X A_{jj}.$$

Equation (1.4.6) can be represented as

$$\frac{dP_{ij}}{dt} + F_{ij} P_{ij} = -\frac{\eta_i}{\eta_j} P_{ii} A_{ij}(t) - \frac{\eta_j}{\eta_i} A_{ji}^T(t) P_{ii}, \quad i \neq j.$$

Consider the homogeneous equations

(1.4.7) $$\frac{dP_{ij}}{dt} + F_{ij}P_{ij} = 0,$$

whose general solution is presented as

$$P_{ij}(t) = \exp\{-F_{ij}t\}C_{ij},$$

where C_{ij} is a constant $n_i \times n_j$ matrix and $\exp\{-F_{ij}t\} = \sum_{k=0}^{\infty} \frac{(-1)^k F_{ij}^k t^k}{k!}$ is an operator exponent.

To find the solution of equation (1.4.6) the method of variation of a constant is applied. Solution of equation (1.4.6) is presented in the form

(1.4.8) $$P_{ij}(t) = \exp\{-F_{ij}t\}C_{ij}(t),$$

where $C_{ij} \in C^1(R, R^{n_1 \times n_2})$ and $C_{ij}(0) = 0$. Substituting (1.4.8) into (1.4.6) yields

$$\frac{dC_{ij}}{dt} = -\exp\{F_{ij}t\}\left(\frac{\eta_i}{\eta_j}P_{ii}A_{ij} + \frac{\eta_j}{\eta_i}A_{ji}^T P_{ii}\right), \quad i \neq j.$$

Integrating the last correlation from 0 to t we determine a partial solution of equation (1.4.6)

(1.4.9.)
$$P_{ij}(t) = -\int_0^t \exp\{-F_{ij}(t-\tau)\}\left(\frac{\eta_i}{\eta_j}P_{ii}A_{ij}(\tau) + \frac{\eta_j}{\eta_i}A_{ji}^T(\tau)P_{jj}\right)d\tau, \quad i \neq j.$$

We establish estimates for the function

$$v(t, x, \eta) = \eta^T U(t, x)\eta = \sum_{i,j=1}^m v_{ij}(t,.)\eta_i\eta_j,$$

where

$$U(t, x) = \begin{pmatrix} v_{11}(x_1) & \cdots & v_{1m}(t, x_1, x_m) \\ \cdots\cdots\cdots\cdots\cdots\cdots\cdots\cdots\cdots\cdots\cdots \\ v_{1m}(t, x_1, x_m) & \cdots & v_{mm}(x_m) \end{pmatrix}.$$

Introduce the designations $\bar{c}_{ii} = \lambda_M(P_{ii})$ and $\underline{c}_{ii} = \lambda_m(P_{ii})$ and assuming $\sup_{t \geq 0} \|P_{ij}(t)\| < \infty$ denote $\bar{c}_{ij} = \sup_{t \geq 0} \|P_{ij}(t)\|$, $\underline{c}_{ij} = -\bar{c}_{ij}$.

Since for the forms (1.4.3) and (1.4.5) the estimates
(1.4.10)
$$\lambda_m(P_{ii})\|x_i\|^2 \le v_{ii}(x_i) \le \lambda_M(P_{ii})\|x_i\|^2, \quad x_i \in R^{n_i};$$
$$-\bar{c}_{ij}\|x_i\|\|x_j\| \le v_{ij}(t, x_i, x_j) \le \bar{c}_{ij}\|x_i\|\|x_j\|, \quad (x_i, x_j) \in R^{n_i} \times R^{n_j},$$

are valid, for the function $v(t, x, \eta) = \eta^{\mathrm{T}} U(t,x) \eta$

(1.4.11) $\qquad w^{\mathrm{T}} H^{\mathrm{T}} \underline{C} H w \le v(t,x,\eta) \le w^{\mathrm{T}} H^{\mathrm{T}} \overline{C} H w \quad \text{for all} \quad x \in R^n,$

where $w = (\|x_1\|, \ldots, \|x_m\|)^{\mathrm{T}}$, $H = \mathrm{diag}\,(\eta_1, \eta_2, \ldots, \eta_m)$, $\overline{C} = [\bar{c}_{ij}]_{i,j=1}^m$, $\underline{C} = [\underline{c}_{ij}]_{i,j=1}^m$.

In order to estimate the derivative of function $v(t, x, \eta)$ along solutions of system (1.4.1) we calculate the constants from Assumption 1.3.3

$$\alpha_{ij}^1 = \alpha_{ij}^2 = 0, \quad \alpha_{ij}^3(t) = \lambda_M(A_{ij}^{\mathrm{T}}(t) P_{ij}(t) + P_{ij}^{\mathrm{T}}(t) A_{ij}(t)),$$
$$\nu_{ki}^1 = \nu_{ijk}^2 = 0, \quad \nu_{ijk}^1(t) = \lambda_M^{1/2}[(P_{ij}^{\mathrm{T}}(t) A_{ik}(t))(P_{ij}^{\mathrm{T}}(t) A_{ik}(t))], \quad \mu_{ijk}^2 = 0.$$

Therefore the elements σ_{ij} of matrix $M(t)$ in estimate (1.3.8) for system (1.4.1) have the structure

$$\sigma_{ii}(t) = -\eta_i^2 \lambda_m(G_{ii}) + 2 \sum_{j=1, j \ne i}^m \eta_i \eta_j \alpha_{ij}^3, \quad i = 1, \ldots, m,$$

$$\sigma_{ij}(t) = \sum_{k=1, k \ne i, k \ne j}^m (\eta_k \eta_j \nu_{ijk}^1 + \eta_i \eta_k \nu_{kij}^1), \quad i, j = 1, \ldots, m, \quad i \ne j.$$

Consequently, the variation of function $Dv(t, x, \eta)$ along solutions of system (1.4.1) is estimated by the inequality

(1.4.12) $\qquad Dv(t,x,\eta)\Big|_{(1.4.1)} \le w^{\mathrm{T}} M(t) w$

for all $(x_1, \ldots, x_m) \in R^{n_1} \times \cdots \times R^{n_m}$.

Remark 1.4.1 In the partial case when matrices A_{ij} and A_{ji} do not depend on t, it is reasonable to choose $P_{ij}(t) = \mathrm{const}$. Then equation (1.4.6) becomes

(1.4.13) $\qquad A_{ii} P_{ij} + P_{ij} A_{jj} = -\dfrac{\eta_i}{\eta_j} P_{ii} A_{ij} - \dfrac{\eta_j}{\eta_i} A_{ji}^{\mathrm{T}} P_{jj}$

or in the operator form

$$F_{ij}P_{ij} = -\frac{\eta_i}{\eta_j} P_{ii}A_{ij} - \frac{\eta_j}{\eta_i} A_{ji}^{\mathrm{T}}P_{jj}.$$

Therefore for the equation (1.4.13) to have a unique solution it is necessary and sufficient that the operator F_{ij} be nondegenerate.

It is known (see Daletskii and Krene [1]) that the set of eigenvalues of the operator F_{ij} consists of the numbers $\lambda_k(A_{ii}) + \lambda_l(A_{jj})$, where $\lambda_k(\cdot)$ is an eigenvalue of the corresponding matrix. Based on these speculations one can formulate the following result.

For the equation (1.4.13) to have a unique solution it is necessary and sufficient that

$$\lambda_k(A_{ii}) + \lambda_l(A_{jj}) \neq 0 \quad \text{for all} \quad k, l,$$

and this solution can be presented as

$$P_{ij} = -F_{ij}^{-1}\left(\frac{\eta_i}{\eta_j} P_{ii}A_{ij} + \frac{\eta_j}{\eta_i} A_{ji}^{\mathrm{T}}P_{jj}\right).$$

This result is summed up as follows.

Theorem 1.4.1 *Assume that for system (1.4.1) the following conditions are satisfied*

(1) *the sign-definite matrices P_{ii}, $i = 1, 2, 3$, are the solution of algebraic equations (1.4.4);*
(2) *the bounded matrices $P_{ij}(t)$ for all $i, j = 1, 2, \ldots, m$, $i \neq j$, are the solution of matrix differential equations (1.4.6);*
(3) *matrices $\overline{C}$ and $\underline{C}$ in estimate (1.4.11) are positive definite;*
(4) *there is a constant $m \times m$ matrix $\widetilde{M} \geq M(t)$ for all $t \geq t_0$ in estimate (1.4.12) which is negative semi-definite (negative definite).*

Then the equilibrium state $x = 0$ of system (1.4.1) is uniformly stable in the whole (uniformly asymptotically stable in the whole).

Example 1.4.1 Consider the motion of two non-autonomously connected oscillators whose behaviour is described by the equations

(1.4.14)
$$\frac{dx_1}{dt} = \gamma_1 x_2 + v\cos\omega t\, y_1 - v\sin\omega t\, y_2,$$
$$\frac{dx_2}{dt} = -\gamma_1 x_1 + v\sin\omega t\, y_1 + v\cos\omega t\, y_2,$$
$$\frac{dy_1}{dt} = \gamma_2 y_2 + v\cos\omega t\, x_1 + v\sin\omega t\, x_2,$$
$$\frac{dy_2}{dt} = -\gamma_2 y_2 + v\cos\omega t\, x_2 - v\sin\omega t\, x_1,$$

where γ_1, γ_2, v, ω, $\omega + \gamma_1 - \gamma_2 \neq 0$ are some constants.

It is easy to prove that the motion of nonautonomously connected oscillators (1.4.14) is uniformly stable in the whole, if

$$|v| < \frac{1}{2}|\omega + \gamma_1 - \gamma_2|.$$

For the details see Martynyuk [16], and Martynyuk and Slyn'ko [1].

1.4.2 Time invariant linear systems
Assume that in the system

(1.4.15)
$$\frac{dx_1}{dt} = A_{11}x_1 + A_{12}x_2 + A_{13}x_3,$$
$$\frac{dx_2}{dt} = A_{21}x_1 + A_{22}x_2 + A_{23}x_3,$$
$$\frac{dx_3}{dt} = A_{31}x_1 + A_{32}x_2 + A_{33}x_3,$$

the state vectors $x_i \in R^{n_i}$, $i = 1, 2, 3$, and $A_{ij} \in R^{n_i \times n_j}$ are constant matrices for all $i, j = 1, 2, 3$.

For the independent systems

(1.4.16)
$$\frac{dx_i}{dt} = A_{ii}x_i, \quad i = 1, 2, 3,$$

we construct auxiliary functions $v_{ii}(x_i)$ as the quadratic forms

(1.4.17)
$$v_{ii}(x_i) = x_i^T P_{ii} x_i, \quad i = 1, 2, 3,$$

whose matrices P_{ii} are determined by

(1.4.18)
$$A_{ii}^T P_{ii} + P_{ii} A_{ii} = -G_{ii}, \quad i = 1, 2, 3,$$

where G_{ii} are prescribed matrices of definite sign.

In order to construct non-diagonal elements $v_{ij}(x_i, x_j)$ of the matrix-valued function $U(x)$ we employ equation (1.3.5). Note that for system (1.4.15)

$$f_i(x_i) = A_{ii}x_i, \quad f_j(x_j) = A_{jj}x_j,$$
$$g_{ij}(x_i, x_j) = A_{ij}x_j, \quad G_i(t, x) = 0, \quad i = 1, 2, 3.$$

Since for the bilinear forms

(1.4.19) $$v_{ij}(x_i, x_j) = v_{ji}(x_j, x_i) = x_i^T P_{ij} x_j$$

the correlations

$$D_{x_i} v_{ij}(x_i, x_j) = x_j^T P_{ij}^T, \quad D_{x_j} v_{ij}(x_i, x_j) = x_i^T P_{ij}$$

are true, equation (1.3.5) becomes

$$x_i^T \left(A_{ii}^T P_{ij} + P_{ij} A_{jj} + \frac{\eta_i}{\eta_j} P_{ii} A_{ij} + \frac{\eta_j}{\eta_i} A_{ji}^T P_{ii} \right) x_j = 0.$$

From this correlation for determining matrices P_{ij} we get the system of algebraic equations

(1.4.20)
$$A_{ii} P_{ij} + P_{ij} A_{jj} = -\frac{\eta_i}{\eta_j} P_{ii} A_{ij} - \frac{\eta_j}{\eta_i} A_{ji}^T P_{ii},$$
$$i \neq j, \quad i, j = 1, 2, 3.$$

Since for (1.4.17) and (1.4.19) the estimates

$$v_{ii}(x_i) \geq \lambda_m(P_{ii}) \|x_i\|^2, \quad x_i \in R^{n_i};$$
$$v_{ij}(x_i, x_j) \geq -\lambda_M^{1/2}(P_{ij} P_{ij}^T) \|x_i\| \|x_j\|, \quad (x_i, x_j) \in R^{n_i} \times R^{n_j},$$

hold true, for function $v(x, \eta) = \eta^T U(x) \eta$ the inequality

(1.4.21) $$w^T H^T C H w \leq v(x, \eta)$$

is satisfied for all $x \in R^n$, $w = (\|x_1\|, \|x_2\|, \|x_3\|)^T$ and the matrix

$$C = \begin{pmatrix} \lambda_m(P_{11}) & -\lambda_M^{1/2}(P_{12} P_{12}^T) & -\lambda_M^{1/2}(P_{13} P_{13}^T) \\ -\lambda_M^{1/2}(P_{12} P_{12}^T) & \lambda_m(P_{22}) & -\lambda_M^{1/2}(P_{23} P_{23}^T) \\ -\lambda_M^{1/2}(P_{13} P_{13}^T) & -\lambda_M^{1/2}(P_{23} P_{23}^T) & \lambda_m(P_{33}) \end{pmatrix}.$$

For system (1.4.15) the constants from Assumption 1.3.3 are:

$$\alpha_{ij}^1 = \alpha_{ij}^2 = 0; \quad \alpha_{ij}^3 = \lambda_M(A_{ij}^T P_{ij} + P_{ij}^T A_{ij}),$$
$$\nu_{ki}^1 = \nu_{ijk}^2 = 0; \nu_{ijk}^1 = \lambda_M^{1/2}[(P_{ij}^T A_{ik})(P_{ij}^T A_{ik})], \quad \mu_{ijk}^2 = 0.$$

Therefore the elements σ_{ij} of matrix M in (1.4.12) for system (1.4.15) have the structure

$$\sigma_{ii} = -\eta_i^2 \lambda_m(G_{ii}) + 2 \sum_{j=1, j \neq i}^{3} \eta_i \eta_j \alpha_{ij}^3, \quad i = 1, 2, 3,$$

$$\sigma_{ij} = \sum_{\substack{k=1, k \neq i, \\ k \neq j}}^{3} (\eta_k \eta_j \nu_{ijk}^1 + \eta_i \eta_k \nu_{kij}^1), \quad i, j = 1, 2, 3, \quad i \neq j.$$

Consequently, the function $Dv(x, \eta)$ variation along solutions of system (1.4.15) is estimated by the inequality

(1.4.22) $$Dv(x, \eta)|_{(1.5.14)} \leq w^T M w$$

for all $(x_1, x_2, x_3) \in R^{n_1} \times R^{n_2} \times R^{n_3}$.

We summarize our presentation as follows.

Corollary 1.4.1 Assume for system (1.4.15) the folowing conditions are satisfied:

(1) algebraic equations (1.4.18) have the sign-definite matrices P_{ii}, $i = 1, 2, 3$, as their solutions;
(2) algebraic equations (1.4.20) have constant matrices P_{ij}, for all $i, j = 1, 2, 3$, $i \neq j$, as their solutions;
(3) matrix C in (1.4.21) is positive definite;
(4) matrix M in (1.4.22) is negative semi-definite (negative definite).

Then the equilibrium state $x = 0$ of system (1.4.15) is uniformly stable (uniformly asymptotically stable).

This corollary follows from Theorem 1.3.1 and hence its proof is obvious.

Example 1.4.2 Partial case of system (1.4.15) is the system

(1.4.23)
$$\frac{dx_1}{dt} = Ax_1 + C_{12}x_2$$
$$\frac{dx_2}{dt} = Bx_2 + C_{21}x_1,$$

where $x_1 \in R^{n_1}$, $x_2 \in R^{n_2}$, and A, B, C_{12} and C_{21} are constant matrices of corresponding dimensions.

Let the matrices from system (1.4.24) be of the form

(1.4.24) $$A = \begin{pmatrix} -2 & 1 \\ 3 & -2 \end{pmatrix}, \quad B = \begin{pmatrix} -4 & 1 \\ 2 & -1 \end{pmatrix},$$

(1.4.25) $$C_{12} = \begin{pmatrix} -0.5 & -0.5 \\ 0.8 & -0.7 \end{pmatrix}, \quad C_{21} = \begin{pmatrix} 1.1 & 0.5 \\ -0.6 & -0.3 \end{pmatrix}.$$

It is easy to prove that the zero solution of system (1.4.24) with matrices (1.4.25), (1.4.26) is uniform asymptotic stability in the whole.

It can be proved that stability of system (1.2.23) with matrices (1.4.24), (1.4.25) can not be studied in terms of the Bailey [1] theorem because the Bailey theorem requires that

$$\|C_{12}\| \|C_{21}\| < 0.0184$$

whereas for system (1.4.23) with (1.4.24), and (1.4.25) we have

$$\|C_{12}\| \|C_{21}\| = 1.75.$$

For the details see Martynyuk and Slyn'ko [1], and Martynyuk [11].

1.5 Polystability of Motion Analysis

One of the applications of the method of the matrix-valued function is the problem on polystability of nonlinear systems with separable motions. In this Section this problem is studied with the aim of establishing various sufficiency conditions for the corresponding motions. Some results are illustrated by examples.

1.5.1 The problem of exponential polystability
Consider a system of differential equations of perturbed motion

(1.5.1) $$\frac{dx}{dt} = f(t, x), \quad x(t_0) = x_0,$$

where $x \in R^n$, $f \in C(R_+ \times \mathcal{D}, R^n)$, $\mathcal{D} \subseteq R^n$, and, hence, $f(t, x) = 0$ for all $t \in R_+$ iff $x = 0$. We assume that this equilibrium state is unique for system (1.5.1).

1. CONTINUOUS SYSTEMS

Let us decompose a vector $x \in R^n$ into two subvectors $x_i \in R^{n_i}$, $i = 1, 2$, $n_1 + n_2 = n$, and rewrite system (1.5.1) as follows:

$$\text{(1.5.2)} \qquad \frac{dx_i}{dt} = f_i(t, x_1, x_2), \qquad x_i(t_0) = x_{i0},$$

where $f \in C(R_+ \times R^{n_i}, R^{n_i})$, $i = 1, 2$.

We use the following notation for norms of vectors:

$$\|x_i\| = \left(\sum_{k=1}^{n_i} x_k^2\right)^{1/2}, \quad \|x\| = \left(\sum_{s=1}^{n} x_s^2\right)^{1/2} = \left(\sum_{j=1}^{2} x_j^2\right)^{1/2}, \quad i = 1, 2.$$

Assume that the right-hand side of system (1.5.1) is continuous in the region $R_+ \times \mathcal{D}$, where $\mathcal{D} = \{x \colon \|x_1\| + \|x_2\| \leq H < +\infty\}$, and the right-hand side of system (1.5.2) is continuous in $R_+ \times \mathcal{D}^*$, where $\mathcal{D}^* = \{x \colon \|x_1\| \leq H, \ 0 < \|x_2\| < +\infty\}$.

If system (1.5.2) is considered in the region $R_+ \times \mathcal{D}^*$, we assume that its solution $x(t; t_0, x_0)$ is x_2-extendable.

Below, we present some definitions, taking into account the results of He and Wang [1], and Martynyuk [7, 8, 16].

Definition 1.5.1 The *equilibrium state* $x = 0$ of system (1.5.1) is called *exponentially x_1-stable (in the small)*, if there exists $\lambda > 0$ and, for any $\varepsilon > 0$, one can find $\delta(\varepsilon) > 0$ such that

$$\text{(1.5.3)} \qquad \|x_1(t; t_0, x_0)\| \leq \varepsilon \exp[-\lambda(t - t_0)] \quad \text{for all} \quad t \geq t_0$$

if $\|x_0\| < \delta(\varepsilon)$.

Definition 1.5.2 The *equilibrium state* $x = 0$ of system (1.5.1) is called *globally exponentially x_1-stable*, if there exists $\lambda > 0$ and, for any Δ, $0 < \Delta < +\infty$, one can find $K(\Delta) > 0$ such that

$$\|x_1(t; t_0, x_0)\| \leq \varepsilon \exp[-\lambda(t - t_0)] \quad \text{for all} \quad t \geq t_0$$

if $\|x_0\| < \Delta$.

Definition 1.5.3 The *equilibrium state* $x = 0$ of system (1.5.1) is called *exponentially polystable (in the small)*, if for positive constants r_1 and r_2 and any $\varepsilon > 0$, there exists $\lambda > 0$ and $\Delta(\varepsilon)$, such that

$$\text{(1.5.4)} \qquad \|x_1(t; t_0, x_0)\|^{2r_1} + \|x_2(t; t_0, x_0)\|^{2r_2} \leq \varepsilon \exp[-\lambda(t - t_0)]$$
$$\text{for all} \quad t \geq t_0$$

if $\|x_0\| < \Delta$.

Definition 1.5.4 The *equilibrium state* $x = 0$ of system (1.5.1) is called *globally exponentially polystable*, if there exists $\lambda > 0$ and, for any Δ, one can find $R(\Delta) > 0$ such that

$$\|x_1(t; t_0, x_0)\|^{2r_1} + \|x_2(t; t_0, x_0)\|^{2r_2} \leq R(\Delta)\exp[-\lambda(t - t_0)] \quad \text{for all} \quad t \geq t_0$$

if $\|x_0\| < \Delta$, $t_0 > 0$.

We study exponential properties of the solution $x = 0$ in the following cases:

Case 1. We study the exponential stability of the solution $x = 0$ with respect to the vector x_1, i.e. the exponential x_1-stability.

Case 2. We study the exponential polystability of the solution $x = 0$.

Remark 1.5.1 The informative part of the notion of polystability in Definitions 1.5.3 and 1.5.4 is, in fact, the difference between the rates of decrease of components of the solution $x(t; t_0, x_0)$ of system (1.5.2).

1.5.2 Approaches to the solution of the problem We investigate the exponential properties of the solution $x = 0$ of system (1.5.1) in Cases 1 and 2 by using scalar and matrix-valued Liapunov functions, respectively.

First, consider Case 1. Suppose that a scalar function $v(t, x) \in C(R_+ \times D^*, R_+)$ is associated with system (1.5.1) and $v(t, x_1, x_2) = 0$ for all $t \in R_+$ if $x_1 = 0$.

Theorem 1.5.1 *Assume that the vector function f in system (1.5.1) is continuous in $R_+ \times D^*$ and there exist*

(1) *functions $v(t, x) \in C(R_+ \times D^*, R_+)$ and functions $\varphi_1, \varphi_2 \in K$ of the same order of magnitude;*

(2) *positive constants c and γ_1 such that*

(1.5.5) $$c\|x_1\|^{\gamma_1} \leq v(t, x_1, x_2) \leq \varphi_1(\|x\|),$$

(1.5.6) $$D^+ v(t, x_1, x_2)\big|_{(1.5.2)} \leq -\varphi_2(\|x\|).$$

Then the equilibrium state $x = 0$ of system (1.5.1) is exponentially x_1-stable in the small.

Proof For functions φ_1 and φ_2 satisfying the conditions of Theorem 1.5.1, there exist constants α_1 and β_1 such that

(1.5.7) $$\alpha_1 \varphi_1(r) \leq \varphi_2(r) \leq \beta_1 \varphi_1(r).$$

1. CONTINUOUS SYSTEMS

In view of (1.5.7), it follows from inequalities (1.5.5) and (1.5.6)

$$D^+v(t, x_1, x_2)\big|_{(1.5.2)} \leq -\alpha_1 v(t, x_1, x_2)$$

and, further,

$$v(t, x(t)) \leq v(t_0, x_0) \exp[-\alpha_1(t - t_0)] \quad \text{for all} \quad t_0 \geq 0.$$

By using the lower bound for the function $v(t, x)$ and inequality (1.5.5), we obtain

$$\|x_1(t; t_0, x_0)\| \leq c^{-1/\gamma_1} \varphi_1^{1/\gamma_1}(\|x_0\|) \exp\left[-\frac{\alpha_1}{\gamma_1}(t - t_0)\right], \quad t \geq t_0.$$

Denote $\lambda = \alpha_1/\gamma_1$. For any $\varepsilon > 0$, we choose $\delta(\varepsilon) = \varphi_1^{-1}(c\varepsilon^{\gamma_1})$. Then we arrive at estimate (1.5.3) if $\|x_0\| < \delta(\varepsilon)$, $t_0 \geq 0$. The theorem is proved.

Theorem 1.5.2 *Suppose that the vector function f in system (1.5.1) is continuous in $R_+ \times R^n$ and there exist*

(1) *functions $v(t, x) \in C(R_+ \times R^n, R_+)$ and functions $\varphi_1, \varphi_2 \in KR$ of the same order of magnitude;*

(2) *positive constants c and γ_2 such that*

(1.5.8)
$$d\|x_1\|^{\gamma_2} \leq v(t, x_1, x_2) \leq \varphi_1(\|x\|),$$
$$D^+v(t, x_1, x_2)\big|_{(1.5.2)} \leq -\varphi_2(\|x\|).$$

Then the equilibrium state $x = 0$ of system (1.5.1) is globally exponentially x_1-stable.

Proof As in the proof of Theorem 1.5.1, we obtain the estimate

$$\|x_1(t; t_0, x_0)\| \leq d^{-1/\gamma_2} \varphi_1^{1/\gamma_2}(\|x_0\|) \exp\left[-\frac{\alpha_2}{\gamma_2}(t - t_0)\right], \quad t \geq t_0.$$

Denote $\lambda = \alpha_2/\gamma_2$. For any $0 < \Delta < +\infty$, we find $K(\Delta) = d^{-1/\gamma_2} \varphi_1^{1/\gamma_2}(\Delta)$. Then

$$\|x(t; t_0, x_0)\| \leq K(\Delta) \exp[-\lambda(t - t_0)], \quad t \geq t_0,$$

for $\|x_0\| < \Delta$, $t_0 \geq 0$.

Consider Case 2. For system (1.5.2), we consider the matrix-valued function

(1.5.9) $$U(t, x) = [v_{ij}(t, x)], \quad i, j = 1, 2$$

the element $v_{ij}(t, x)$ of which satisfies special conditions.

Assumption 1.5.1 There exist

(1) functions $\varphi_1, \varphi_2 \in K(KR)$ of the same order of magnitude;
(2) the matrix function (1.5.9) whose elements satisfy the following estimates:
   (a) $\underline{c}_{11}\|x_1\|^{2r_1} \leq v_{11}(t, x_1) \leq \bar{c}_{11}\varphi_1^2(\|x_1\|)$ for all $(t, x) \in R_+ \times \mathcal{D}$ (for all $(t, x) \in R_+ \times R^n$);
   (b) $\underline{c}_{22}\|x_2\|^{2r_2} \leq v_{22}(t, x_2) \leq \bar{c}_{22}\varphi_2^2(\|x_2\|)$ for all $(t, x) \in R_+ \times \mathcal{D}$ (for all $(t, x) \in R_+ \times R^n$), here, $\underline{c}_{ii} > 0$ and $\bar{c}_{ii} > 0$, $i = 1, 2$;
   (c) $\underline{c}_{12}\|x_1\|^{r_1}\|x_2\|^{r_2} \leq v_{12}(t, x_1, x_2) \leq \bar{c}_{12}\varphi_1(\|x_1\|)\varphi_2(\|x_2\|)$;
   (d) $v_{12}(t, x_1, x_2) = v_{21}(t, x_1, x_2)$ for all $(t, x) \in R_+ \times \mathcal{D}$ (for all $(t, x) \in R_+ \times R^n$), here, $\underline{c}_{ij} = \underline{c}_{ji}$, $\bar{c}_{ij} = \bar{c}_{ji}$, $i \neq j$, and $r_i > 0$, $i, j = 1, 2$.

Proposition 1.5.1 Suppose that all conditions of Assumption 1.5.1 are satisfied. Then the function

$$v(t, x, \eta) = \eta^T U(t, x)\eta$$

with $\eta \in R_+^2$ satisfies the bilateral inequality

(1.5.10) $$u_1^T A_1 u_1 \leq v(t, x, \eta) \leq u_2^T A_2 u_2$$

for all $(t, x) \in R_+ \times \mathcal{D}$ (for all $(t, x) \in R_+ \times R^n$). Here,

$$u_1^T = (\|x_1\|^{r_1}, \|x_2\|^{r_2}), \quad u_2^T = (\varphi_1(\|x_1\|), \varphi_2(\|x_2\|)),$$
$$A_1 = H^T C_1 H, \quad A_2 = H^T C_2 H, \quad H = \operatorname{diag}(\eta_1, \eta_2),$$

$$C_1 = \begin{pmatrix} \underline{c}_{11} & \underline{c}_{12} \\ \underline{c}_{21} & \underline{c}_{22} \end{pmatrix}, \quad C_2 = \begin{pmatrix} \bar{c}_{11} & \bar{c}_{12} \\ \bar{c}_{21} & \bar{c}_{22} \end{pmatrix}.$$

Proof By substituting inequalities (a)–(c) from Assumption 1.5.1 into the expression

$$v(t, x, \eta) = \sum_{i,j=1}^{2} \eta_i \eta_j v_{ij}(t, \cdot),$$

we get estimate (1.5.10).

Assumption 1.5.2 There exist
(1) functions $\psi_1, \psi_2 \in K(KR)$ of the same order of magnitude
(2) functions $\mu_{ij}(t)$, $i = 1, 2$, $j = 1, 2, \ldots, 10$, continuous on any finite interval and such that

(a) $D_t^+ v_{ii}(t, x_i) + (D_{x_i}^+ v_{ii})^T f_i(t, x_i) \leq \mu_{i1}(t)\psi_i^2(\|x_i\|) + r_{i1}(t, \psi)$
for all $(t, x) \in R_+ \times D$ (for all $(t, x) \in R_+ \times R^n$);

(b) $(D_{x_i}^+ v_{ii})^T g_i(t, x_1, x_2) \leq \mu_{i2}(t)\psi_i^2(\|x_i\|)$
$+ \mu_{i3}(t)\psi_1(\|x_1\|)\psi_2(\|x_2\|) + \mu_{i4}(t)\psi_2^2(\|x_2\|) + r_{i2}(t, \psi)$
for all $(t, x) \in R_+ \times D$ (for all $(t, x) \in R_+ \times R^n$);

(c) $D_t^+ v_{12}(t, x_1, x_2) + (D_{x_i}^+ v_{12})^T f_i(t, x_i) \leq \mu_{i5}(t)\psi_i^2(\|x_i\|)$
$+ \mu_{i6}(t)\psi_1(\|x_1\|)\psi_2(\|x_2\|) + \mu_{i7}(t)\psi_2^2(\|x_2\|) + r_{i3}(t, \psi)$
for all $(t, x) \in R_+ \times D$ (for all $(t, x) \in R_+ \times R^n$);

(d) $(D_{x_i}^+ v_{12})^T g_i(t, x_1, x_2) \leq \mu_{i8}(t)\psi_i^2(\|x_i\|)$
$+ \mu_{i9}(t)\psi_1(\|x_1\|)\psi_2(\|x_2\|) + \mu_{i10}(t)\psi_2^2(\|x_2\|) + r_{i4}(t, \psi)$
for all $(t, x) \in R_+ \times D$ (for all $(t, x) \in R_+ \times R^n$).

Here, $f_i(t, x_i) = f_i(t, x_i, x_j)$ for $x_j = 0$, $j = 1, 2$, $g_i(t, x_i, x_j) = f_i(t, x_i, x_j) - f_i(t, x_i)$, $i, j = 1, 2$, and $r_{ik}(t, \psi)$, $i = 1, 2$, $k = 1, 2, 3, 4$, are polynomials in ψ_i, $i = 1, 2$, of degree higher than two.

Proposition 1.5.2 If all conditions of Assumption 1.5.2 are satisfied, then the following estimate is true for the function $D^+v(t, x, \eta)$ for all $(t, x) \in R_+ \times D$ (for all $(t, x) \in R_+ \times R^n$):

(1.5.11) $\qquad \eta^T D^+ U(t,x)\eta \leq u_3^T(\|x\|) A_3(t) u_3(\|x\|) + R(t, \psi).$

Here, $u_3^T(\|x\|) = (\psi_1(\|x_1\|), \psi_2(\|x_2\|))$,

$$R(t, \psi) = \eta_1^2(r_{11}(t, \psi) + r_{12}(t, \psi)) + \eta_2^2(r_{21}(t, \psi) + r_{22}(t, \psi))$$
$$+ 2\eta_1\eta_2(r_{13}(t, \psi) + r_{14}(t, \psi) + r_{23}(t, \psi) + r_{24}(t, \psi)),$$

and $A_3(t)$ is a 2×2 matrix continuous on every finite interval with elements defined as follows:

$$a_{11}(t) = \eta_1^2(\mu_{11}(t) + \mu_{12}(t)) + \eta_2^2\mu_{22}(t)$$
$$+ 2\eta_1\eta_2\big(\mu_{15}(t) + \mu_{18}(t) + \mu_{25}(t) + \mu_{28}(t)\big),$$

$$a_{22}(t) = \eta_2^2(\mu_{21}(t) + \mu_{22}(t)) + \eta_1^2\mu_{14}(t)$$
$$+ 2\eta_1\eta_2\big(\mu_{17}(t) + \mu_{110}(t) + \mu_{27}(t) + \mu_{210}(t)\big),$$

$$a_{12}(t) = \frac{1}{2}\left(\eta_1^2(\mu_{13}(t) + \eta_2^2\mu_{23}(t)\right)$$
$$+ \eta_1\eta_2\big(\mu_{16}(t) + \mu_{19}(t) + \mu_{26}(t) + \mu_{29}(t)\big).$$

Assume that the matrix $A_3(t)$ is negative definite for all $t \in R_+ = [0, +\infty)$. Then, for any $\mu \in (0,1)$, there exists $H(\mu) > 0$ and $\alpha > 0$ such that, for $x \in \Omega(H) \subseteq \mathcal{D}$ and $\Omega(H) = \{x \colon \|x\| < H(\mu)\}$, the estimate

$$(1.5.12) \qquad u_3^T(\|x\|)^T A_3(t) u_3(\|x\|) + |R(t,\psi)| < -\alpha(1-\mu) v(t,x,\eta)$$

is true for all $t \in R_+$, and estimate (1.5.11) takes the form

$$(1.5.13) \qquad D^+ v(t,x,\eta) \leq -\alpha(1-\mu) v(t,x,\eta)$$

in the region $(t,x) \in R_+ \times \Omega$.

Let $\|u\| = (u^T u)^{1/2}$ be the Euclidean norm of a vector u in the cone $K = \{u \colon u \geq 0\}$.

The proof of Proposition 1.5.2 is based on the direct application of the estimates from Assumption 1.5.2 to the function $D^+ v(t,x,\eta)$.

Proposition 1.5.3 The following estimates are true for the quadratic forms $u_1^T A_1 u_1$ and $u_2^T A_2 u_2$:

$$(1.5.14) \qquad \lambda_m(A_1) u_1^T u_1 \leq u_1^T A_1 u_1 \leq \lambda_M(A_1) u_1^T u_1,$$

$$(1.5.15) \qquad \lambda_m(A_2) u_{20}^T u_{20} \leq u_{20}^T A_{20} u_{20} \leq \lambda_M(A_2) u_{20}^T u_{20},$$

where $u_{20} = (\varphi_1(\|x_{10}\|), \varphi_2(\|x_{20}\|))$.

Proposition 1.5.3 can be proved by standard methods of theory of quadratic forms.

Theorem 1.5.3 *Suppose that the vector function f of system (1.5.1) is continuous in $R_+ \times \Omega$ and*

(1) *the conditions of Assumptions 1.5.1 and 1.5.2 are satisfied;*
(2) *the matrices A_1 and A_2 are positive definite;*
(3) *the matrix $A_3(t)$ is negative definite for all $t \in R_+$.*

Then the equilibrium state $x = 0$ of system (1.5.2) is exponentially stable in the small.

Proof It follows from (1.5.13) that

$$(1.5.16) \qquad v(t,x(t),\eta) \leq v(t_0,x_0,\eta) \exp[-\alpha(1-\mu)(t-t_0)], \quad t \geq t_0.$$

By virtue of Propositions 1.5.1 and 1.5.3, we have

$$u_1^T(t) A_1 u_1(t) \leq u_{20}^T A_2 u_{20} \exp[-\alpha(1-\mu)(t-t_0)], \quad t \geq t_0,$$

and, further,

(1.5.17) $$\lambda_m(A_1)u_1^T(t)u_1(t) \leq \lambda_M(A_2)u_{20}^T u_{20}\exp[-\alpha(1-\mu)(t-t_0)], \quad t \geq t_0.$$

Denoting $a = \lambda_m^{-1}(A_1)\lambda_M(A_2)$, we rewrite estimate (1.5.17) as follows:

(1.5.18) $$\|x_1(t)\|^{2r_1} + \|x_2(t)\|^{2r_2} \leq a(\varphi_1^2(\|x_{10}\|) + \varphi_2^2(\|x_{20}\|))\exp[-\alpha(1-\mu)(t-t_0)], \quad t \geq t_0.$$

Since the functions $\varphi_1, \varphi_2 \in K$ have the same order of magnitude (see condition (i) in Assumption 1.5.1), there exists a function $\varphi \in K$ such that

(1.5.19) $$\varphi_1^2(\|x_{10}\|) + \varphi_2^2(\|x_{20}\|) \leq \varphi^2(\|x_0\|).$$

Inequality (1.5.18) holds if the following inequality is satisfied:

(1.5.20) $$\|x_1(t)\|^{2r_1} + \|x_2(t)\|^{2r_2} \leq a\varphi^2(\|x_0\|)\exp[-\alpha(1-\mu)(t-t_0)], \quad t \geq t_0.$$

For any $\varepsilon > 0$, we choose $\delta(\varepsilon) = \min\left(H(\mu), \varphi_1^{-1}(a^{-1/2}\varepsilon^{1/2})\right)$ and denote $\lambda = \alpha(1-\mu)$, $0 < \mu < 1$. Then it follows from inequality (1.5.20) that if $\|x_0\| < \delta(\varepsilon)$, $t_0 \geq 0$, then

$$\|x_1(t)\|^{2r_1} + \|x_2(t)\|^{2r_2} \leq \varepsilon\exp[-\alpha(1-\mu)(t-t_0)], \quad t \geq t_0,$$

i.e., the separable motions of system (1.5.2) are exponentially polystable. The theorem is proved.

Theorem 1.5.4 *Suppose that the vector function f of system (1.5.1) is continuous in $R_+ \times R^n$ and*

(1) *the conditions of Assumptions 1.5.1 and 1.5.2 with functions $\varphi_1, \varphi_2 \in KR$ and $\psi_1, \psi_2 \in KR$, respectively, are satisfied;*
(2) *for any $\mu \in (0,1)$, inequality (1.5.12) holds for $(t,x) \in R_+ \times R^n$;*
(3) *conditions (2) and (3) of Theorem 1.5.3 are satisfied.*

Then the equilibrium state $x = 0$ of system (1.5.2) is globally exponentially stable.

Proof By analogy with the proof of Theorem 1.5.3, we obtain inequality (1.5.20) with the function $\varphi(\|x_0\|) \in KR$. As above, we denote $\lambda =$

$\alpha(1-\mu)$, $0 < \mu < 1$, and, for any $0 < \Delta < +\infty$, choose $R(\Delta)$ in the form $R(\Delta) = a\varphi^2(\Delta)$. Then the following estimate is true for $\|x_0\| < \Delta$, $t_0 \geq 0$:

$$\|x_1(t)\|^{2r_1} + \|x_2(t)\|^{2r_2} \leq R(\Delta)\exp[-\alpha(1-\mu)(t-t_0)], \quad t \geq t_0.$$

Theorem 1.5.4 is proved.

The statement below establishes the relationship between the global exponentially x_1-stability of the solution $x = 0$ and other types of stability of this solution.

Theorem 1.5.5 *The equilibrium state $x = 0$ of system (1.5.2) is globally exponentially x_1-stable if and only if it is exponentially x_1-stable in the small and globally uniformly asymptotically x_1-stable.*

Proof Necessity If the equilibrium state $x = 0$ of system (1.5.2) is globally exponentially x_1-stable, then it is exponentially x_1-stable in the small. Definition 1.5.2 implies that

(1.5.21) $\qquad \|x_1(t;t_0,x_0)\| < M(\Delta)$ for all $t \geq t_0$ and $\|x_0\| < \Delta$,

where $M(\Delta) = K(\Delta)\Delta$. Inequality (1.5.21) follows from the fact that the global uniform asymptotic x_1-stability implies the uniform x_1-boundedness of the solution $x = 0$. If $\|x_0\| < \delta(\varepsilon)$ for $t \geq t_0$, where $\delta(\varepsilon) = \varepsilon$, then estimate (1.5.3) yields

$$\|x_1(t;t_0,x_0)\| < \varepsilon \quad \text{for all} \quad t \geq t_0$$

because the equilibrium state $x = 0$ is uniformly x_1-stable.

It is easy to show that, for any $\Delta > 0$, $\varepsilon > 0$, and $t_0 \in r_+$, there exists

$$T(\varepsilon, \Delta) = (1/\lambda)\ln(M(\Delta)/\varepsilon)$$

such that

(1.5.22) $\qquad \|x_1(t;t_0,x_0)\| < \varepsilon \quad \text{for all} \quad t \geq t_0 + T(\delta,\Delta)$

whenever $\|x_0\| < \Delta$ and $t_0 \geq 0$. Thus, the equilibrium state $x = 0$ of system (1.5.2) is globally uniformly x_1-stable.

Sufficiency It follows from the exponential x_1-stability of the solution $x = 0$ in the small that one can find $\lambda > 0$ for any $\delta > 0$, $0 < \delta \leq r_0 < 1$, and $a > 0$ such that the condition $\|x_0\| \leq \beta$, $t_0 > 0$ implies the estimate

(1.5.23) $\qquad \|x_1(t;t_0,x_0)\| \leq a\|x_0\|\exp(-\lambda(t-t_0)) \quad \text{for all} \quad t \geq t_0.$

For any $\varepsilon > 0$, we choose $\delta(\varepsilon) = \varepsilon/2$. Then, for $\|x_0\| < \delta(\varepsilon)$, inequality (1.5.23) yields

$$\|x_1(t; t_0, x_0)\| \leq \varepsilon \exp(-\lambda(t - t_0)) \quad \text{for all} \quad t \geq t_0.$$

Here, $0 < \delta \leq \varepsilon$, $a \geq 1 \geq \varepsilon$, and $S(r_0) = \{x \colon \|x_1\| < r_0, \ 0 < \|x_2\| < \infty\}$. It follows from the condition of global uniform asymptotic x_1-stability in Theorem 1.5.5 that, for any $\Delta > 0$, there exists $M(\Delta) > 0$ such that

(1.5.24) $\qquad \|x_1(t; t_0, x_0)\| < M(\Delta) \quad \text{for all} \quad t \geq t_0$

whenever $\|x_0\| < \Delta$. Furthermore, for any $\Delta > 0$, $\varepsilon > 0$, and $t_0 \in R_+$, one can find $T = T(\varepsilon, \Delta) > 0$ such that the condition $\|x_0\| \leq \Delta$ implies the estimate

(1.5.25) $\qquad \|x_1(t; t_0, x_0)\| < \delta(\varepsilon) \quad \text{for all} \quad t \geq t_0 + T(\varepsilon, \Delta).$

Let

$$R(\Delta) = \max(M(\Delta) \exp(\lambda T(\varepsilon, \Delta), a).$$

Let us estimate the solution $x_1(t; t_0, x_0)$ for $t_0 \leq t \leq t_0 + T(\varepsilon, \Delta)$ and $t \leq t_0 + T(\varepsilon, \Delta)$, respectively. Assume that $t_0 \leq t \leq t_0 + T(\varepsilon, \Delta)$. Since

$$R(\Delta) \exp[-\lambda(t - t_0)] \geq R(\Delta) \exp[-\lambda T(\varepsilon, \Delta)] = M(\Delta),$$

we have

(1.5.26) $\qquad \|x_1(t; t_0, x_0)\| \leq R(\Delta) \exp(-\lambda(t - t_0)), \quad t \geq t_0$

for $\|x_0\| < \Delta$. Let $t \geq t_0 + T(\varepsilon, \Delta)$. Denote $\tilde{x} = x(t_1; t_0, x_0)$. In this case, we have $\|x_1\| < \delta(\varepsilon)$. Estimate (1.5.23) yields

(1.5.27) $\qquad \|x_1(t; t_0, x_0)\| \leq \varepsilon \exp(-\lambda(t - t_0)), \quad t \geq t_1.$

Note that, by virtue of the continuity and uniqueness of solutions of system (1.5.2), the following relation is true:

$$x_1(t; t_0, \tilde{x}) = x_1(t; t_1, x(t_1; t_0, x_0)) = x_1(t; t_1, \tilde{x}), \quad t \geq t_1.$$

It is now easy to show that there exists $\lambda > 0$ and, for any $\beta > 0$, one can find $R(\Delta) > 0$ such that

(1.5.28) $\qquad \|x_1(t; t_0, x_0)\| \leq R(\Delta) \exp(-\lambda(t - t_0)), \quad t \geq t_0,$

whenever $\|x_0\| < \Delta$ and $t_0 \geq 0$. For $\|x_0\| \leq r_0$, we have estimate (1.5.26). Hence, it remains to consider the case $r_0 \leq \|x_0\| \leq \Delta < +\infty$. For $\|x_0\|/r_0 \geq 1$, we get $K(\Delta) = R(\Delta)/r_0$, and inequality (1.5.28) implies the following estimate:

$$\|x_1(t;t_0,x_0)\| \leq K(\Delta)\|x_0\|\exp(-\lambda(t-t_0)), \quad t \geq t_0.$$

This completes the proof of the theorem.

1.5.3 Autonomous system

Consider the perturbed motion equation

(1.5.29)
$$\frac{dx_1}{dt} = f_1(x_1) + g_1(x_1, x_2),$$
$$\frac{dx_2}{dt} = f_2(x_2) + g_2(x_1, x_2),$$

where $x_1 \in R^{n_1}$, $x_2 \in R^{n_2}$, $x = (x_1^T, x_2^T)^T \in R^n$, $f_1 \in C(\mathcal{D}_1, R^{n_1})$, $f_2 \in C(\mathcal{D}_2, R^{n_2})$, $g_1 \in C(\mathcal{D}_1 \times \mathcal{D}_2, R^{n_1})$, $g_2 \in C(\mathcal{D}_1 \times \mathcal{D}_2, R^{n_2})$. Here $\mathcal{D}_1 = \{x \in R^{n_1}: 0 < \|x_1\| < h_1\}$, $\mathcal{D}_2 = \{x \in R^{n_2}: 0 < \|x_2\| < h_2\}$, $h_1, h_2 = \text{const} > 0$.

Suppose that system (1.5.29) has a continuous solution $x(t, x_0)$ in open neighborhood $\mathcal{S} \subseteq \mathcal{D}_1 \times \mathcal{D}_2$ of the unique equilibrium state $x = 0$ for any $x_0 \in \mathcal{S}$ and its motions are definite and continuous in $(t, x_0) \in I_0 \times \mathcal{S}$, $I_0 \subseteq R_+$, $I_0 \neq 0$, $I_0 = I_0(x_0)$. We shall establish exponential polystability conditions for system (2.6.29) in the sense of Definition 1.5.1 the method of constructive application of the matrix-valued Liapunov function.

We shall formulate some assumptions which are the basis of the proposed method of analysis of exponential polystability of motion.

Assumption 1.5.3 There exists

(1) open connected neighborhood $\mathcal{S}$ of equilibrium state $x = 0$ of system (1.5.29);
(2) matrix-valued function $U(x) = [v_{ij}(\cdot)]$, $i, j = 1, 2$, with elements $v_{ii} \in C(\mathcal{D}_i, R_+)$, $v_{ij} \in C(\mathcal{D}_1 \times \mathcal{D}_2, R)$, $i \neq j$;
(3) real constants $\bar{c}_{ii} > 0$, $\underline{c}_{ii} > 0$, $\bar{c}_{12}, \underline{c}_{12} \in R$, and
(4) comparison functions $\varphi_1, \varphi_2 \in K$ such that

$$\underline{c}_{11}\|x_1\|^{2r_1} \leq v_{11}(x_1) \leq \bar{c}_{11}\varphi_1^2(\|x_1\|)$$

for all $x_1 \in \mathcal{D}_1$,

$$\underline{c}_{22}\|x_2\|^{2r_2} \leq v_{22}(x_2) \leq \bar{c}_{22}\varphi_2^2(\|x_2\|)$$

for all $x_2 \in \mathcal{D}_2$,
$$\underline{c}_{12}\|x_1\|^{r_1}\|x_2\|^{r_2} \leq v_{12}(x_1,x_2) \leq \bar{c}_{12}\varphi_1(\|x_1\|)\varphi_2(\|x_2\|)$$
for all $(x_1, x_2) \in \mathcal{D}_1 \times \mathcal{D}_2$, where r_1 and r_2 are positive constants.

Proposition 1.5.4 If all conditions of Assumption 1.5.3, are satisfied, then for function

(1.5.30) $\qquad v(x,\eta) = \eta^{\mathrm{T}} U(x)\eta, \quad \eta \in R_+^2, \quad \eta > 0$

the bilateral inequality

(1.5.31) $\qquad u_1^{\mathrm{T}} H^{\mathrm{T}} \underline{C} H u_1 \leq v(x,\eta) \leq u_2^{\mathrm{T}} H^{\mathrm{T}} \overline{C} H u_2$

holds true for all $(x_1, x_2) \in \mathcal{D}_1 \times \mathcal{D}_2$.

Here

$$u_1 = (\|x_1\|^{r_1}, \|x_2\|^{r_2})^{\mathrm{T}}, \qquad u_2 = (\varphi_1(\|x_1\|), \varphi_2(\|x_2\|))^{\mathrm{T}},$$

$$H = \mathrm{diag}\,(\eta_1, \eta_2),$$

$$\underline{C} = \begin{pmatrix} \underline{c}_{11} & \underline{c}_{12} \\ \underline{c}_{21} & \underline{c}_{22} \end{pmatrix}, \quad \overline{C} = \begin{pmatrix} \bar{c}_{11} & \bar{c}_{12} \\ \bar{c}_{21} & \bar{c}_{22,} \end{pmatrix},$$

$$\bar{c}_{12} = \bar{c}_{21}, \quad \underline{c}_{12} = \underline{c}_{21}.$$

Assumption 1.5.4 Assume that
(1) conditions (1), (2) and (4) of Assumption 1.5.3 are satisfied;
(2) there exist constants α_{ij}, $i = 1,2$, $j = 1,2,3$, β_{ij}, $i = 1,2$, $j = 1,\ldots,5$ such that
  (a) $(D_{x_1} v_{11})^{\mathrm{T}} f_1(x_1) \leq \alpha_{11} \psi_1^2(\|x_1\|)$;
  (b) $(D_{x_1} v_{11})^{\mathrm{T}} g_1(x_1,x_2) \leq \alpha_{12} \psi_1(\|x_1\|)\psi_2(\|x_2\|) + \alpha_{13}\psi_1^2(\|x_1\|)$;
  (c) $(D_{x_2} v_{22})^{\mathrm{T}} f_2(x_2) \leq \alpha_{21} \psi_2^2(\|x_2\|)$;
  (d) $(D_{x_2} v_{22})^{\mathrm{T}} g_2(x_1,x_2) \leq \alpha_{22} \psi_1(\|x_1\|)\psi_2(\|x_2\|) + \alpha_{23}\psi_2^2(\|x_2\|)$;
  (e) $(D_{x_1} v_{12})^{\mathrm{T}} f_1(x_1) \leq \beta_{11} \psi_1^2(\|x_1\|) + \beta_{12}\psi_1(\|x_1\|)\psi_2(\|x_2\|)$;
  (f) $(D_{x_1} v_{12})^{\mathrm{T}} g_1(x_1,x_2) \leq \beta_{13} \psi_1^2(\|x_1\|) + \beta_{14}\psi_1(\|x_1\|)\psi_2(\|x_2\|) + \beta_{22}\psi_2^2(\|x_2\|)$;
  (g) $(D_{x_1} v_{12})^{\mathrm{T}} f_2(x_2) \leq \beta_{21} \psi_1(\|x_1\|)\psi_2(\|x_2\|) + \beta_{22}\psi_2^2(\|x_2\|)$;
  (h) $(D_{x_2} v_{12})^{\mathrm{T}} g_2(x_1,x_2) \leq \beta_{23} \psi_1^2(\|x_1\|) + \beta_{24}\psi_1(\|x_1\|)\psi_2(\|x_2\|) + \beta_{25}\psi_2^2(\|x_2\|)$
  for all $(x_1, x_2) \in \mathcal{D}_1 \times \mathcal{D}_2$.

Proposition 1.5.5 If all conditions of Assumption 1.5.4 are satisfied, then for the total derivative of function (1.5.30) along solutions of system (1.5.29) the inequality

(1.5.32) $$Dv(x,\eta)\big|_{(1.5.29)} \leq u^{\mathrm{T}} S u$$

holds true for all $x \in \mathcal{D}_1 \times \mathcal{D}_2$.

Here $u = (\psi(\|x_1\|), \psi(\|x_2\|))^{\mathrm{T}}$, and S is 2×2 matrix with elements

$$\sigma_{11} = \eta_1^2(\alpha_{11} + \alpha_{13}) + 2\eta_1\eta_2(\beta_{11} + \beta_{13} + \beta_{23}),$$

$$\sigma_{22} = \eta_2^2(\alpha_{21} + \alpha_{23}) + 2\eta_1\eta_2(\beta_{15} + \beta_{22} + \beta_{25}),$$

$$\sigma_{12} = \sigma_{21} = \eta_1^2\alpha_{12} + \eta_2^2\alpha_{22} + 2\eta_1\eta_2(\beta_{12} + \beta_{14} + \beta_{21} + \beta_{24}).$$

Proof We omit the proofs of Propositions 1.5.3 and 1.5.4 because they are similar to the known ones (see Martynyuk and Miladzhanov [1], and Djordjević [2]).

Estimates (1.5.31), (1.5.32) are sufficient for formulation of a new test for the presence of exponential polystability of separable motion in system (1.5.29).

Theorem 1.5.6 *Assume that differential equations of perturbed motion (1.5.29) are such that all conditions of Assumptions 1.5.3 and 1.5.4 are satisfied and moreover:*

(1) $\inf \dfrac{\psi_1^2(r) + \psi_2^2(r)}{\varphi_1^2(r) + \varphi_2^2(r)} = \alpha > 0$ *for all* $r \in [0, a)$;

(2) *in estimate (2.6.31) matrices* $\underline{C}$ *and* $\overline{C}$ *are positive definite;*

(3) *in inequality (1.5.32) matrix* S *is negative definite.*

Then the equilibrium state $x = 0$ *of system (1.5.29) is exponentially polystable in the small.*

Proof Designate $\lambda_1 = \lambda_m(H^{\mathrm{T}}\overline{C}H)$, $\lambda_2 = \lambda_M(H^{\mathrm{T}}\underline{C}H)$ and $\gamma = \lambda_M(S)$, $\gamma < 0$. By condition (1) of Theorem 1.5.6 $\|u\|^2 \geq \|u_2\|^2$ and therefore the sequence of inequalities

(1.5.33) $$-\|u\|^2 \leq -\alpha\|u_2\|^2 \leq -\dfrac{\alpha}{\lambda_1} v(x,\eta)$$

is satisfied. According to (1.5.33) inequality (1.5.32) becomes

(1.5.34) $$Dv(x,\eta) \leq -\Delta v(x,\eta),$$

1. CONTINUOUS SYSTEMS

where $\Delta = -\dfrac{\alpha\gamma}{\lambda_1} > 0$. From inequality (1.5.34) it is easy to find

$$
\begin{aligned}
\lambda_2 \|u_1\|^2 &= \lambda_2(\|x_1\|^{2r_1} + \|x_2\|^{2r_2}) \\
&\leq v(x_0,\eta)\exp[-\Delta(t-t_0)] \\
&\leq \lambda_1(\varphi_1^2(\|x_{10}\|) + \varphi_2^2(\|x_{20}\|))\exp[-\Delta(t-t_0)].
\end{aligned}
\tag{1.5.35}
$$

Since the functions $(\varphi_1,\varphi_2) \in K$, the fact that $\|x_{10}\| \leq \|x_0\|$ and $\|x_{20}\| \leq \|x_0\|$ implies $\varphi_1(\|x_{10}\|) \leq \varphi_1(\|x_0\|)$ and $\varphi_1(\|x_{10}\|) \leq \varphi_1(\|x_0\|)$. Consequently we get from (1.5.35)

$$
\begin{aligned}
\|x_1(t,x_0)\|^{2r_1} + \|x_2(t,x_0)\|^{2r_2} &\leq \frac{\lambda_1}{\lambda_2}(\varphi_1^2(\|x_{10}\|) + \varphi_2^2(\|x_{20}\|)) \\
\times \exp[-\Delta(t-t_0)] &\leq \frac{\lambda_1}{\lambda_2}(\varphi_1^2(\|x_0\|) + \varphi_2^2(\|x_0\|))\exp[-\Delta(t-t_0)]
\end{aligned}
\tag{1.5.36}
$$

for all $t \geq t_0$.

For arbitrary $\varepsilon > 0$ we take $\delta = \delta(\varepsilon) > 0$ according to the formula

$$
\delta(\varepsilon) = \min\left\{\varphi_1^{-1}\left[\left(\frac{\varepsilon\lambda_2}{2\lambda_1}\right)^{1/2}\right],\ \varphi_2^{-1}\left[\left(\frac{\varepsilon\lambda_2}{2\lambda_1}\right)^{1/2}\right]\right\}.
$$

Besides, from (1.5.36) we get the estimate of separable motions

$$
\|x_1(t,x_0)\|^{2r_1} + \|x_2(t,x_0)\|^{2r_2} \leq \varepsilon\exp[-\Delta(t-t_0)]
\tag{1.5.37}
$$

for all $t \geq t_0$ whenever $\|x_0\| < \delta$. This proves the theorem.

Example 1.5.1 Let perturbed motion equations be

$$
\begin{aligned}
\frac{dx_1}{dt} &= A_1 x_1 + B_1 x_1 \|x_1\|^{-r_1} \|x_2\|^{r_2}, \quad 0 < r_1 < 1, \\
\frac{dx_2}{dt} &= A_2 x_2 + B_2 x_2 \|x_1\|^{r_1} \|x_2\|^{-r_2}, \quad 0 < r_2 < 1,
\end{aligned}
\tag{1.5.38}
$$

where $x_1 \in R^{n_1}$, $x_2 \in R^{n_2}$, A_i, B_i, $i = 1,2$, are matrices of corresponding dimensions. In order to use Theorem 1.5.6 we construct the matrix-valued function $U(x) = [v_{ij}(\cdot)]$, $i,j = 1,2$, with elements:

$$
\begin{aligned}
v_{11}(x_1) &= (x_1^T x_1)^{r_1},\quad v_{22}(x_2) = (x_2^T x_2)^{r_2}, \\
v_{12}(x_1,x_2) &= v_{21}(x_1,x_2) = \alpha(x_1^T x_1)^{r_1/2}(x_2^T x_2)^{r_2/2},
\end{aligned}
\tag{1.5.39}
$$

where $\alpha = \text{const}$, $|\alpha| < 1$.

Denote $\lambda_1 = \lambda_M(A_1 + A_1^T)$, $\beta_1 = \lambda_M(B_1 + B_1^T)$, and $\lambda_2 = \lambda_M(A_2 + A_2^T)$, $\beta_2 = \lambda_M(B_2 + B_2^T)$ are maximal eigenvalues of the corresponding matrices.

It is easy to show that in region $D_1 \times D_2$

(1.5.40)
$$Dv_{11}(x_1)\big|_{(1.5.38)} \leq r_1 \lambda_1 \|x_1\|^{2r_1} + r_2 \beta_1 \|x_1\|^{r_1} \|x_2\|^{r_2},$$
$$Dv_{22}(x_2)\big|_{(1.5.38)} \leq r_2 \lambda_2 \|x_2\|^{2r_2} + r_1 \beta_1 \|x_1\|^{r_1} \|x_2\|^{r_2},$$
$$Dv_{12}(x_2)\big|_{(1.5.38)} \leq \frac{\alpha r_1}{2} \lambda_1 \|x_1\|^{r_1} \|x_2\|^{r_2} + \frac{\alpha r_1}{2} \lambda_1 \|x_2\|^{2r_2}$$
$$+ \frac{\alpha r_2}{2} \lambda_2 \|x_1\|^{r_1} \|x_2\|^{r_2} + \frac{\alpha r_2}{2} \lambda_2 \|x_1\|^{2r_1}.$$

Hence it follows that the variation of total derivative of function $v(x, \eta)$ by virtue of system (1.5.38) is estimated by the inequality

(1.5.41)
$$Dv(x, \eta)\big|_{(1.5.38)} \leq \psi^T S \psi,$$

where $\psi = (\|x_1\|^{r_1}, \|x_2\|^{r_2})^T$ and 2×2 matrix S has the elements

$$\sigma_{11} = \eta_1^2 r_1 \lambda_1 + \eta_1 \eta_2 \alpha r_2 \beta_2;$$
$$\sigma_{22} = \eta_2^2 r_2 \lambda_2 + \eta_1 \eta_2 \alpha r_1 \beta_1;$$
$$\sigma_{12} = \sigma_{21} = \eta_1^2 r_1 \beta_1 + \eta_2^2 r_2 \beta_2 + \alpha r_1 \eta_1 \eta_2 \lambda_1 + \alpha r_2 \eta_1 \eta_2 \lambda_2.$$

It is easy to check that all conditions of Theorem 1.5.6 are fulfilled for the function $U(x)$ with elements (1.5.39) if

(1.5.42) $\qquad \sigma_{11} < 0, \quad \sigma_{22} < 0, \quad \sigma_{11}\sigma_{22} - \sigma_{12}^2 > 0.$

Conditions (1.5.42) are sufficient for exponential polystability of system (1.5.35) motions.

1.5.4 Polystability by the first order approximations

System (1.5.2) is represented as two groups of equations

(1.5.43)
$$\frac{dx_1}{dt} = A(t)x_1 + B(t)x_2 + Y(t, x_1, x_2),$$
$$\frac{dx_2}{dt} = C(t)x_1 + D(t)x_2 + Z(t, x_1, x_2).$$

Here A, B, C and D are matrix functions of t continuous for all $t \in R_+$, the dimensions of which are coordinated with the dimensions of vectors $x_1 \in R^{n_1}$ and $x_2 \in R^{n_2}$, $n_1 + n_2 = n$. Vector functions Y and Z contain variables x_1, and x_2 in power higher than two, and together with linear approximation satisfy existence conditions for solutions of system (1.5.43).

1. CONTINUOUS SYSTEMS

Definition 1.5.5 State $x = (x_1^T, x_2^T)^T = 0$ of system (1.5.43) is *polystable*, if it is uniformly Liapunov stable and (simultaneously) exponentially x_1-stable, i.e. for any $\varepsilon > 0$ and $t_0 > 0$ one can find numbers $\delta(\varepsilon) > 0$ and $\gamma > 0$ such that for $\|x_0\| < \delta$ the inequalities

$$\|x(t; t_0, x_0)\| < \varepsilon, \quad \|x_1(t; t_0, x_0)\| < \varepsilon \exp[-\gamma(t - t_0)]$$

hold true for all $t \geq t_0$.

Theorem 1.5.7 *Suppose that the perturbed motion equations (1.5.43) are such that:*

(1) *the equilibrium state* $x = (x_1^T, x_2^T)^T$ *of system*

(1.5.44)
$$\frac{dx_1}{dt} = A(t)x_1 + B(t)x_2,$$
$$\frac{dx_2}{dt} = C(t)x_1 + D(t)x_2$$

is polystable in the sense of Definition 1.5.5;

(2) *vector functions Y and Z satisfy the conditions*

$$Y(t, 0, 0) = Y(t, 0, x_2) = 0, \quad Z(t, 0, 0) = Z(t, 0, x_2) = 0,$$

$$\frac{\|Y(t, x_1 x_2)\| + \|Z(t, x_1, x_2)\|}{\|x_1\|} \to 0$$

for $\|x_1\| + \|x_2\| \to 0$ *uniformly in t.*

Then the equilibrium state $x = (x_1^T, x_2^T)^T = 0$ of system (1.5.43) is polystable in the sense of Definition 1.5.5.

Proof If condition (1) of Theorem 1.5.7 is satisfied, it is possible to construct for system (1.5.44) the matrix-valued function $U(t, x)$ and to find vector $\eta \in R_+$ such that the function $v(t, x, \eta) = \eta^T U(t, x)\eta$ for all $t \geq 0$, $\|x\| < +\infty$, will satisfy the conditions

(a) $\|x_1\| \leq v(t, x, \eta) \leq M\|x\|$, $M = \text{const} > 0$.
(b) $|v(t, x', \eta) - v(t, x'', \eta)| \leq M\|x' - x''\|$,
(c) $Dv(t, x, \eta)\big|_{(1.5.44)} \leq -\alpha v(t, x, \eta)$.

It is easy to see that for the function $Dv(t, x, \eta)\big|_{(1.5.43)}$ the estimate

(1.5.45) $$Dv(t, x, \eta)\big|_{(1.5.43)} \leq -\alpha v(t, x, \eta) + H(t, x),$$

where $H(t,x) = (\text{grad } v, X(t,x))$, $X(t,x) = (Y^T, Z^T)^T$.

For $H(t,x)$ in estimate (1.5.45)

$$|H(t,x)| \leq \varepsilon M v(t,x,\eta), \quad (1.5.46)$$

where $\varepsilon \to 0$ as $\|x\| \to 0$, because of conditions (a), (b) imposed on function $v(t,x,\eta)$ and due to condition (2) of Theorem 1.5.7. If inequality (1.5.46) is true, there exists a β $(0 < \beta < d < +\infty)$ such that in the domain $t \geq 0$, $\|x\| \leq \beta$ estimate (1.5.45) becomes

$$Dv(t,x,\eta)\big|_{(1.5.44)} \leq -\alpha_1 v(t,x,\eta), \quad (1.5.47)$$

where $\alpha_1 = \text{const} > 0$. Note that for arbitrary solution $x(t;t_0,x_0)$ of system (1.5.44) with the initial conditions $t \geq 0$, $\|x_0\| \leq \delta$ $(0 < \delta < \beta)$ estimate $\|x(t;t_0,x_0)\| \leq \beta$ holds true at least on some interval (t_0, t^*). Therefore due to condition (a) imposed on function $v(t,x,\eta)$ we get from inequality (1.5.47)

$$\|x_1(t;t_0,x_0)\| \leq v(t, x(t;t_0,x_0)) \leq M\|x_0\| \exp[-\alpha_1(t-t_0)]. \quad (1.5.48)$$

Condition (2) of Theorem 1.5.7 and inequality (1.5.48) imply that there exists a constant $\alpha_2 = \text{const} \to 0$ as $\|x\| \to 0$ such that

$$\|X(t, x(t;t_0,x_0))\| \leq \alpha_2 \|x_0\| \exp[-\alpha_1(t-t_0)], \quad (1.5.49)$$

for all $t \in (t_0, t^*)$.

Let $K(t,\tau)$ be the Cauchy matrix of linear system (1.5.44). It is known that solution $x(t;t_0,x_0)$ of system (1.5.43) can be represented as

$$x(t;t_0,x_0) = K(t,t_0)x_0 + \int_{t_0}^{t} K(t,\tau) X(\tau, x(\tau;t_0,x_0))\, d\tau, \quad (1.5.50)$$

for all $t \geq t_0$. Since the state $x = (x_1^T, x_2^T)^T = 0$ of system (1.5.44) is uniformly Liapunov stable, there exists a constant $N > 0$ such that $\|K(t,t_0)\| \leq N$ for all $t \geq t_0$, $t_0 \geq 0$. In view of this fact and estimating (1.5.49) from (1.5.50) we get

$$\|x(t;t_0,x_0)\| \leq N(1 + \alpha_1^{-1}\alpha_2)\|x_0\| \quad (1.5.51)$$

for all $t \in (t_0, t^*)$.

1. CONTINUOUS SYSTEMS

Let ε be arbitrary small, $0 < \varepsilon < \beta$ so that $\delta < \min\{M^{-1}, [N(1 + \alpha_1^{-1}\alpha_2)]^{-1}\}\varepsilon$. Moreover, estimates (1.5.48) and (1.5.45) yield

(1.5.52)
$$\|x_1(t;t_0,x_0)\| \le \varepsilon \exp[-\alpha_1(t-t_0)],$$
$$\|x(t;t_0,x_0)\| < \varepsilon$$

for all $t \in (t_0, t^*)$.

Inequalities (1.5.52) hold for all values of time for which estimate (1.5.51) takes place. According to the choice ε, $\varepsilon < \beta$, estimate (1.5.52) is fulfilled for all $t \ge t_0$. This proves Theorem 1.5.7.

Note that if in Theorem 1.5.7 condition (2) is replaced by

(2)' in domain $t \ge 0$, $\|x\| < d < +\infty$, for given function $v(t,x,\eta)$ the inequality

$$\|Y(t,x_1,x_2)\| + \|Z(t,x_1,x_2)\| \le \gamma v(t,x,\eta),$$

where $\gamma = \text{const} > 0$, sufficiently small,

then the Theorem 1.5.7 remains valid.

Theorem 1.5.7 may be extended to systems more general than (1.5.43). In particular, consider the perturbed motion equations in the form

(1.5.53)
$$\frac{dx_1}{dt} = A(t)x_1 + B(t)x_2 + Y(t,x_1,x_2,x_3),$$
$$\frac{dx_2}{dt} = C(t)x_1 + D(t)x_2 + Z(t,x_1,x_2,x_3),$$
$$\frac{dx_3}{dt} = W(t,x_1,x_2,x_3).$$

In domain $\mathcal{D}_2 = \{t \ge t_0,\ \|x\| \le d < +\infty,\ \|x_3\| < +\infty\}$ we assume that the existence and uniqueness conditions are fulfilled for solutions of system (1.5.53) and other solutions of system (1.5.53) for which x_3 is extendable, i.e. definite for all $t \ge 0$ for which $\|x\| \le d$.

Definition 1.5.6 *The equilibrium state* $y = (x_1^T, x_2^T, x_3^T)^T = 0$ *of system (1.5.53) is polystable with respect to a part of the variables, if it is uniformly (x_1^T, x_2^T)-stable and (simultaneously) exponentially x_1-stable, i.e. for any values of ε, $t \ge t_0$, there exist numbers $\delta(\varepsilon) > 0$ and $\gamma > 0$ such that for $\|x_0\| + \|x_{30}\| < \delta$ the inequalities*

$$\|x(t;t_0,x_0,x_{30})\| < \varepsilon, \quad \|x_1(t;t_0,x_0,x_{30})\| < \varepsilon \exp[-\gamma(t-t_0)]$$

are fulfilled for all $t \ge t_0$.

The following assertion is proved in the same way as Theorem 1.5.7.

Theorem 1.5.8 *Assume that*

(1) *the equilibrium state* $x = (x_1^T, x_2^T)^T = 0$ *of system (1.5.44) is uniformly Liapunov stable and (simultaneously) exponentially x_1-stable;*

(2) *in domain $\mathcal{D}_2$ the conditions*

$$Y(t, 0, 0, 0) = Y(t, 0, x_2, x_3) = 0,$$
$$Z(t, 0, 0, 0) = Z(t, 0, x_2, x_3) = 0,$$
$$W(t, 0, 0, 0) = 0,$$
$$\frac{\|Y(t, x_1, x_2, x_3)\| + \|Z(t, x_1, x_2, x_3)\|}{\|x_1\|} \to 0$$

are fulfilled for $\|x_1\| + \|x_2\| \to 0$.

Then the equilibrium state $y = (x_1^T, x_2^T, x_3^T)^T$ of system (1.5.53) is polystable with respect to a part of the variables.

Example 1.5.2 Angular motion of a solid with respect to the mass center subjected to the linear moments of forces is described by the equation

(1.5.54) $$\frac{dx}{dt} = L(t)x + X(x),$$

where $x = (x_1, x_2, x_3) \in R^3$, L is a matrix 3×3 whose elements are continuous for all $t \in R_+$ functions characterizing the action of linear moments, dissipative and accelerating forces, and vector $X(x)$ is of the form

$$X(x) = ((I_2 - I_3)I_1^{-1}x_2 x_3, \ (I_3 - I_1)I_2^{-1}x_2 x_3, \ (I_1 - I_2)I_3^{-1}x_2 x_3)^T.$$

Here x_1, x_2, and x_3 are projections of the x-angular velocity vector on main central axes of inertia, I_i are main central moments of inertia.

Assume that the equilibrium state $x = (x_1, x_2, x_3) = 0$ of the linear system

(1.5.55) $$\frac{dx}{dt} = L(t)x, \quad x \in R^3,$$

is uniformly Liapunov stable and (simultaneously) exponentially (x_1, x_2)-stable. It is easy to verify that for the vector function $X(x)$ the condition (2) of Theorem 1.5.7 is satisfied and therefore, the equilibrium state $x =$

$(x_1, x_2, x_3)^T = 0$ of system (1.5.54) possesses the same properties as the linear approximation (1.5.55).

1.5.5 Polystability by pseudolinear approximations Consider the nonlinear system of differential equations

(1.5.56)
$$\frac{dx}{dt} = A(t,x)x + B(t,y)y + F(t,x,y),$$
$$\frac{dy}{dt} = D(x)x + G(t,x,y),$$

where $x \in R^{n_1}$, $y \in R^{n_2}$. Assume that functions $A(t,x)$, $B(t,y)$, $D(x)$, $F(t,x,y)$ and $G(t,x,y)$ are definite and continuous in the domain

$$D = \{(t,x,y) \in R_+ \times R^{n_1} \times R^{n_2} : t \geq 0, \; \|x\| \leq h, \; \|y\| \leq h\},$$
$$h = \text{const} > 0,$$

and functions $F(t,x,y)$ and $G(t,x,y)$ satisfy the inequalities

$$\|F(t,x,y)\| \leq c_1(x,y)\|x\|^{\gamma_1}, \quad \|G(t,x,y)\| \leq c_2\|x\|^{\gamma_2} \quad \text{for all} \quad (t,x,y) \in D.$$

Here function $c_1(x,y) \to 0$ as $\|x\| + \|y\| \to 0$, $F(t,0,0) = 0$, $G(t,0,0) = 0$ for all $t \in J_t^+$, $c_2 > 0$, $\gamma_1, \gamma_2 > 0$. We introduce the following definition.

Definition 1.5.7 The equilibrium state $x = 0$ of system (1.5.56) is called

(i) *x-polystable*, iff it is stable and asymptotically x-stable;
(ii) *uniformly x-polystable*, if it is uniformly stable and uniformly asymptotically x-stable;

Assumption 1.5.5 The pseudo-linear system

(1.5.57)
$$\frac{dx}{dt} = A(t,x)x$$

satisfies the following conditions

(1) the equilibrium state $x = 0$ of system (1.5.57) is uniformly asymptotically stable;

(2) there exists a function $v(t,x)$ continuously differentiable in the domain $H = \{(t,x)\colon t \geq 0,\ \|x\| \leq h\}$, positive definite and such that

$$\underline{c}(\|x\|)\|x\|^2 \leq v(t,x) \leq \bar{c}(\|x\|)\|x\|^2,$$

$$\left.\frac{dv}{dt}\right|_{(1.5.57)} \leq -\alpha(\|x\|)\|x\|^2,$$

$$\left\|\frac{\partial v}{\partial x}\right\| \leq \rho\|x\|^\alpha, \quad \rho > 0 \quad \alpha > 0,$$

where $\underline{c}, \bar{c}, \alpha \in C(R_+, R_+)$.

Consider a pseudo-linear approximation of system (1.5.56)

$$\frac{dx}{dt} = A(x)x + B(t,y)y,$$
$$\frac{dy}{dt} = D(x)x,$$

and construct a matrix-valued function $U(t,x,y)$. The diagonal elements of this function are taken as

$$v_{11}(x) = v(t,x), \quad v_{22}(y) = y^\mathrm{T} y.$$

To construct the non-diagonal elements $v_{12}(t,x,y)$ of the matrix-valued function we consider the equation
(1.5.58)
$$D_t v_{12} + (D_x v_{12})^\mathrm{T} A(t,x)x = -\frac{\eta_1}{2\eta_2}(D_x v(x))^\mathrm{T} B(t,y)y - \frac{\eta_2}{\eta_1} y^\mathrm{T} D(x)x$$

for some $\eta = (\eta_1, \eta_2)^\mathrm{T}$. Applying function $U(t,x,y)$ and vector η we construct a scalar function $v(t,x,y) = \eta^\mathrm{T} U(t,x,y)\eta$.

Theorem 1.5.9 *Assume that the perturbed motion equations are such that*

(1) *all conditions of Assumption 1.5.5 are satisfied;*
(2) *equation (1.5.58) has a solution in the form of a continuously differentiable function $v_{12}(t,x,y)$ admitting the estimates*

$$\underline{c}_{12}(x,y)\|x\|\|y\| \leq v_{12}(t,x,y) \leq \bar{c}_{12}(x,y)\|x\|\|y\|$$
$$\|D_x v_{12}\| \leq \rho_1 \|x\|^{\alpha_1}\|y\|^{\beta_1}; \quad \|D_x v_{12}\| \leq \rho_2 \|x\|^{\alpha_2}\|y\|^{\beta_2}, \quad \rho_1, \rho_2 > 0,$$

1. CONTINUOUS SYSTEMS 75

where $\underline{c}_{12} \in C(R^{n_1} \times R^{n_2}, R)$, $\bar{c}_{12} \in C(R^{n_1} \times R^{n_2}, R)$;
(3) matrices

$$\underline{C}(x,y) = \begin{pmatrix} \underline{c}_{11}(x) & \underline{c}_{12}(x,y) \\ \underline{c}_{12}(x,y) & 1 \end{pmatrix}, \quad \underline{c}_{11}(x) = \underline{c}(\|x\|),$$

$$\overline{C}(x,y) = \begin{pmatrix} \bar{c}_{11}(x) & \bar{c}_{12}(x,y) \\ \bar{c}_{12}(x,y) & 1 \end{pmatrix}, \quad \bar{c}_{11}(x) = \bar{c}(\|x\|),$$

satisfy in the domain $D = \{(x,y)\colon \|x\| \leq h,\ \|y\| \leq h\}$ the generalized Silvester conditions;
(4) there exists a constant $\varkappa > 0$ such that

$$-a(\|x\|)\eta_1^2 + \sup_{\|x\|=1} \frac{(D_y v_{12})^\mathrm{T} D(x)x + x^\mathrm{T} D^\mathrm{T}(x) D_y v_{12}}{\|x\|^2} < -\varkappa$$

for all $(t,x,y) \in D$;

(5)
$$\sup_{\|y\|=1} \frac{(D_x v_{12})^\mathrm{T} B(t,y)y + y^\mathrm{T} B^\mathrm{T}(t,y) D_x v_{12}}{\|y\|^2} \leq 0$$

for all $t \geq 0$ and $\|x\| \leq h$;
(6) $\alpha + \gamma_1 \geq 2$, $\alpha_1 + \gamma_1 \geq 2$, $\alpha_2 + \gamma_2 \geq 2$, $\beta_1 > 0$, $\beta_2 > 0$.

Then the equilibrium state $x = y = 0$ of system (1.5.56) is uniformly x-polystable.

Proof Conditions (2) from Assumption 1.5.5 and Theorem 1.5.9 for the components of matrix-valued function $U(t,x,y)$ allow us to estimate the scalar function $v(t,x,y,\eta) = \eta^\mathrm{T} U(t,x,y)\eta$ as

$$u^\mathrm{T} H^\mathrm{T} \underline{C}(x,y) H u \leq v(t,x,y,\eta) \leq u^\mathrm{T} H^\mathrm{T} \overline{C}(x,y) H u,$$

where $u = (\|x\|, \|y\|)^\mathrm{T}$, $H = \mathrm{diag}\,(\eta_1, \eta_2)$. Under condition (3) the function $v(t,x,y,\eta)$ is positive definite and decreasing. We shall estimate the total derivative of function $v(t,x,y,\eta)$ along solutions of system (1.5.56) taking into account conditions (4) and (5)

$$\begin{aligned}
\frac{dv}{dt}\bigg|_{(1.5.56)} &\leq -\varkappa\|x\|^2 + \eta_1^2 (D_x v)^\mathrm{T} F(t,x,y) + \eta_2^2 y^\mathrm{T} G(t,x,y) \\
&\quad + 2\eta_1\eta_2 (D_x v_{12})^\mathrm{T} F(t,x,y) + 2\eta_1\eta_2 (D_y v_{12})^\mathrm{T} G(t,x,y) \\
&\leq -\varkappa\|x\|^2 + \rho\eta_1^2 \|x\|^\alpha c_1(x,y)\|x\|^{\gamma_1} + \eta_2^2 \rho_2 c_2 \|x\|^{\alpha_2} \|y\|^{\beta_2} \|x\|^{\gamma_2} \\
&\quad + 2\eta_1\eta_2 \rho_1 c_1(x,y) \|x\|^{\alpha_1} \|y\|^{\beta_1} \|x\|^{\gamma_1} + 2\eta_1\eta_2 \rho_2 c_2 \|x\|^{\alpha_2} \|y\|^{\beta_2} \|x\|^{\gamma_2} \\
&\leq -\varkappa\|x\|^2 + c(x,y)\|x\|^2,
\end{aligned}$$

where the function $c(x,y) \to 0$ as $\|x\| + \|y\| \to 0$. Therefore there exists a magnitude $h_1 \leq h$ such that $c(x,y) < \varkappa/2$ for $\|x\|+\|y\| \leq h_1$. Thus in the domain $\widetilde{D} = \{(t,x,y): t \geq 0, \|x\| + \|y\| \leq h_1\}$ the derivative of function $v(t,x,y)$ along solutions of system (1.4.1) is estimated by the inequality

$$\left.\frac{dv}{dt}\right|_{(1.5.56)} \leq -\frac{\varkappa}{2}\|x\|^2.$$

From Theorem 2.5.2 from Martynyuk [9] we conclude on the uniform asymptotic stability of the equilibrium state $x = y = 0$ of system (1.5.56). The assertion on uniform asymptotic x-stability follows from Theorem 2.6.1 by Martynyuk [9] (see also Theorem 6.1 from Rumyantsev and Oziraner [1]).

1.6 Stability of Dynamical Systems in Metric Space

This section presents an approach to stability analysis of dynamical systems determined in metric space. The method of analysis of invariant sets of dynamical systems was proposed by Zubov on the basis of the generalized direct Liapunov method. In our approach a generalized comparison principle is used together with the idea of multicomponent mapping (matrix-valued Liapunov functional).

1.6.1 Basic concepts and definitions Let X be a set of elements (no matter of what nature) and a measure $\rho(x,y)$ be defined for $x, y \in X$.

Definition 1.6.1 (X, ρ) is a metric space if the following conditions are fulfilled for any $x, y, z \in X$:

(i) $\rho(x,y) \geq 0$,
(ii) $\rho(x,y) = 0 \Leftrightarrow x = y$,
(iii) $\rho(x,y) = \rho(y,x)$,
(iv) $\rho(x,y) \leq \rho(x,z) + \rho(y,z)$,

and, additionally, for any $X_0 \subseteq X$, $\rho(x, X_0) = \inf\limits_{y \in X_0} \rho(x,y)$.

Definition 1.6.2 A metric space (T, ρ) is called a temporal space if:
(i) T is completely ordered by the ordering "<";
(ii) T has a minimum element $t_{\min} \in T$, i.e. $t_{\min} < t$ for any $t \in T$, such that $t \neq t_{\min}$;
(iii) for any $t_1, t_2, t_3 \in T$ such that $t_1 < t_2 < t_3$ it holds that

$$\rho(t_1, t_3) = \rho(t_1, t_2) + \rho(t_2, t_3);$$

(iv) T is unbounded from above; i.e., for any $M > 0$, there exists $t \in T$ such that $\rho(t, t_{\min}) > M$.

Definition 1.6.3 Let (X, ρ) be a metric space with a subset $A \subseteq X$ and let (T, ρ) be a temporal space with subset $T \subseteq R_+$. A mapping $p(\cdot, a, \tau_0) \colon T_{a,\tau_0} \to X$ is called *motion* if $p(\tau_0, a, \tau_0) = a$, where $a \in A$, $\tau_0 \in T$ and $T_{a,\tau_0} = [\tau_0, \tau_1) \cap T$ for $\tau_1 > \tau_0$, with τ_1 being a finite value or infinity.

Definition 1.6.4 Let $T_{a,\tau_0} \times \{a\} \times \{\tau_0\} \to X$ denote the set of mappings of $T_{a,\tau_0} \times \{a\} \times \{\tau_0\}$ into X, $\Lambda = \bigcup_{(a,\tau_0) \in A \times T}(T_{a,\tau_0} \times \{a\} \times \{\tau_0\} \to X)$ and S be a family of motions; i.e.,

$$S \subseteq \{p(\cdot, a, \tau_0) \in \Lambda \colon p(\tau_0, a, \tau_0) = a\}.$$

Then the tuple (T, X, A, S) of sets and spaces is called *a dynamical system*.

Note that Definition 1.6.4 possesses some generality. Specifically,

(i) if X is a normed linear space and every motion $p(\tau, a, \tau_0)$ is assumed to be continuous with respect to τ, a and τ_0, then Definition 1.6.4 corresponds to the concept of a family of motions in Hahn [1];

(ii) under some additional conditions imposed on $p(\tau, a, \tau_0)$ (see Zubov [1] P.183–184), Definition 1.6.4 reduces to the concept of a general system introduced by Zubov.

In what follows, we consider dynamical systems satisfying the standard semigroup property

$$p(\tau_2, p(\tau_1, a, \tau_0)) = p(\tau_2 + \tau_1, a, \tau_0)$$

for all $a \in A$ and any $\tau_1, \tau_2 \in R_+$.

Definition 1.6.5 A dynamical system (R_+, X, A, S) is called *continuous* if any of its motions $p \in S$ is continuous; i.e., any mapping $p(\cdot, a, \tau_0) \colon T_{a,\tau_0} \to X$ is continuous.

Let (X_1, ρ_1) and (X_2, ρ_2) be metric spaces, and let (R_+, X_1, A_1, S_1) be a continuous dynamical system. We assume that the space X_1 is a Decart product of spaces $X_{11}, X_{12}, \ldots, X_{1m}$, on which the multicomponent mapping

(1.6.1) $\qquad U(t, x) \colon T \times X_{11} \times X_{12} \times \ldots \times X_{1m} \to X_2$

is acting.

It is assumed that the mapping $U\colon R_+ \times X_{11} \times X_{12} \times \ldots \times X_{1m} \to X_2$ has the following properties: for any motion $p(\,\cdot\,,a,t_0) \in S_1$, the function $q(\,\cdot\,,b,t_0) = U(\,\cdot\,, p(\,\cdot\,,a,t_0),\,)$ with initial value $b = U(t_0,a)$ is another motion for which $T_{a,t_0} = T_{b,t_0}$, $b \in A_2 \subset X_2$.

Let S_2 denote the set of motions q determined by initial values $a \in A_1$ and $t_0 \in R_+$. Then (R_+, X_2, A_2, S_2) is a continuous dynamical system.

The mapping given by (1.6.1) induces a mapping of S_1 into S_2, denoted by $\mathfrak{M}$; i.e., $S_2 = \mathfrak{M}(S_1)$. Moreover, we denote by $M_1 \subset A_1$ and $M_2 \subset A_2$ some sets invariant under S_1 and S_2, respectively. The set M_2 is then defined by the formula

$$(1.6.2) \quad M_2 = U(R_+ \times M_1) = \{x_2 \in X_2 \colon x_2 = U(t', x_1) \\ \text{for some } x_1 \in M_1 \text{ and } t' \in R_+,\}.$$

In what follows, we consider continuous dynamical systems (R_+, X_1, A_1, S_1) and (R_+, X_2, A_2, S_2) with invariant sets $M_1 \subset A_1$ and $M_2 \subset A_2$, respectively.

Definition 1.6.7 Multicomponent *mapping* (1.6.1)

$$(1.6.3) \quad V\colon R_+ \times X_{11} \times X_{12} \times \ldots \times X_{1m} \to X_2$$

preserves some type of stability of a continuous dynamical system if the sets

$$(1.6.4) \quad S_2 = \mathfrak{M}(S_1) \triangleq \{q(\,\cdot\,,b,t_0)\colon q(t,b,t_0) = U(t, p(t,a,t_0)), \\ p(\,\cdot\,,a,t_0) \in S_1, \quad \eta \in R^m, \quad b = U(t_0,a), \\ T_{b,t_0} = T_{a,t_0}, \quad a \in A_1, \quad t_0 \in R_+\}$$

and M_2 (see formula (1.6.2)) satisfy the following conditions:

(i) the invariance of (S_1, M_1) is equivalent to the invariance of (S_2, M_2);
(ii) some type of stability of (S_1, M_1) is equivalent to the same type of stability of (S_2, M_2).

1.6.2 Sufficient conditions for the stability of a dynamical system

Note that the mapping U induces a mapping $\mathfrak{M}\colon S_1 \to S_2$, that preserves some types of stability of (S_1, M_1) and $(S_2, U(R_+ \times M_1))$.

1. CONTINUOUS SYSTEMS

Theorem 1.6.1 *Let a dynamical system (R_+, X_1, A_1, S_1) be assigned a comparison system (R_+, X_2, A_2, S_2) by means of a multicomponent mapping $U(t, p)\colon R_+ \times X_1 \to X_2$. Suppose that there exist closed sets $M_i \subset A_i$, $i = 1, 2$, and the following conditions are fulfilled:*

(1) for $\mathfrak{M}(S_1)$ and S_2, $\mathfrak{M}(S_1) = S_2$;

(2) there exist constant $m \times m$ matrix A_i, $i = 1, 2$, and comparison functions $\psi_1, \psi_2 \in K$ such that

$$(1.6.5) \qquad \psi_1^{\mathrm{T}} A_1 \psi_1 \leq \rho_2(U(t, p), M_2) \leq \psi_2^{\mathrm{T}} A_2 \psi_2$$

for all $p \in X_1$ and $t \in R_+$, where

$$\psi_1 = (\psi_{11}(\rho_1(p, M_1)), \ldots, \psi_{1m}(\rho_1(p, M_1)))^{\mathrm{T}},$$
$$\psi_2 = (\psi_{21}(\rho_1(p, M_1)), \ldots, \psi_{2m}(\rho_1(p, M_1)))^{\mathrm{T}}.$$

Here, ρ_1 and ρ_2 are metrics defined on X_1 and X_2, respectively.

If the matrices A_i, $i = 1, 2$, are positive definite, then the following is true:

(1) the invariance of (S_2, M_2) implies the invariance of (S_1, M_1);

(2) the stability, uniform stability, asymptotic stability, or uniform asymptotic stability of (S_2, M_2) implies the respective type of stability of (S_1, M_1);

(3) if in estimate (1.6.5) $\psi_1^{\mathrm{T}} A_1 \psi_1 = a(\rho_1(p, M_1))^b$, where $a > 0$ and $b > 0$, then the exponential stability of (S_2, M_2) implies the exponential stability of (S_1, M_1).

Proof of item (1) Let (S_2, M_2) be an invariant pair. Then, for any $a \in M_1$ and any motion $p(\,\cdot\,; a, t_0) \in S_1$, we find that $q(\,\cdot\,; b, t_0) = U(t, p(\,\cdot\,; a, t_0)) \in S_2$, where $b = U(t_0, a)$. This follows from condition (1) in Theorem 1.6.2 and from the definition of $\mathfrak{M}(S_1)$ by formula (1.6.4). Moreover, the invariance of (S_2, M_2) implies that $q(t; b, t_0) = U(t, p(t; a, t_0)) \in M_2$ for all $t \in T_{b, t_0} = T_{a, t_0}$. Since M_1 and M_2 are closed and the matrices A_1 and A_2 are positive definite and satisfy (1.6.5), we conclude that $p(t; a, t_0) \in M_1$ for all $t \in T_{a, t_0}$. This implies the invariance of (S_1, M_1).

Proof of item (2) Assume that (S_2, M_2) is stable. Then, by the definition of stability, for any $\varepsilon_2 > 0$ and $t_0 \in R_+$, there exists $\delta_2 = \delta_2(t_0, \varepsilon_2) >$

0 such that $\rho_2(q(t;b,t_0), M_2) < \varepsilon_2$ for all $q(\,\cdot\,;b,t_0) \in S_2$ and all $t \in T_{b,t_0}$ whenever $\rho_2(b, M_2) < \delta_2(t_0, \varepsilon_2)$. Estimates (1.6.5) can be transformed into

(1.6.6) $\quad \lambda_m(A_1)\widetilde{\psi}_1(\rho_1(p, M_1)) \leq \rho_2(U(t,p), M_2) \leq \lambda_M(A_2)\widetilde{\psi}_2(\rho_1(p, M_1)).$

Here $\lambda_m(A_1) > 0$ and $\lambda_M(A_2) > 0$ are the minimum and maximum eigenvalues of the positive definite matrices A_1 and A_2, and $\widetilde{\psi}_1, \widetilde{\psi}_2 \in K$ are such that

$$\psi_1^{\mathrm{T}}(\rho_1(p, M_1))\psi_1(\rho_1(p, M_1)) \geq \widetilde{\psi}_1(\rho_1(p, M_1))$$

and

$$\psi_2^{\mathrm{T}}(\rho_1(p, M_1))\psi_2(\rho_1(p, M_1)) \geq \widetilde{\psi}_2(\rho_1(p, M_1)).$$

Since (S_2, M_2) is stable, for any $\varepsilon > 0$ and any $t_0 \in R_+$, we choose $\varepsilon_2 = \lambda_m(A_1)\widetilde{\psi}_1(\varepsilon)$ and $\delta_1 = \lambda_M^{-1}(A_2)\widetilde{\psi}_2^{-1}(\delta_2)$. Assuming that $\rho_1(a, M_1) < \delta_1$ and taking into account (1.6.6), we obtain

$$\rho_2(b, M_2) \leq \lambda_M(A_2)\widetilde{\psi}_2(\rho_1(a, M_1)) < \lambda_M(A_2)\widetilde{\psi}_2(\delta_1)$$
$$= \lambda_M(A_2)\widetilde{\psi}_2(\lambda_M^{-1}(A_2)\widetilde{\psi}_2^{-1}(\delta_2)) = \delta_2.$$

It follows that, for all motions $q(\,\cdot\,;b,t_0) \in S_2$, the estimate $\rho_2(q(t;b,t_0), M_2) < \varepsilon_2$ holds for all $t \in T_{b,t_0}$. Returning to estimates (1.6.6), we find that, for all $p(\,\cdot\,;a,t_0) \in S_1$ and all $t \in T_{a,t_0} = T_{b,t_0}$, where $b = U(t_0, a)$, we have

$$\rho_1(p(t;a,t_0), M_1) \leq \lambda_m^{-1}(A_1)\widetilde{\psi}_1^{-1}(\rho_2(V(p(t;a,t_0)), M_2))$$
$$\leq \lambda_m^{-1}(A_1)\widetilde{\psi}_1^{-1}(\rho_2(q(t;b,t_0), M_2)) \leq \lambda_m^{-1}(A_1)\widetilde{\psi}_1^{-1}(\lambda_m(A_1)\widetilde{\psi}_1(\varepsilon)) = \varepsilon,$$

whenever $\rho_1(a, M_1) < \delta_1$. It follows that (S_1, M_1) is stable.

It is well known that a system motion is asymptotically stable if it is stable and attracting. Assume that (S_2, M_2) is attracting. Then, for any $t_0 \in R_+$ there exists $\Delta_2 = \Delta_2(t_0) > 0$ such that, for all $q(\,\cdot\,;b,t_0) \in S_2$, the limit relation

$$\lim_{t \to \infty} \rho_2(q(t;b,t_0), M_2) = 0,$$

holds true whenever $\rho_2(b, M_2) < \Delta_2$. In other words, for any $\varepsilon_2 > 0$, there exists $\tau = \tau(\varepsilon_2, t_0, q) > 0$ with $q = q(\,\cdot\,;b,t_0) \in S_2$ such that $\rho_2(q(t;b,t_0), M_2) < \varepsilon_2$ for all $t \in T_{b,t_0+\tau}$, whenever $\rho_2(b, M_2) < \Delta_2$. According to condition (1) in Theorem 1.6.1, for any motion $p(\,\cdot\,;a,t_0) \in S_1$, we set $b = U(t_0, a)$. Then $q(\,\cdot\,; b, t_0) = U(p(\,\cdot\,; a, t_0)) \in S_2$. Furthermore, for any $\varepsilon_1 > 0$, we choose $\varepsilon_2 = \lambda_m(A_1)\widetilde{\psi}_1(\varepsilon_1)$ and set $\Delta_1 = \lambda_M^{-1}(A_2)\widetilde{\psi}_2^{-1}(A_2)$. For any motion $p(\,\cdot\,; a, t_0) \in S_1$, we then have

$$\rho_2(b, M_2) \leq \lambda_M(A_2)\widetilde{\psi}_2(\rho_1(a, M_1)) < \lambda_M(A_2)\widetilde{\psi}_2(\Delta_1) = \Delta_2$$

whenever $\rho_1(a, M_1) < \Delta_1$ and $t \in T_{a,t_0+\tau} = T_{b,t_0+\tau}$. Hence, $\rho_2(q(t; a, t_0), M_2) < \varepsilon_2 = \lambda_m(A_1)\widetilde{\psi}_1(\varepsilon_1)$ for all $t \in T_{a,t_0+\tau}$. Returning to estimate (1.6.2), we find that

$$\rho_1(p(t; a, t_0), M_1) \leq \lambda_m^{-1}(A_1)\widetilde{\psi}_1^{-1}(\rho_2(q(t; a, t_0), M_2)) < \lambda_m^{-1}(A_1)\widetilde{\psi}_1^{-1}(\varepsilon_1),$$

i.e., (S_1, M_1) is an attractive pair. Thus, if (S_2, M_2) is asymptotically stable, then (S_1, M_1) is asymptotically stable as well.

The statement on uniform stability and uniform asymptotic stability are proved following the same scheme, but δ_2 and Δ_2 are chosen to be independent of $t_0 \in R_+$.

Let us prove statement (3) of theorem. Assume that (S_2, M_2) is exponentially stable. Then there exists $\alpha_2 > 0$ and, for any $\varepsilon_2 > 0$, there exists $\delta_2 = \delta_2(\varepsilon_2) > 0$ such that for any motion $q(\,\cdot\,; b, t_0) \in S_2$ and all $t \in T_{b,t_0}$

$$\rho_2(q(t; b, t_0), M_2) < \varepsilon_2 e^{-\alpha_2(t-t_0)}$$

whenever $\rho_2(b, M_2) < \delta_2$. According to condition (1) in Theorem 1.6.1, for any motion $p(\,\cdot\,; a, t_0) \in S_1$, there exists a motion $q(\,\cdot\,; b, t_0) = U(p(\,\cdot\,; a, t_0)) \in S_2$, where $b = U(t_0, a)$. Furthermore, for any $\varepsilon_1 > 0$, we choose $\varepsilon_2 = a\varepsilon_1^b$. Let $\alpha_1 = \alpha_2/b$ and $\delta_1 = \lambda_M^{-1}(A_2)\psi_2^{-1}(\delta_2)$. For $p(t; a, t_0) \in M_1$ with $\rho_1(a, M_1) < \delta_1$, in view of (1.6.6), we obtain

$$\rho_2(b, M_2) \leq \lambda_M(A_2)\widetilde{\psi}_2(\rho_1(a, M_1)) < \lambda_M(A_2)\widetilde{\psi}_2(\delta_1) = \delta_2.$$

Consequently,

$$\rho_2(q(t; b, t_0), M_2) < \varepsilon_2 e^{-\alpha_2(t-t_0)}$$

for all $t \in T_{b,t_0}$.

According to the hypothesis of Theorem 1.6.1, we have to set

$$\psi_1^T A_1 \psi_1 = a(\rho_1(p, M_1))^b$$

in (1.6.6). It is easy to see that

$$\rho_1(p(t; a, t_0), M_1) < \left(\frac{\varepsilon_2}{a}\right)^{1/b} e^{-\frac{\alpha_2}{b}(t-t_0)} = \varepsilon_1 e^{-\alpha_1(t-t_0)}$$

for all $t \in T_{a,t_0}$. Thus, (S_1, M_1) is exponentially stable.

1.6.3 Stability analysis of hybrid systems Many physical and technical problems of the real world are modelled by mixed systems of equations and correlations. For example, in motion control theory the feedback consists of several interconnected blocks. These blocks are described by equations of different types. Such systems are called *hybrid*. Under certain assumptions real hybrid system σ can correspond to the dynamical system (T, X, A, S) in metric space.

Assume that (X, ρ) and (X_i, ρ_i), $i = 1, 2, \ldots, m$, are metric spaces. Let $X = X_1 \times X_2 \times \ldots \times X_m$ and there exist constants $a_1, a_2 > 0$ such that

$$a_1 \rho(x, y) \leq \sum_{i=1}^{m} \rho_i(x_i, y_i) \leq a_2 \rho(x, y) \tag{1.6.7}$$

for all $x, y \in X$, where $x = (x_1, \ldots, x_m)^{\mathrm{T}}$, $y = (y_1, \ldots, y_m)^{\mathrm{T}}$, $x_i \in X_i$, $y_i \in X_i$, $i = 1, 2, \ldots, m$. Further on we will assume that

$$\rho(x, y) = \sum_{i=1}^{m} \rho_i(x_i, y_i). \tag{1.6.8}$$

Definition 1.6.8 (cf. Michel, Wang *et al.* [1]) Dynamical system (T, X, A, S) is *hybrid*, if its metric space (X, ρ) consists of metric spaces (X_i, ρ_i), $i = 1, 2, \ldots, m$, where X_i are nontrivial unsplit with metrics $\rho_i(x_i, y_i)$, and if there exist at least two metric spaces X_i and X_j, $1 \leq i \neq j \leq m$, which are not isometric.

The proposition below is necessary when the multicomponent mapping is made by matrix-valued functional.

Proposition 1.6.1 Let multicomponent mapping $U(t, x) \colon T \times X \to X_2$ be performed by matrix-valued functional $U(t, x) = [v_{ij}(t, x)]$, $i, j = 1, 2, \ldots, m$, for the elements of which:

(1) $v_{ii} \in C(R_+ \times X, R_+)$, $i = 1, 2, \ldots, m$, $v_{ij} \in C(R_+ \times X, R)$ for all $i \neq j$ and for all $x \in X$ and $t \in R_+$;

(2) there exist comparison functions $\varphi_{i1}, \varphi_{i2}$ of class K, positive constants $\underline{c}_{ii} > 0$, $\bar{c}_{ii} > 0$ and arbitrary constants $\underline{c}_{ij} \in R$, $\bar{c}_{ij} \in R$ for $i \neq j$ such that

$$\begin{aligned}
\underline{c}_{ii} \varphi_{i1}^2(\rho_i(x_i, M_i)) &\leq v_{ii}(t, x) \leq \bar{c}_{ii} \varphi_{i2}^2(\rho_i(x_i, M_i)), \\
\underline{c}_{ij} \varphi_{i1}(\rho_i(x_i, M_i)) \varphi_{j1}(\rho_j(x_j, M_j)) &\leq v_{ij}(t, x) \\
&\leq \bar{c}_{ij} \varphi_{i2}(\rho_i(x_i, M_i)) \varphi_{j2}(\rho_j(x_j, M_j))
\end{aligned} \tag{1.6.9}$$

for all $x_i \in X_i$, $x \in X$ and $t \in R_+$.

Then for the functional

$$v(t,x,\eta) = \eta^{\mathrm{T}} U(t,x)\eta, \quad \eta \in R^m_+, \quad \eta_i > 0,$$

the bilaterel inequality

(1.6.10)
$$\begin{aligned}u_1^{\mathrm{T}}(\rho(x,M))H^{\mathrm{T}}\underline{C}Hu_1(\rho(x,M)) &\le v(t,x,\eta) \\ &\le u_2^{\mathrm{T}}(\rho(x,M))H^{\mathrm{T}}\overline{C}Hu_2(\rho(x,M))\end{aligned}$$

holds for all $x \in X$ and $t \in R_+$, where

$$H = \mathrm{diag}\,(\eta_1, \eta_2, \ldots, \eta_m),$$
$$\underline{C} = [\underline{c}_{ij}], \quad \overline{C} = [\overline{c}_{ij}], \quad i,j = 1,2,\ldots,m,$$
$$u_1(\cdot) = (\varphi_{i1}(\rho_1(x_1,M_1)), \ldots, \varphi_{m1}(\rho_m(x_m,M_m)))^{\mathrm{T}},$$
$$u_2(\cdot) = (\varphi_{i2}(\rho_1(x_1,M_1)), \ldots, \varphi_{m2}(\rho_m(x_m,M_m)))^{\mathrm{T}}.$$

Proof Estimate (1.6.10) is obtained by direct substitution by estimates (2) of Proposition 1.6.1 in the expression

$$v(t,x,\eta) = \sum_{i=1}^m \sum_{j=1}^m v_{ij}(t,x)\eta_i\eta_j.$$

Theorem 1.6.2 *Assume that behaviour of the hybrid system Σ is correctly described by the dynamical system (T, X, A, S), where $T = R_+$, $X = X_1 \times \ldots \times X_m$ and X_i are subspaces with metrics ρ_i, $i = 1, 2, \ldots, m$. Let $M_i \subset X_i$ and $M = M_1 \times M_2 \times \ldots \times M_m$ be an invariant set. If*
(1) *there exist functionals $v_{ij}(t,x)$ mentioned in Proposition 1.6.1;*
(2) *given functionals $v_{ij}(t,x)$ and a vector $\eta \in R^m_+$, $\eta > 0$, there exist bounded for all $x \in X$ functions $\Phi_{ij}(x,\eta)$, $i,j = 1,2,\ldots,m$, and comparison functions φ_{i3} of class K such that*

$$D^+ v(t,x,\eta)|_{(S)} \le u_3^{\mathrm{T}} \Phi(x,\eta) u_3$$

on the system of motions S for all $x \in X$ and $t \in R_+$, where

$$u_3(\rho(x,M)) = (\varphi_{13}(\rho_1(x_1,M_1)), \ldots, \varphi_{m3}(\rho_m(x_m,M_m)))^{\mathrm{T}}.$$

Then

(a) *if matrices* $B_1 = H^\mathrm{T}\underline{C}H$, $B_2 = H^\mathrm{T}\overline{C}H$ *are positive definite and constant* $m \times m$ *matrix* $\overline{\Phi} \geq \dfrac{1}{2}(\Phi^\mathrm{T}(x,\eta)+\Phi(x,\eta))$ *for all* $x \in X$ *is negative semi-definite, then the couple* (S,M) *is uniformly stable;*

(b) *if matrices* B_1 *and* B_2 *are positive definite and matrix* $\overline{\Phi}$ *is negative definite, then the couple* (S,M) *is uniformly asymptotically stable;*

(c) *if matrices* B_1 *and* B_2 *are positive definite, matrix* $\overline{\Phi}$ *is negative semidefinite, the set* M *is bounded and the comparison functions* $\varphi_{i1}, \varphi_{i2} \in KR$ *class* $i = 1, 2, \ldots, m$, *then the family of motion* S *is uniformly bounded;*

(d) *if in condition (c) the matrix* $\overline{\Phi}$ *is negative definite, then the family of motions* S *is uniformly limitedly bounded and the couple* (S,M) *is uniformly asymptotically stable in the whole;*

(e) *if there exist constants* a_1, a_2, b, c *such that*

$$a_1 r^b \leq u_1^\mathrm{T}(\rho(x,M))H^\mathrm{T}\underline{C}u_1(\rho(x,M)),$$
$$u_2^\mathrm{T}(\rho(x,M))H^\mathrm{T}\overline{C}u_2(\rho(x,M)) \leq a_2 r^b,$$
$$\varphi_3^\mathrm{T}\overline{\Phi}\varphi_3 \geq cr^b$$

for all $r \in R_+$, *then the couple* (S,M) *is exponentially stable in the whole.*

Proof Let us prove statement (a) of Theorem 1.6.2. Under condition (1) of Theorem 1.6.2 the functional $v(t,x,\eta)$ is positive definite and decreasing because matrices B_1 and B_2 are positive definite. Under condition (2) of Theorem 1.6.2 the functional $D^+v(t,x,\eta)$ on the system of motions S is negative semidefinite due to restrictions on matrix $\overline{\Phi}$. In this case the functional $v(t,x,\eta)$ is nonincreasing for all $t \geq 0$ along the system of motions S. Further, given $\varepsilon > 0$ we compute $\lambda = \inf\limits_{t \geq 0} v(t,x,\alpha)$ for $\rho(x,M) = \varepsilon$. Because of estimate (1.6.9) we can find by value λ the value $\delta > 0$ such that for $\rho(x,M) < \delta$ the estimate $v(t,x,\alpha) < \lambda$ holds for all $t \geq 0$. Now we show that the obtained value $\delta > 0$ corresponds to the given $\varepsilon > 0$, i.e. for $\rho(x,M) < \delta$ the inequality

$$\rho(q(t;a,t_0), M) < \varepsilon$$

holds for all $t \geq 0$. Assume on the contrary, let there exist a motion $q(t;a,t_0) \in S$ such that for some value $t^* \in R_+$ the inequality $\rho(q(t^*;a,t_0),$

$M) = \varepsilon$ take place. Then we get

$$v(t, q(t^*; a, t_0), \alpha) \geq \lambda,$$

but due to condition (a) of Theorem 1.6.2 the functional $v(t, x, \alpha)$ is nonincreascent along the system of motions S. Therefore

$$v(t, q(t; a, t_0), \alpha) \leq v(t, x, \alpha) < \lambda$$

for any $q(t; a, t_0) \in S$.

The contradiction obtained shows that the system of motions S of the hybrid system Σ is uniformly (S, M) stable.

The proof of statements (b) – (e) of Theorem 1.6.2 is similar to that of statement (a) following the Liapunov (ε, δ)-technique.

1.6.4 Stability of the two-component systems analysis We consider a hybrid two-component system

(1.6.11) $$\frac{dx}{dt} = X(t, x(t)) + g_1(t, z, x(t), w(t, z)), \quad x(t_0) = x_0,$$

(1.6.12) $$\frac{\partial w}{\partial t} = L(t, x, \partial/\partial z)w + g_2(t, z, x(t), w(t, z)),$$

where

$$w(t_0, z) = w^0(z), \quad M(t, z, \partial/\partial z)w|_{\partial \Omega} = w^1(t, s), \quad s \in \partial\Omega, \quad \Omega \subset R^k,$$
$$X \colon T_0 \times U \to R^n, \quad L \colon B_1 \to B_2, \quad M \colon B_1 \to B_3, \quad w^0 \in B_4,$$

L, M are some differential operators and $B_1, \ldots, B_4$ are Banach spaces.

A hybrid system (1.6.11), (1.6.12) consists of the independent subsystems

(1.6.13) $$\frac{dx}{dt} = X(t, x(t)),$$

(1.6.14) $$\frac{\partial w}{\partial t} = L(t, z, \partial/\partial z)w$$

and interconnection functions between them

$$g_1 = g_1(t, z, x, w) \colon T_0 \times \Omega \times H \times Q \to R^n,$$
$$g_2 = g_2(t, z, x, w) \colon T_0 \times \Omega \times H \times Q \to R^m.$$

Let us introduce assumptions on subsystems (1.6.13), (1.6.14) and interconnection functions between them.

Assumption 1.6.1 There exist functions $v_{ij} \in C(R_+ \times H \times Q, R)$, $i,j = 1,2$, $v_{ij}(t,x,w)$ is locally Lipshitzian in x and w, functions of comparison $\varphi_i, \psi_i \in K$, $i = 1,2$, and positive constants $\underline{\alpha}_{ii}, \overline{\alpha}_{ii} > 0$, $i = 1,2$, and arbitrary constants $\underline{\alpha}_{12}, \overline{\alpha}_{12}$ such that

$$\underline{\alpha}_{11}\varphi_1^2(\|x\|) \leq v_{11}(t,x,w) \leq \overline{\alpha}_{11}\varphi_2^2(\|x\|);$$
$$\underline{\alpha}_{22}\psi_1^2(\|x\|) \leq v_{22}(t,x,w) \leq \overline{\alpha}_{22}\psi_2^2(\|x\|);$$
$$\underline{\alpha}_{12}\varphi_1(\|x\|)\psi_1(\|x\|) \leq v_{12}(t,x,w) \leq \overline{\alpha}_{12}\varphi_2(\|x\|)\psi_2(\|x\|)$$

for all $x \in H$, $w \in Q$ and $t \geq 0$.

Lemma 1.6.1 *If all the conditions of Assumption 1.6.1 are fulfilled and matrices*

$$A_1 = \begin{pmatrix} \underline{\alpha}_{11} & \underline{\alpha}_{12} \\ \underline{\alpha}_{21} & \underline{\alpha}_{22} \end{pmatrix}, \quad \underline{\alpha}_{12} = \underline{\alpha}_{21},$$

$$A_2 = \begin{pmatrix} \overline{\alpha}_{11} & \overline{\alpha}_{12} \\ \overline{\alpha}_{21} & \overline{\alpha}_{22} \end{pmatrix}, \quad \overline{\alpha}_{12} = \overline{\alpha}_{21},$$

are positive definite, then the function

(1.6.15) $$v(t,x,w) = \eta^{\mathrm{T}} U(t,x,w)\eta,$$

where $\eta = (\eta_1, \eta_2)^{\mathrm{T}}$, $\eta_i > 0$, is positive definite and decreasing.

Proof We introduce the notations

$$r = (\varphi_1(\|x\|), \psi_1(\|w\|))^{\mathrm{T}}, \quad q = (\varphi_2(\|x\|), \psi_2(\|w\|))^{\mathrm{T}}, \quad B = \begin{pmatrix} \eta_1 & 0 \\ 0 & \eta_2 \end{pmatrix}.$$

Having fulfilled the conditions of Assumption 1.6.1 for the function (1.6.15) such a bilateral estimation

(1.6.16) $$r^{\mathrm{T}} B^{\mathrm{T}} A_1 B r \leq \eta^{\mathrm{T}} U(t,x,w)\eta \leq q^{\mathrm{T}} B^{\mathrm{T}} A_2 B q$$

holds.

By virtue of conditions of the Lemma 1.6.1 it follows from the estimation (1.6.16) that the function $v(t,x,w)$ is positive definite and decreasing.

Assumption 1.6.2 There exist:
(1) functions $v_{11}(t,x)$, $v_{22}(t,w)$ and functions $v_{12}(t,x,w) = v_{21}(t,x,w)$;
(2) constants β_{ik}, $i = 1,2$, $k = 1,\ldots,8$, and functions $\xi_1 = \xi_1(\|x\|)$ and $\xi_2 = \xi_2(\|w\|)$ of the K-class such that
  (a) $D_t^+ v_{11}(t,x) + D_x^+ v_{11}(t,x)|_X \leq \beta_{11}\xi_1^2$;
  (b) $D_x^+ v_{11}(t,x)|_{g_1} \leq \beta_{12}\xi_1^2 + \beta_{13}\xi_1\xi_3$;
  (c) $D_t^+ v_{22}(t,w) + D_w^+ v_{22}(t,w)|_W \leq \beta_{21}\xi_2^2$;
  (d) $D_w^+ v_{22}(t,w)|_{g_2} \leq \beta_{22}\xi_2^2 + \beta_{23}\xi_1\xi_2$;
  (e) $D_t^+ v_{12}(t,x,w) + D_x^+ v_{12}(t,x,w)|_X \leq \beta_{14}\xi_1^2 + \beta_{15}\xi_1\xi_2$;
  (f) $D_w^+ v_{12}(t,x,w)|_L \leq \beta_{24}\xi_1^2 + \beta_{25}\xi_1\xi_2$;
  (g) $D_x^+ v_{12}(t,x,w)|_{g_1} \leq \beta_{16}\xi_1^2 + \beta_{17}\xi_1\xi_2 + \beta_{18}\xi_2^2$;
  (h) $D_w^+ v_{12}(t,x,w)|_{g_2} \leq \beta_{26}\xi_1^2 + \beta_{27}\xi_1\xi_2 + \beta_{28}\xi_2^2$.

Lemma 1.6.2 *If all conditions of the Assumption 1.6.2 are fulfilled and the matrix*

$$C = \begin{pmatrix} c_{11} & c_{12} \\ c_{21} & c_{22} \end{pmatrix}, \quad c_{12} = c_{21},$$

with the elements

$$c_{11} = \eta_1^2(\beta_{11} + \beta_{12}) + 2\eta_1\eta_2(\beta_{14} + \beta_{16} + \beta_{26}),$$
$$c_{22} = \eta_2^2(\beta_{21} + \beta_{22}) + 2\eta_1\eta_2(\beta_{18} + \beta_{24} + \beta_{28}),$$
$$c_{12} = \frac{1}{2}(\eta_1^2\beta_{13} + \eta_2^2\beta_{23}) + \eta_1\eta_2(\beta_{15} + \beta_{25} + \beta_{17} + \beta_{27})$$

is negative definite, then the derivative

$$D^+ v(t,x,w) = \eta^T D^+ U(t,x,w)\eta$$

of the function $v(t,x,w)$ is a negative definite function by virtue of the system (1.6.11), (1.6.12).

Proof In virtue of the estimation (a)–(d) of the Assumption 1.6.2 the estimation

$$D^+ v(t,x,w) \leq p^T C p$$

holds, where $p = (\xi_1(\|x\|), \xi_2(\|x\|))^T$.

The definite negativeness of the derivative follows from the condition of the Lemma 1.6.2.

Theorem 1.6.3 *If the two-component system (1.6.11), (1.6.12) is such that all conditions of Lemmas 1.6.1 and 1.6.2 are fulfilled, then the state of equilibrium $x = 0$, $w = 0$ of the system is uniform asymptotically stable.*

If in the Assumption 1.6.1 $N_x = R^k$, $N_w = Q$, functions φ_i, ψ_i, ξ_i belong to the KR-class and conditions of the Lemmas 1.6.1, 1.6.2 are fulfilled, then the state of equilibrium $x = 0$, $w = 0$ of the system (1.6.11), (1.6.12) is uniform asymptotically stable in the whole.

Proof Having fulfilled the enumerated conditions the function $v(t, x, w)$ and its full derivative satisfies all conditions of the Theorem 1.6.2. It proves the statement of the Theorem 1.6.3.

Remark 1.6.3 If in estimations (a)–(d) of Assumption 1.6.2 we change the sign of the inequality for the opposite one and leave in the inequalities of Assumption 1.6.1 only estimation below, then it isn't difficult to define conditions of instability of the state $x = 0$, $w = 0$ of the system (1.6.11), (1.6.12).

1.7 Problems for Investigation

1.7.1 To obtain existence conditions for solutions to system (1.3.5) which satisfy bilinear estimates (condition (2) of Assumption 1.3.2) or other similar conditions allowing us to establish algebraic conditions of sign-definiteness and decrease (radial unboundedness) of function (1.3.6).

1.7.2 To construct an algorithm of approximate solution of system (1.3.5) in terms of the method of perturbed nonlinear mechanics.

1.7.3 To obtain the criterion for exponential stability of system (1.3.3) in terms of function (1.3.6) provided that the independent subsystems (1.3.4) are not exponentially stable.

1.7.4 To investigate other than stability in the sense of Liapunov dynamical properties of system (1.3.3) or its partial cases such as stability, boundedness, and uniform boundedness in terms of two measures.

1.7.5 In terms of Liapunov function (1.3.6) to construct algorithms for estimation of domains of stability, attraction and asymptotic stability of system (1.3.3) and its partial cases in the phase space or/and in the parameter space.

Hint. For the initial definitions of the corresponding domains of stability, attraction and asymptotic stability see Grujić, *et al.* [1], Krasovskii [1], and Martynyuk [16].

1.7.6 For the class of autonomous systems

$$(1.7.1) \qquad \frac{dx}{dt} = X(x(t)), \quad x(t_0) = x_0,$$

where $x \in R^n$, $X \in C^1(R^n, R^n)$, $X(0) = 0$, admitting decomposition to (1.3.3) form, to establish conditions of global asymptotic stability under condition that the origin for system (1.7.1) is an asymptotic attractor.

1.8 Notes and References

Section 1.2 Nonlinear dynamics of continuous systems is a traditional domain of intensive investigations starting with the works by Galileo, Newton, Euler, Lagrange, etc. The problem of motion stability arises whenever the engineering or physical problem is formulated as a mathematical problem of qualitative analysis of equations. Poincaré [1] and Liapunov [1] developed a background for the method of auxiliary functions for continuous systems which do not allow the integration of the motion equations for their qualitative analysis. The ideas of Poincare and Liapunov were further developed and applied in many branches of modern natural sciences.

A definition of stability (of motion) was given in a precise form by Liapunov [1] (see also Comments on Liapunov's original definition by Grujić *et al.* [1]). The concept of stability of equilibrium was previously discussed by Lagrange and Dirichlet (see Appel [1]). Lagrange gave also a criteria for stability of equilibrium points based upon the consideration of the potential energy (see Chetaev [1], Matrosov [1] and references therin). It is interesting to notice that the energy is a particular case of a Liapunov function.

The results of Liapunov [1], Chetaev [1], Persidskii [1], Malkin [1], Ascoli [1], Barbasin and Krasovskii [1], Massera [1], and Zubov [1], were the basis of the Definitions 1.2.1–1.2.3 (ad hoc see Grujić, *et al.* [1], pp. 8–12) and cf. Rao Mohana Rao [1], Yoshizawa [1], Rouche, *et al.* [1], Antosiewicz [1], Lakshmikantham and Leela [1], Hahn [1], etc.). For the Definitions 1.2.4–1.2.7 and 1.2.13 see Hahn [1], and Martynyuk [9]. Definitions 1.2.8–1.2.12 are based on some results by Liapunov [1], Hahn [1] (see and cf. Djordjević [1], Grujić [2], Martynyuk [3–6]). The proofs of Proposalls 1.2.1–1.2.5 are in Hahn [1], Kuz'min [1], Martynyuk [9], Zubov [2], etc.

Theorems 1.2.1–1.2.7 are set out according to Martynyuk [11, 12] (see also Martynyuk [13, 14]). For the proof of Corollary 1.2.1 see Liapunov [1], and Chetaev [1]; for the proof of Corollary 1.2.2 see Barbashin and Krasovskii [1]; for the proof of Corollary 1.2.3–1.2.4 see Liapunov [1], Massera [1], Yoshizava [1], Halanay [1], etc; for the proof of Corollary 1.2.5–1.2.6 see He and Wang [1], Krasovskii [1], and Hahn [2]; for the proof of Corollary 1.2.8 see Chetaev [1], Rouche, et al. [1]; and for the proof of Corollary 1.2.9–1.2.10 see Liapunov [1], and Rouche, et al. [1]. For the Theorems 1.2.9–1.2.17 see Martynyuk [19] and references therin. The Example 1.2.6 became known to the author from the paper by Alexandrov and Platonov [1].

Further results obtained via Liapunov's methods can be found in Burton [1], Galperin [1], Gruyitch [1], Hale [1], Ioss and Joseph [1], Lakshmikantham, Leela and Martynyuk [1], Rama Mohana Rao [1], Coppel [1], Cesari [1], Lakshmikantham and Leela [2], Martynyuk [15], Skowronskii [1], Vorotnikov [1], Zubov [3] (see also CD ROM by Kramer and Hofmann [1] and Sell [1] for references), etc.

Section 1.3 The problem of constructing Liapunov functions for nonlinear nonautonomous systems of general type remains still unsolved though its more than one-hundred existence. Meanwhile the efforts of many mathematicians and mechanical scientists have resulted in an efficient approach of constructing the appropriate auxiliary functions for specific classes of systems of equations with reference to many applications (see Alexandrov and Platonov [1], Antosiewicz [1], Barbashin [1], Wang, et al. [1], etc.)

The approach proposed in this section is based on the idea of matrix-valued function as an appropriate medium for Liapunov function construction. This approach has been developed since 1984 and some of the obtained results are published and summarized by Martynyuk [10, 11, 16], and Kats and Martynyuk [1].

Actually, the problem of constructing Liapunov functions for the class of nonlinear systems of (1.3.3) type is reduced to the solution of systems of first order partial equations (1.3.5) which are more simple than the Liapunov equation for the initial system proposed in 1892 in his famous dissertation paper.

This section is based on some results by Martynyuk and Slyn'ko [1–3], and Slyn'ko [2]. Besides, some results by Djordjević [1, 2], Hahn [1], Krasovskii [1], Lankaster [1], etc. are used.

Section 1.4 Linear nonautonomous system of (1.4.1) type or autonomous system (1.4.14) is of essential interest in the context of the problem of constructing the Liapunov function since this allows us to investigate stability of the equilibrium state of some quasilinear systems. In spite of the seeming simplicity of linear systems the problem of constructing the appropriate Liapunov function remains open in this case es well (see, e.g. Barbashin [1], Zhang [1], etc.).

In this section for the above-mentioned systems we adopt the algorithm of Liapunov function construction presented in Section 1.3. Since in this case systems of equations (1.3.5) turn out to be linear differential or algebraic, their exact solutions can sometimes be found. The section is based on the results by Martynyuk and Slyn'ko [1–3].

The Bellman–Bailey approach (see Bellman [1] and Bailey [1]) to stability investigation of large-scale systems has been developed considerably in many papers. In the books by Barbashin [1], Martynyuk [1], Michel and Miller [1], Šiljak [1, 2], Grujic, *et al.* [1], Voronov and Matrosov [1], Lakshmikantham, *et al.* [1], Borne, *et al.* [1], etc. alongside the original results the results of many investigations of dynamics of linear and nonlinear systems in terms of vector Liapunov functions are summarized. An essential deficiency of this approach is the supersufficiency of stability conditions for the systems of motion equations under consideration (see Piontkovskii and Rutkovskaya [1], and Martynyuk and Slyn'ko [1]).

The application of the matrix-valued Liapunov function for the same classes of systems of equations provides wider conditions of stability. The reasons for this were scrutinized by Martynyuk [9, 11]. In this section by the example of a linear system it is shown how supersufficient the stability conditions obtained via the Bellman–Bailey approach are as compared with those obtained via the application of matrix-valued function.

Section 1.5 The phenomenon of motion polystability has been investigated in nonlinear dynamics since 1987. As noticed by Aminov and Sirazetdinov [1], and Martynyuk [7] this phenomenon was discovered while developing the notion of stability with respect to a part of variables. In monographs by Martynyuk [11, 16] some results are presented obtained in the development of the theory of motion polystability including sufficient conditions for exponential polystability in the first approximation (see also Martynyuk [13, 14], and Slyn'ko [1]).

Section 1.6 The results of this section are new. Similar to the Theorem 1.6.1 in the paper by Martynyuk [17] was proved for discontinuous dynamical system (see also Martynyuk [19]). The trends in the stability theory in metric space are based on results by Zubov [1], Nemytskii and Stepanov [1], Birkhoff [1], Bhatia and Szego [1], etc. The mappings preserving stability in metric space were first considered by Thomas [1] and Hahn [1]. In the papers by Michel and Miller [1] and Michel, et al. [1] and other mappings of the type were studied in the stability analysis of large-scale systems.

The application of multicomponent mapping $U(t,p)\colon R_+ \times X_1 \to X_2$ adds more flexibility to the approach to stability analysis of dynamical system in metric space, because this mapping admits a wider class of components for its elements $v_{ij}(t,p)$.

Section 1.7 The problems set out in this section are addressed first of all to young researchers in the area of motion stability theory and its application. Solution of any of the problems will be not only a subject of a significant paper but an essential contribution to the development of the method of matrix Liapunov functions which was called by Prof. V. Lakshmikantham (in Moscow, 2001) one of three outstanding achievements of stability theory in the 20th century.

References

Alexandrov, A.Yu. and Platonov, A.V.
[1] Constructing of Lyapunov function for a class of nonlinear systems. *Nonlinear Dynamics and Systems Theory*, (to appear).

Aminov, A.B. and Sirazetdinov, T.K.
[1] The method of Liapunov functions in problems of multistability of motion. *Prikl. Mat. Mekh.* **51** (1987) 553–558. [Russian]

Antosiewicz, H.A.
[1] A survey of Lyapunov's second method. *Contributions to the Theory of Nonlinear Oscillations* (4) (1958) 141–166.

Ascoli, G.
[1] Osservazioni sopora alcune questioni di stabilita 1. *Atti Accad. Naz. Lincei Rend Cl. Sci. Fis. Mat. Nat.* **9** (1950) 129–134.

Bailey, F.N.
[1] The application of Liapunov's second method to interconnected systems. *J. Soc. Ind. Appl. Math. Ser. A* **3** (1965) 443–462.

Barbashin, Ye.A.
[1] *The Liapunov Functions.* Moscow: Nauka, 1970. [Russian]

Barbashin, Ye.A. and Krasovskii, N.N.
[1] On the stability of motion in the large. *Dokl. Akad. Nauk SSSR* **86** (1952) 453–456.

Bellman, R.
[1] Vector Liapunov functions. *SIAM J. Contr. Ser. A* **1** (1962) 32–34.

Bhatia, N.P. and Szego, G.P.
[1] *Stability Theory of Dynamical Systems*. Berlin: Springer-Verlag, 2002.

Birkhoff, G.G.
[1] *Dynamical Systems*. Moscow-Leningrad: GITTL, 1941. [Russian]

Borne, P., Dambrine, M., Perruquetti, W. and Richard, J.P.
[1] Vector Lyapunov functions: Time-varing, ordinary and functional differential equations. In: *Advances in Stability Theory at the End of XXth Century*. (Ed.: A.A. Martynyuk), London: Taylor and Francis, 2002, P.89–112.

Burton, T.A.
[1] *Stability and Periodic Solutions of Ordinary and Functional Differential Equations*. Orlando: Academic Press, 1985.

Bylov, B.F., Vinograd, R.E., Grobman, D.M. and Nemytskii, V.V.
[1] *Lyapunov Characteristic Exponents Theory and Its Applications to Stability Issues*. Moscow: Nauka, 1966. [Russian]

Cesari, L.
[1] *Asymptotic Behaviour and Stability Problems in Ordinary Differential Equations*. 2nd edn. Berlin: Springer-Verlag, 1963.

Chetaev, N.G.
[1] *Stability of Motion*. Moscow: Nauka, 1990. [Russian]

Conti, R.
[1] Sulla prolungabilita delle soluzioni di un sistema di equazioni differentiali ordinarie. *Boll. Unione Math. Ital. Ser. 3* **2** (4) (1956) 510–514.

Coppel, W.A.
[1] *Stability and Asymptotic Behaviour of Differential Equations*. Boston: Heath, 1965.

Corduneanu, C.
[1] Applications of differential inequalities to stability theory, *An. Sti. Univ. "Al i Cuza", Iasi Sect. 1, Mat.* **6**(1) (1960) 47–58. [Russian]
[2] Supplement to the paper "Applications of differential inequalities to stability theory", *An. Sti. Univ. "Al i Cuza", Iasi Sect. 1 Mat.* **7**(2) (1961) 247–252. [Russian]

Daletskii, Yu.L. and Krein, M.G.
[1] *Stability of Solutions of Differential Equations in the Banach Space*. Moscow: Nauka, 1986. [Russian]

Djordjević, M.Z.
[1] Stability analysis of large scale systems whose subsystems may be unstable. *Large Scale Systems* **5** (1983) 252–262.
[2] Stability analysis of interconnected systems with possibly unstable subsystems. *Systems and Control Letters* **3**(3) (1983) 165–169.

Galperin, E.A.
[1] Some generalizations of Lyapunov's approach to stability and control. *Nonlinear Dynamics and Systems Theory* **2**(1) (2002) 1–23.

Gruyitch, Lj.T.
[1] Consistent Lyapunov methodology: Non-differentiable non-linear systems. *Nonlinear Dynamics and Systems Theory* **1**(1) (2001) 1–22.
[2] On large-scale systems stability. *Proc. 12th World Congress IMACS*, Vol. 1, 1988, P.224–229.

Grujić, Lj.T., Martynyuk, A.A. and Ribbens-Pavella, M.
[1] *Large-Scale Systems Stability under Structural and Singular Perturbations.* Berlin: Springer-Verlag, 1987.

Hahn, W.
[1] *Stability of Motion.* Berlin: Springer-Verlag, 1967.
[2] Uber Stabilitatserhaltende Abbildungen and Ljapunovsche Funktionen. *J. fur Reine und Andewandte Math.* **228** (1967) 189–192.

Halanay, A.
[1] *Differential Equations.* New York: Academic Press, 1966.

Hale, J.K.
[1] *Asymptotic Behaviour of Dissipative Systems.* Math. Surveys and Monographs, Providence: Amer. Math. Soc., Vol. 25, 1988.

Hatvani, L.
[1] On application of differential inequalities in the theory of stability, *Vestnik Moscow Univ. Ser. Math. Mekh.* **3** (1975) 83–88. [Russian]

He, J.X. and Wang, M.S.
[1] Remarks on exponential stability by comparison functions of the same order magnitude. *Ann. of Diff. Eqs.* **7**(4) (1991) 409–414.

Iooss, G. and Joseph, D.D.
[1] *Elementary Stability and Bifurcation Theory.* New York: Springer-Verlag, 1990.

Kats, I.Ya. and Martynyuk, A.A.
[1] *Stability and Stabilization of Nonlinear Systems with Random Structures.* London and New York: Taylor and Francis, 2002.

Kramer and Hofman G.E.I.
[1] Zentralblatt MATH CD-ROM, Volumes 926–962, Berlin: Springer, 2000.

Krasovskii, N.N.
[1] *Stability of Motion.* Stanford, California: Stanford Univ. Press, 1963.

Kuz'min, P.A.
[1] *Small Oscillations and Stability of Motion.* Moscow: Nauka, 1973. [Russian]

Lakshmikantham, V. and Leela, S.
[1] *Differential and Integral Inequalities. Theory and Applications.* New York: Academic Press, 1969.

Lakshmikanthan, V., Leela, S. and Martynyuk, A.A.
[1] *Stability Analysis of Nonlinear Systems.* New York: Marcel Dekker, 1989.

Lakshmikantham, V., Matrosov, V.M. and Sivasundaram, S.
[1] *Vector Lyapunov Function and Stability Analysis of Nonlinear Systems.* Dordrecht, etc.: Kluwer Academic Publishers, 1991.

Lankaster, P.
[1] *Matrix Theory*. Moscow: Nauka, 1978. [Russian]

Lefschetz, S.
[1] *Differential Equations: Geometric Theory*. New York: Interscience Publishers, 1957.

Lyapunov, A.M.
[1] *The General Problem of Stability of Motion*. Kharkov: Kharkov Mathematical Society, 1892. [Russian] (see also French translation in: *Ann. Fac Toulouse* **9** (1907) 203–474; and English translation in: *Ann. of Mathematical Study*, No. 17, Princeton: Princeton University Press, 1949, and London: Taylor and Francis, 1992).

Malkin, I.G.
[1] To the question of inverse Liapunov theorem on asymptotic stability. *Prikl. Mat. Mekh.* **18**(2) (1954) 129–138. [Russian]

Martynyuk, A.A.
[1] *Stability of Motion of Complex Systems*. Kiev: Naukova Dumka, 1975. [Russian]
[2] The Lyapunov matrix function. *Nonlin. Anal.* **8** (1984) 1223–1226.
[3] Extension of the state space of dynamical systems and the problem of stability. *Colloquia Mathematica Societaties Janas Bolyai 47. Differential Equations: Qualitative Theory*, Szeged (Hungary), 1984, P.711–749.
[4] The matrix-function of Liapunov and hybrid systems stability. *Int. Appl. Mech.* **21**(4) (1985) 89–96.
[5] The Lyapunov matrix function and stability of hybrid systems. *Nonlin. Anal.* **10** (1986) 1449–1457.
[6] Stability analysis of nonlinear systems via Liapunov matrix function. *Prikl. Mekh.* **27**(8) (1991) 3–15. [Russian]
[7] A theorem on polystability. *Dokl. Akad. Nauk SSSR* **318** (4) (1991) 808–811.
[8] On polystability with respect to a part of variables. *Dokl. Akad. Nauk Russia* **324** (1) (1992) 39–41.
[9] Matrix method of comparison in the theory of the stability of motion. *Int. Appl. Mech.* **29** (1993) 861–867.
[10] *Stability Analysis: Nonlinear Mechanics Equations*. New York: Gordon and Breach Science Publishers, 1995.
[11] *Stability by Liapunov's Matrix Function Method with Applications*. New York: Marcel Dekker, 1998.
[12] Stability and Liapunov's matrix functions method in dynamical systems. *Prikl. Mekh.* **34**(10) (1998) 144–152. [English]
[13] A survey of some classical and modern developments of stability theory. *Nonlin. Anal.* **40** (2000) 483–496.
[14] Some results of developing classical and modern theories of stability of motion. *Int. Appl. Mech.* **37**(9) (2001) 1142–1157.
[15] Stability analysis of continuous systems with structural perturbations. *Int. Appl. Mech.* **38**(7) (2002) 783–805.
[16] *Qualitative Methods in Nonlinear Dynamics: Novel Approaches to Liapunov Matrix Functions*. New York: Marcel Dekker, 2002.
[17] On stability of motion of discontinuous dynamical systems. *Dokl. Acad. Nauk* **397** (3) (2004) 308–312. [Russian]
[18] Stability of dynamical systems in metric space. *Nonlinear Dynamics and Systems Theory* **5**(2) (2005) 157–167.
[19] Stability analysis by comparison technique. *Nonlin. Anal.* **62** (2005) 629–641.

Martynyuk, A.A. (Ed.)
[1] *Advances in Stability Theory at the End of the 20th Century.* London and New York: Taylor and Francis, 2003.

Martynyuk, A.A. and Gutowski, R.
[1] *Integral Inequalities and Stability of Motion.* Kiev: Naukova Dumka, 1979. [Russian]

Martynyuk, A.A., Lakshmikantham, V. and Leela, S.
[1] *Stability of Motion: The Method of Integral Inequalities.* Kiev: Naukova Dumka, 1989. [Russian]

Martynyuk, A.A. and Miladzhanov, V.G.
[1] *Stability Analysis in the Whole of Dynamical System via Matrix Liapunov Functions.* Preprint 87.62, Institute of Mathematics, Kiev, 1987. [Russian]

Martynyuk, A.A. and Obolenskii, A.Yu.
[1] Stability of autonomous Wazewski systems. *Diff. Uravn.* **16** (1980) 1392,-1407. [Russian]

Martynyuk, A.A. and Slyn'ko, V.I.
[1] Solution of the problem of constructing Liapunov matrux function for a class of large scale system. *Nonlinear Dynamics and Systems Theory* **1**(2) (2001) 193–203.
[2] On stability of large-scale system with non-autonomous connected subsystems. *Int. Appl. Mech.* **37**(10) (2001) 1341–1351.
[3] On stability of motion of large-scale system. *Diff. Uravn.* **39**(6) (2003) 754–758.

Massera, J.L.
[1] Contributions to stability theory. *Ann. of Math.* **64** (1956) 182–206.

Matrosov, V.M.
[1] *Vector Lyapunov Functions Method: Nonlinear Analysis of Dymnamical Properties,* Moscow: Fizmatlit, 2001. [Russian]

Michel, A.N. and Miller, R.K.
[1] *Qualitative Analysis of Large Scale Dynamical Systems.* New York: Academic Press, 1977.

Michel, A.N., Miller, R.K. and Tang, W.
[1] Lyapunov stability of interconnected systems: decomposition into strongly connected subsystems. *IEEE Trans. on Circuits and Systems* **CAS−25** (1978) 799–810.

Michel, A.N., Wang, K. and Hu, B.
[1] *Qualitative Theory of Dynamical Systems. The Role of Stability Preserving Mappings.* New York: Marcel Dekker, 2001.

Nemytskii, V.V. and Stepanov, V.V.
[1] *Qualitative Theory of Differential Equations.* Moscow and Leningrad: GITTL, 1949. [Russian]

Olech, C. and Opial, Z.
[1] Sur une inegalité differentielle. *Ann. Polon. Math.* **7** (1960) 241–254.

Peiffer, K. and Rouche, N.
[1] Liapunov's second method applied to partial stability. *J. Mécanique* **8** (2) (1969) 323–334.

Persidskii, K.P.
[1] *Stability Theory of Solutions of Differential Equations. Theory Probability.* Selected works in two volumes, Vol. 1. Alma-Ata: Nauka, 1976. [Russian]

Piontkovskii, A.A. and Rutkovskaya, L.D.
[1] Investiagation of certain stability theory problems by the vector Lyapunov function method. *Avtomatika i Telemekhanika* **10** (1967) 23–31. [Russian]

Poincaré, H.
[1] *On the Curves Defined by Equations Differential.* Moscow and Leningrad: GITTL, 1947. [Russian]

Rama Mohana Rao, M.
[1] *Ordinary Differential Equations. Theory and Applications.* New Delhi and Madras: Affiliated East-West Press Pvt Ltd, 1980.

Rouche, N., Habets, P. and Laloy, M.
[1] *Stability Theory by Liapunov's Direct Method.* New York: Springer-Verlag, 1977.

Rumiantsev, V.V. and Oziraner, A.S.
[1] *Stability and stabilization of motion with respect to a part of variables.* Moscow: Nauka, 1987. [Russian]

Sell, G.R.
[1] References on dynamical systems. *University of Minnesota.* Preprint 91–129, 1993.

Šiljak, D.D.
[1] *Large-Scale Dynamic Systems: Stability and Structure.* New York: North Holland, 1978.
[2] *Decentralized Control of Complex Systems.* Boston: Academic Press, 1991.

Schroder, J.
[1] *Operator Inequalities.* New York: Academic Press, 1980.

Skowronski, J.M.
[1] *Nonlinear Liapunov Dynamics.* Singapore, etc.: World Scientific, 1990.

Slyn'ko, V.I.
[1] To the problem on polystability of motion. *Int. Appl. Mech.* **37**(12) (2001) 1624–1628.
[2] On constructing of non-diagonal elements of the Liapunov's matrix function. *Dop. Nats. Akad. Nauk of Ukr.* **4** (2001) 58–62.

Thomas, J.
[1] Uber die Invarianz der Stabilitat bei einem Phasenraum-Homoomorphismus. *J. fur Reine und Angewandte Math.* **213** (1964) 147–150.

Voronov, A.A. and Matrosov, V.M. (Eds.)
[1] *Method of Vector Lyapunov Functions in Stability Theory.* Moscow: Nauka, 1987. [Russian]

Vorotnikov, V.I.
[1] *Stability of dynamical systems with respect to part of variables.* Moscow: Nauka, 1991. [Russian]

Walter, W.
[1] *Differential and Integral Inequalities.* Berline: Springer-Verlag, 1970.

Wang, Y., Shi, Y. and Sasaki, H.
[1] A criterion for stability of nonlinear time-varing dynamic system. *Nonlinear Dynamics and Systems Theory* **4**(2) (2004) 217–229.

Yoshizawa, T.
[1] *Stability Theory by Liapunov's Second Method.* Tokyo: The Math. Soc. of Japan, 1966.

Zhang, Bo
[1] Formulas of Liapunov functions for systems of linear ordinary and delay differential equations. *Funkcial. Ekvac.* **44**(2) (2001) 253–278.

Zubov, V.I.
[1] *Methods of A. M. Liapunov and its Applications.* Leningrad: Izdat. Leningrad. Gos. Universitet, 1957. [Russian]
[2] *Mathematical Methods of Investigations of Automatic Control Systems.* Leningrad: Mashinostroeniye, 1979. [Russian]
[3] *Control Processes and Stability.* St-Petersburg: St-Petersburg Gos. Universitet, 1999. [Russian]

2

STABILITY ANALYSIS OF DISCRETE-TIME SYSTEMS

2.1 Introduction

In this Chapter, we present some approaches to stability investigation of systems modelled by difference equations. These approaches are based on the development of the main idea of the Liapunov direct method in terms of multicomponent auxiliary functions.

This Chapter is arranged as follows.

In Section 2.2 a statement of problems of qualitative analysis of motions modelled by a system of difference equations is discussed.

Section 2.3 sets out the main results of stability analysis of time-discrete systems. These results are obtained via the direct Liapunov method in terms of auxiliary matrix-valued and vector function. As for continuous systems, the application of the above mentioned class of functions provides more versatility for the direct Liapunov method for the class of system under consideration.

Section 2.4 presents a method of constructing the matrix-valued Liapunov function for linear difference system. Also, the application examples are cited for the proposed approach.

Section 2.5 describes the method of stability analysis of a discrete-time system based on the idea of the hierarchical Liapunov function. Also, a problem on connected stability for discrete-time system is discussed.

Section 2.6 establishes new stability conditions for solutions of system given on a time scale. The conditions are obtained by the direct Liapunov method encorporating matrix-valued auxiliary functions.

In the final Section 2.7, some unsolved problems are indicated, which are of interest for both the general theory of discrete-time systems and applications.

2.2 Large-Scale Discrete-Time Systems

Consider a system with a finite number of degrees of freedom described by the system of difference equations in the form

(2.2.1) $$x(\tau + 1) = f(\tau, x(\tau)),$$

where $\tau \in N_\tau^+ \triangleq \{\tau_0, \ldots, \tau_0 + k, \ldots\}$, $\tau_0 \geq 0$, $k = 1, 2, \ldots$, $x \in R^n$, $f: N_\tau^+ \times R^n \to R^n$, $f(\tau, x)$ is continuous in x. Let solution $x(\tau; \tau_0, x_0)$ of system (2.2.1) be definite for all $\tau \in N_\tau^+$ and $x(\tau_0; \tau_0, x_0) = x_0$. Assume that $f(\tau, x) = x$ for all $\tau \in N_\tau^+$ iff $x = 0$. Besides, system (2.2.1) admits zero solution $x = 0$ and it corresponds to the unique equilibrium state of system (2.2.1).

The definitions of the dynamical properties of solutions of system (2.2.1) are obtained by replacing the independent variable $t \in R$ by $\tau \in N_\tau^+$ in Definitions 1.2.1–1.2.3 and so are omitted.

Stability (instability) of the equilibrium state $x = 0$ of system (2.2.1) is sometimes studied by means of reducing this system to the form

(2.2.2) $$x(\tau + 1) = Ax(\tau) + g(\tau, x(\tau)),$$

where A is a constant $n \times n$–matrix, $g: N_\tau^+ \times R^n \to R^n$ is a vector-function continuous in x and satisfies certain conditions of smallness.

In this case, under some additional restrictions on the properties of matrix A, stability (instability) of the state $x = 0$ of system (2.2.2) can be studied in terms of the first order approximation equations.

Of essential interest is the case when the order of system (2.2.1) is rather high, or when this system is a composition of more simple subsystems. In this case the finite-dimensional system of equations of the type of

(2.2.3) $$\begin{aligned} x_i(\tau + 1) &= f_i(\tau, x_i(\tau)) + g_i(\tau, x_1(\tau), \ldots, x_m(\tau)), \\ i &= 1, 2, \ldots, m, \end{aligned}$$

is considered, where $x_i \in R^{n_i}$, $f_i: N_\tau^+ \times R^{n_i}$, $g_i: N_\tau^+ \times R^{n_1} \times \cdots \times R^{n_m} \to R^{n_i}$.

Designate

$$n = \sum_{i=1}^{m} n_i, \quad x = (x_1^\mathrm{T}, \ldots, x_m^\mathrm{T})^\mathrm{T},$$

$$f(\tau, x) = (f_1^\mathrm{T}(\tau, x_1), \ldots, f_m^\mathrm{T}(\tau, x_m))^\mathrm{T},$$

and
$$g(\tau, x) = (g_1^T(\tau, x_1, \ldots, x_m), \ldots, g_m^T(\tau, x_1, \ldots, x_m))^T.$$

Now system (2.2.3) can be presented in the vector form

(2.2.4) $\quad x(\tau + 1) = f(\tau, x(\tau)) + g(\tau, x(\tau)) \triangleq H(\tau, x(\tau)),$

where $x \in R^n$, $H \colon N_\tau^+ \times R^n \to R^n$. Formally system (2.2.4) coincides in form with system (2.2.1). However, if $g(\tau, x(\tau)) \equiv 0$, this system falls into the independent subsystems

(2.2.5) $\quad x_i(\tau + 1) = f_i(\tau, x_i(\tau)), \quad i = 1, 2, \ldots, m,$

each of the latter can possess the same degree of complexity of the solution behavior as the system (2.2.1).

2.3 Tests for Stability Analysis

2.3.1 Liapunov's like theorems via matrix-valued functions

Qualitative analysis of stability (instability) of the equilibrium state $x = 0$ of system (2.2.1) can be carried out efficiently if an appropriate auxiliary Liapunov function is available.

Assume that for system (2.2.1) a matrix-valued function

(2.3.1) $\quad U(\tau, x(\tau)) = [v_{ij}(\tau, x(\tau))], \quad i, j = 1, 2, \ldots, m.$

is constructed where $v_{ii} \colon N_\tau^+ \times R^n \to R_+$ and $v_{ij} \colon N_\tau^+ \times R^n \to R$ for all $i, j = 1, 2, \ldots m$. Let $x(\tau; \tau_0, x_0)$ be the solution of system (2.2.1) with the initial conditions $x(\tau_0) = x_0$, so that $x(\tau_0; \tau_0, x_0) = x_0$. For any matrix-valued function (2.3.1) we define the first forward difference of U with respect to τ along the solution of (2.2.1) by

(2.3.2) $\quad \begin{aligned} \Delta U(\tau, x(\tau))|_{(2.2.1)} &= U(\tau + 1, x(\tau + 1)) - U(\tau, x(\tau)) \\ &= U(\tau + 1, f(\tau, x(\tau))) - U(\tau, x(\tau)). \end{aligned}$

Here the first forward difference of U along solutions of system (2.2.1) is calculated element-wise.

By means of vector $\eta \in R_+^m$, $\eta > 0$, and the matrix-valued function $U(\tau, x)$ we construct the scalar function

(2.3.3) $\quad v(\tau, x, \eta) = \eta^T U(\tau, x) \eta.$

It is clear that

(2.3.4) $$\Delta v(\tau, x, \eta)|_{(2.2.1)} = \eta^T \Delta U(\tau, x(\tau)) \eta.$$

If there exists a function $w \colon N_\tau^+ \times R_+ \to R$ such that

(2.3.5) $$\Delta v(\tau, x, \eta)|_{(2.2.1)} \leq \omega(\tau, v(\tau, x, \eta))$$

we consider the inequality

(2.3.6) $$\begin{aligned} v(\tau+1,\, x(\tau+1),\, \eta) &\leq v(\tau, x(\tau), \eta) + \omega(\tau, v(\tau, x(\tau), \eta)) \\ &\triangleq g(\tau,\, v(\tau, x(\tau), \eta)) \end{aligned}$$

and the corresponding comparison equation

(2.3.7) $$u(\tau+1) = g(\tau, u(\tau)) = u(\tau) + \omega(\tau, u(\tau)).$$

Further we assume that $g(\tau, 0) = 0$ for all $\tau \in N_\tau^+$.

Since the function (2.3.3) is scalar and the comparison equation (2.3.7) is scalar as well, the main stability theorem in terms of the comparison principle for the matrix-valued Liapunov function (2.3.1) remains the same as in Theorem 4.8.1 by Lakshmikantham, Leela, and Martynyuk [1].

Theorem 2.3.1 *Assume that the vector-function f in system (2.2.1) is continuous with respect to the second argument and definite on $N_\tau^+ \times \mathcal{N}$ (on $N_\tau^+ \times R^n$). If there exist*

(1) *an open connected discrete-time invariant neighborhood $\mathcal{G} \subseteq \mathcal{N}$ of the point $x = 0$;*

(2) *matrix-valued function $U \colon N_\tau^+ \times \mathcal{G} \to R^{m \times m}$ ($U \colon N_\tau^+ \times R \to R^{m \times m}$), $U(\tau, 0) = 0$ for all $\tau \in N_\tau^+$ and the vector $\eta \in R_+^m$ such that the function (2.3.3) is positive definite, decreasing, radially unbounded and continuous with respect to the second argument;*

(3) *function $g \colon N_\tau^+ \times R_+ \to R$, $g(\tau, 0) = 0$, $g(\tau, u)$ is nondecreasing in u and such that inequality (2.3.7) holds for all $(\tau, x, \eta) \in N_\tau^+ \times \mathcal{G} \times R_+^m$ (for all $(\tau, x, \eta) \in N_\tau^+ \times R^n \times R_+^m$).*

Then

(1) *stability (in the whole) of zero solution $u(\tau) = 0$ of the discrete equation (2.3.7) implies stability (in the whole) of the equilibrium state $x = 0$ of discrete system (2.2.1);*

(2) *asymptotic stability (in the whole) of zero solution $u(\tau) = 0$ of the discrete equation (2.3.7) implies asymptotic stability (in the whole) of the equilibrium state $x = 0$ of discrete system (2.2.1).*

The proof of this Theorem is similar to the proof of Theorem 4.8.1 by Lakshmikantham, Leela, and Martynyuk [1].

General Theorems 1.2.6–1.2.7 from Section 1.2 on stability of continuous systems are extended for the discrete systems (2.2.1). We shall cite here some of the assertions.

Theorem 2.3.2 *Assume that the vector-function f in system (2.2.1) is continuous with respect to the second argument and definite in the domain $N_\tau^+ \times \mathcal{N}$. If there exist*

(1) *an open connected discrete-time invariant neighborhood $\mathcal{G} \subseteq \mathcal{N}$, $\mathcal{N} \subseteq R^n$ of the point $x = 0$;*

(2) *a matrix-valued function $U \colon N_\tau^+ \times \mathcal{G} \to R^{m \times m}$, $U(\tau, 0) = 0$ for all $\tau \in N_\tau^+$ and a vector $\eta \in R_+^m$ such that the function $v(\tau, x, \eta) = \eta^T U(\tau, x)\eta$ is continuous in the second argument;*

(3) *the comparison functions $\psi_{i1}, \psi_{i2}, \psi_{i3} \in K$, $\widetilde{\psi}_{i2} \in CK$, $i = 1, 2, \ldots, m$ and $m \times m$-matrices $A_j(\eta)$, $j = 1, 2, 3$, $\tilde{A}_2(\eta)$ such that*

(a) $\psi_1^T(\|x\|) A_1(\eta) \psi_1(\|x\|) \leq v(\tau, x, \eta) \leq \widetilde{\psi}_2^T(\tau, \|x\|) \tilde{A}_2(\eta) \widetilde{\psi}_2(\tau, \|x\|)$ *for all $(\tau, x, \eta) \in N_\tau^+ \times \mathcal{G} \times R_+^m$;*

(b) $\psi_1^T(\|x\|) A_1(\eta) \psi_1(\|x\|) \leq v(\tau, x, \eta) \leq \psi_2^T(\|x\|) A_2(\eta) \psi_2(\|x\|)$ *for all $(\tau, x, \eta) \in N_\tau^+ \times \mathcal{G} \times R_+^m$;*

(c) $\Delta v(\tau, x, \eta)|_{(2.2.1)} \leq \psi_3^T(\|x\|) A_3(\eta) \psi_3(\|x\|)$ *for all $(\tau, x, \eta) \in N_\tau^+ \times \mathcal{G} \times R_+^m$.*

Then, if the matrices $A_1(\eta)$, $A_2(\eta)$, $\tilde{A}_2(\eta)$ for all $(t, x, \eta) \in N_\tau^+ \times \mathcal{G} \times R_+^m$ are positive definite and the matrix $A_3(\eta)$ is negative semidefinite, then

(a) *the equilibrium state $x = 0$ of discrete system (2.2.1) is stable under condition (3a);*

(b) *the equilibrium state $x = 0$ of discrete system (2.2.1) is uniformly stable under condition (3b).*

Corollary 2.3.1 *Let there exist at least one couple of indices $(p, q) \in [1, m]$, for which the function $v(\tau, x, e) = e^T U(\tau, x(\tau))e$ is positive definite and continuous with respect to the second argument.*

Then if $\Delta v(\tau, x, e)|_{(2.2.1)}$ is negative semi-definite, the state $x_e = 0$ of system (2.2.1) is stable.

Corollary 2.3.2 *If for the function $v(\tau, x, e)$ from Corollary 2.3.1 the above mentioned conditions are satisfied and moreover, the function*

$v(\tau, x, e)$ is radially unbounded, then the state $x_e = 0$ of system (2.2.1) is stable in the whole.

Having changed a little the conditions of Theorem 2.3.2 one can easily obtain the conditions of asymptotic and uniform asymptotic stability for the equilibrium state $x = 0$ of system (2.2.1).

Corollary 2.3.3 *For the function $v(t, x, e)$, mentioned in Corollary 2.3.1 let*

(1) *there exist functions φ_1, φ_2 of class K (class KR) such that for all $(\tau, x) \in N_\tau^+ \times \mathcal{G}$ (for all $(\tau, x) \in N_\tau^+ \times R^n$)*

$$\varphi_1(\|x\|) \leq v(\tau, x, e) \leq \varphi_2(\|x\|);$$

(2) *for all $(\tau, x) \in N_\tau^+ \times \mathcal{G}$ (for all $(\tau, x) \in N_\tau^+ \times R^n$)*

$$\Delta v(\tau, x, e)|_{(2.2.1)} \leq -\psi_1(\|x\|)$$

for some function ψ_1 of class K.

Then the state $x_e = 0$ of system (2.2.1) is uniformly asymptotically stable (uniformly asymptotically stable in the whole).

Theorem 2.3.3 *Assume that the vector-function f in system (2.2.1) is continuous with respect to the second argument and definite in the domain $N_\tau^+ \times \mathcal{N}$ ($N_\tau^+ \times R^n$). If there exist*

(1) *an open connected discrete-time invariant neighborhood $\mathcal{G} \subseteq \mathcal{N}$, $\mathcal{N} \subseteq R^n$ of the point $x = 0$;*

(2) *a matrix-valued function $U \colon N_\tau^+ \times \mathcal{G} \to R^{m \times m}$ ($U \colon N_\tau^+ \times R^n \to R^{m \times m}$), $U(\tau, 0) = 0$ for all $\tau \in N_\tau^+$ and a vector $\eta \in R_+^m$ such that the function $v(\tau, x, \eta) = \eta^T U(\tau, x)\eta$ is continuous in the second argument;*

(3) *the functions $\sigma_{1i}, \sigma_{2i} \in K$, $i = 1, 2, \ldots m$, a positive real number Δ_1 and positive integer p, $m \times m$ matrices $F_1(\eta), F_2(\eta)$ such that*

(a) $\Delta_1 \|x\|^p \leq v(\tau, x, \eta) \leq \sigma_1^T(\|x\|) F_1(\eta) \sigma_1(\|x\|)$ *for all $(\tau, x, \eta \neq 0)$ $\in N_\tau^+ \times \mathcal{N} \times R_+^m$ (for all $(\tau, x, \eta \neq 0) \in N_\tau^+ \times R^n \times R_+^m$);*

(b) $\Delta v(\tau, x, \eta)|_{(2.2.1)} \leq \sigma_2^T(\|x\|) F_2(\eta) \sigma_2(\|x\|)$ *for all $(\tau, x, \eta \neq 0)$ $\in N_\tau^+ \times \mathcal{N} \times R_+^m$ (for all $(\tau, x, \eta \neq 0) \in N_\tau^+ \times R^n \times R_+^m$).*

Then, provided that the matrix $F_1(\eta)$, $(\eta \neq 0) \in R_+^m$, is positive definite, the matrix $F_2(\eta)$, $(\eta \neq 0) \in R_+^m$, is negative definite and functions σ_{1i},

σ_{2i} are of the same magnitude, then the state $x = 0$ of system (2.2.1) is exponentially stable.

If in the conditions of Theorem 2.3.3 the functions $\sigma_{1i}, \sigma_{2i} \in KR$, $i = 1, 2, \ldots m$, and all its conditions are fulfilled, than the state $x = 0$ of system (2.2.1) is exponentially stable in the whole.

The proof of Theorem 2.3.3 is similar to the proof of Theorem 2.5.5 from Martynyuk [2].

Corollary 2.3.4 *Let there exist at least one couple of indices* $(p, q) \in [1, m]$, *for which the function* $v(\tau, x, e) = e^{\mathrm{T}} U(\tau, x) e$ *is positive definite function,* $v \in C(N_\tau^+ \times \mathcal{N}, R_+)$, *such that*

(a) $\Delta_1 \|x\|^p \leq v(\tau, x, e) \leq \Delta_2 \|x\|^p$ *for all* $(\tau, x, e) \in N_\tau^+ \times \mathcal{N} \times I$, $I = (1, \ldots, 1) \in R^m$;

(b) $\Delta v(\tau, x, e)|_{(2.2.1)} \leq -\Delta_3 \|x\|^p$,

where Δ_1, Δ_2, Δ_3 *and* p *are positive constants, then the state* $x = 0$ *of system (2.2.1) is exponentially stable.*

Theorem 2.3.4 *Assume that the vector-function f in system (2.2.1) is continuous with respect to the second argument and definite in the domain* $N_\tau^+ \times \mathcal{N}$ $(N_\tau^+ \times R^n)$. *If there exist*

(1) *an open connected discrete-time invariant neighborhood* $\mathcal{G} \subseteq \mathcal{N}$, $\mathcal{N} \subseteq R^n$ *of the point $x = 0$;*

(2) *a matrix-valued function* $U \colon N_\tau^+ \times \mathcal{G} \to R^{m \times m}$ $(U \colon N_\tau^+ \times R^n \to R^{m \times m})$, $U(\tau, 0) = 0$ *for all $\tau \in N_\tau^+$ and a vector $\eta \in R_+^m$ such that the function* $v(\tau, x, \eta) = \eta^{\mathrm{T}} U(\tau, x) \eta$ *is continuous in the second argument;*

(3) *the functions* $\psi_{1i}, \psi_{2i}, \psi_{3i} \in K(KR)$, $i = 1, 2, \ldots m$, *and $m \times m$ matrices* $A_j(\eta)$, $j = 1, 2, 3$, *such that*

(a) $\psi_1^{\mathrm{T}}(\|x\|) A_1(\eta) \psi_1(\|x\|) \leq v(\tau, x, \eta) \leq \psi_2^{\mathrm{T}}(\|x\|) A_2(\eta) \psi_2(\|x\|)$ *for all* $(\tau, x, \eta \in N_\tau^+ \times \mathcal{N} \times R_+^m$ *(for all* $(\tau, x, \eta) \in N_\tau^+ \times R^n \times R_+^m$);

(b) $\Delta v(\tau, x, \eta)|_{(2.2.1)} \leq \psi_3^{\mathrm{T}}(\|x\|) A_3(\eta) \psi_3(\|x\|)$ *for all* $(\tau, x, \eta) \in N_\tau^+ \times \mathcal{N} \times R_+^m$ *(for all* $(\tau, x, \eta) \in N_\tau^+ \times R^n \times R_+^m$)

Then, if the matrices $A_1(\eta)$ and $A_2(\eta)$ are positive definite and the matrix $A_3(\eta)$ is negative definite, then the state $x = 0$ of system (2.2.1) is asymptotically stable (asymptotically stable in the whole).

Corollary 2.3.5 *Let there exist at least one couple of indices* $(p, q) \in [1, m]$ *for which the function* $v(\tau, x, e) = e^{\mathrm{T}} U(\tau, x) e$ *is positive definite,* decreasent and radially unbounded, $v \in C(N_\tau^+ \times \mathcal{N} \times I)$ ($v \in C(N_\tau^+ \times R^n \times I)$) such that there exist $\psi_1, \psi_2 \in KR$, and $\psi_3 \in K$ and

(a) $\psi_1(\|x\|) \leq v(\tau, x, e) \leq \psi_2(\|x\|)$ for all $(\tau, x, e) \in N_\tau \times \mathcal{N} \times I$ (for all $(\tau, x, e) \in N_\tau^+ \times R^n \times I$);

(b) $\Delta v(\tau, x, e)|_{(2.2.1)} \leq -\psi_3(\|x\|)$ for all $(\tau, x, e) \in N_\tau \times \mathcal{N} \times I$,

then the state $x = 0$ of system (2.2.1) is uniformly asymptotically stable (uniformly asymptotically stable in the whole).

In the next results we address boundedness properties of the solution of (2.2.1) via matrix-valued Liapunov's function. In these results we let $\mathcal{N} = R^n$.

Theorem 2.3.5 *Assume that the vector-function f in system (2.2.1) is continuous with respect to the second argument and definite in the domain $N_\tau^+ \times R^n$. If there exist*

(1) *a matrix-valued function $U: N_\tau^+ \times R^n \to R^{m \times m}$, $U(\tau, 0) = 0$ for all $\tau \in N_\tau^+$, and a vector $\eta \in R_+^m$, $\eta > 0$, such that the function $v(\tau, x, \eta) = \eta^T U(\tau, x)\eta$ is continuous in the second argument;*

(2) *continuous, strictly increasing, real valued functions ψ_{1i}, ψ_{2i}, $i = 1, 2, \ldots, m$, which are defined on R_+ with $\lim\limits_{r \to +\infty} \psi_{ji}(r) = \infty$, $i = 1, 2, \ldots, m$, $j = 1, 2$, and $m \times m$ matrices $A_k(\eta)$, $k = 1, 2, 3$, and a constant $M > 0$ such that*

(a) $\psi_1^T(\|x\|) A_1(\eta) \psi_1(\|x\|) \leq v(\tau, x, \eta) \leq \psi_2^T(\|x\|) A_2(\eta) \psi_2(\|x\|)$ *for all $\|x\| \geq M$ and $\tau \in N_\tau^+$;*

(b) $\Delta v(\tau, x, \eta)|_{(2.2.1)} \leq \psi_3^T(\|x\|) A_3(\eta) \psi_3(\|x\|)$ *for all $\|x\| \geq M$ and $\tau \in N_\tau^+$, $\psi_{3i} \in K$, $i = 1, 2, \ldots, m$,*

then, if the matrices $A_1(\eta)$, $A_2(\eta)$ are positive definite and the matrix $A_3(\eta)$ is negative semidefinite (negative definite), then

(a) *the set of all solutions of (2.2.1) is uniformly bounded;*

(b) *the set of all solutions of (2.2.1) is uniformly ultimately bounded under $\psi_{3i} \in KR$, $i = 1, 2, \ldots, m$.*

Corollary 2.3.6 *Let there exist at least one couple of indices $(p, q) \in [1, m]$ for which the function $v(\tau, x, e) = e^T U(\tau, x) e$ is such that for all $\|x\| \geq M$ and $\tau \in N_\tau^+$*

(a) *there exist two continuous, strictly increasing, real valued functions ψ_1, ψ_2 which are defined on R_+ with $\lim\limits_{r \to +\infty} \psi_i(r) = +\infty$, $i = 1, 2$,* and a constant $M > 0$ such that

$$\psi_1(\|x\|) \leq v(t, x, e) \leq \psi_2(\|x\|);$$

(b) $\Delta v(\tau, x, e)|_{(2.2.1)} \leq 0$ for all $\|x\| \geq M$ and $\tau \in N_\tau^+$.

Then the set of all solutions of (2.2.1) is uniformly bounded.

If in addition to (a) and (b) there exists a function $\psi_3 \in KR$ such that

$$\Delta v(\tau, x, e)|_{(2.2.1)} \leq -\psi_3(\|x\|)$$

for all $\|x\| \geq M$ and $\tau \in N_\tau^+$, then the set of all solutions of (2.2.1) is uniformly ultimately bounded.

Further we shall set out instability conditions of the equilibrium state $x = 0$ of system (2.2.1).

Theorem 2.3.6 *Assume that the vector function f in system (2.2.1) is continuous with respect to the second argument and is definite in the domain $N_\tau^+ \times \mathcal{N}$. If there exist*

(1) *an open connected discrete-time invariant neighborhood $\mathcal{G} \subseteq \mathcal{N}$, $\mathcal{N} \subseteq R^n$ of the state $x = 0$;*

(2) *matrix-valued function $U \colon N_\tau^+ \times \mathcal{G} \to R^{m \times m}$, $U(\tau, 0) = 0$ for all $\tau \in N_\tau^+$, and the vector $\vartheta \in R^m$ such that the function $v(\tau, x, \vartheta) = \vartheta^T U(\tau, x) \vartheta$ is continuous with respect to the second argument;*

(3) *comparison functions $\eta_{1i}, \eta_{2i}, \eta_{3i} \in K$, $i = 1, 2, \ldots, m$, and matrices $A_j(\vartheta)$, $j = 1, 2, 3$, such that*

(a) $\eta_1^T(\|x\|) A_1(\vartheta) \eta_1(\|x\|) \leq v(\tau, x, \vartheta) \leq \eta_2^T(\|x\|) A_2(\vartheta) \eta_2(\|x\|)$
for all $(\tau, x, \vartheta) \in N_\tau^+ \times \mathcal{G} \times R^m$;

(b) $\Delta v(\tau, x, \vartheta)|_{(2.2.1)} \geq \eta_3^T(\|x\|) A_3(\vartheta) \eta_3(\|x\|)$
for all $(\tau, x, \vartheta) \in N_\tau^+ \times \mathcal{G} \times R^m$;

(4) *point $x = 0$ belongs to $\mathcal{G}$;*

(5) $v(\tau, x, \vartheta) = 0$ on $N_\tau^+ \times (\partial \mathcal{G} \cap B_\Delta)$, where $\overline{B}_\Delta \subset \mathcal{N}$.

Then, if the matrices $A_1(\vartheta)$, $A_2(\vartheta)$ and $A_3(\vartheta)$ for all $(\vartheta \neq 0) \in R^m$ are positive definite, the equilibrium state $x = 0$ of the discrete system (2.2.1) is unstable.

The proofs of Theorems 2.3.4–2.3.6 are similar to the proofs of Theorems 1.2.1–1.2.7.

2.3.2 A trend of application of vector Liapunov's functions

Together with the system of difference equations (2.2.1) we consider the vector-function

(2.3.8) $$V(\tau, x) = (v_1(\tau, x), v_2(\tau, x), \ldots, v_m(\tau, x))^{\mathrm{T}}$$

where $v_i \colon N_\tau^+ \times B(r) \to R_+$ and satisfy the conditions

(2.3.9) $$a_{i1} \psi_{i1}^{1/2}(\|x_i\|) \leq v_i(\tau, x) \leq a_{i2} \psi_{i2}^{1/2}(\|x_i\|), \qquad i = 1, 2, \ldots, m$$

Here a_{i1}, a_{i2} are some positive constants and ψ_{i1}, ψ_{i2} are of class $K(KR)$.

Similarly to the case of continuous system for discrete-time systems there are two approaches in application of vector-function (2.3.8).

2.3.2.1 Scalar approach Further we need the following definition (cf. Ostrovskii [1]).

Definition 2.3.1 *The matrix $A = (a_{ij})_{i,j=1}^m$ is M-matrix, if $a_{ii} > 0$, $a_{ij} \leq 0$ for $i \neq j$, and the successive principal minors of A are all positive.*

In the framework of scalar approach we state the following.

Theorem 2.3.7 *Assume that for the system of equations (2.2.1):*

(1) *there exists a vector-function (2.3.8) whose components are locally Lipschiitzian in x and satisfy (2.3.9);*

(2) *there exist constants $d_{ij} > 0$ for all $i, j = 1, 2, \ldots, m$ and comparison functions $\psi_{3j} \in K$, such that*

$$\Delta V(\tau, x)\big|_{(2.2.1)} \leq D\psi_3(\|x\|)$$

for all $x \in R^n$ and $\tau \in N_\tau^+$, where $D = [d_{ij}]$, $\psi_3(\|x\|) = (\psi_{31}(\|x\|), \ldots, \psi_{3m}(\|x\|))^{\mathrm{T}}$;

(3) *for all $(\tau, x) \in N_\tau^+ \times R^n$ the estimates*

$$(\psi_1^{1/2}(\|x\|))^{\mathrm{T}} B_1 \psi_1^{1/2}(\|x\|) \leq v(\tau, x, \alpha) \leq (\psi_2^{1/2}(\|x\|))^{\mathrm{T}} B_2 \psi_2^{1/2}(\|x\|)$$

are satisfied.

Then, if the matrices B_1, B_2 are positive definite and the matrix D is M-matrix, then the equilibrium state $x = 0$ of system (2.2.1) is uniformly asymptotically stable (in the whole).

Proof We rewrite system (2.2.1) as

(2.3.10) $$x_i(\tau + 1) = f_i(\tau, x_1(\tau), \ldots, x_m(\tau)), \qquad i = 1, 2, \ldots, m$$

and consider the function

(2.3.11) $$v(\tau, x, \alpha) = \alpha^{\mathrm{T}} V(\tau, x), \quad \alpha \in R_+^m, \quad \alpha > 0.$$

Under condition (3) of Theorem 2.3.7 function (2.3.11) is positive definite, decreasing and radially unbounded.

In view of conditions (2) of Theorem 2.3.7

$$\Delta v(\tau, x, \alpha)\big|_{(2.3.10)} = \alpha^{\mathrm{T}} \Delta V(\tau, x)\big|_{(2.3.10)} \leq \alpha^{\mathrm{T}} D \psi_3(\|x\|)$$

for all $x \in R^n$ and $\tau \in \mathcal{N}_\tau^+$. Since D is an M-matrix, there exists $D^{-1} \geq 0$. Besides we take $\alpha = (D^{-1})^{\mathrm{T}} w > 0$ for $w \in R_+^m$, $w > 0$, so that

$$\Delta v(\tau, x, \alpha)\big|_{(2.3.10)} < -w^{\mathrm{T}} \psi(\|x\|) < 0$$

for all $(x \neq 0) \in R^n$ and $\tau \in \mathcal{N}_\tau^+$.

Therefore, $\Delta v(\tau, x, \alpha)\big|_{(2.3.10)}$ is a negative definite function for all $x \in R^n$ and $\tau \in \mathcal{N}_\tau^+$. This proves the assertion of Theorem 2.3.7.

Assume that for the system

(2.3.12) $$x(\tau + 1) = A(\tau, x(\tau)) x(\tau)$$

where $A \colon \mathcal{N}_\tau^+ \times R^n \to R^{n \times n}$, $x \in R^n$, $A(\tau, x(\tau))$ is a bounded $n \times m$ matrix for any τ and $A(\tau, 0) \neq 0$ for all $\tau \in \mathcal{N}_\tau^+$, is a vector-function (2.3.8) constructed in some way.

One can easily verify the following.

Theorem 2.3.8 (cf. Šiljak [1], Theorem 4.11) *The equilibrium state $x = 0$ of system (2.3.12) is exponentially stable in the whole, if for the function $v(\tau, x, \alpha)$, specified according to (2.3.11) there exist constants $c_1 > 0$, $c_2 > 0$, $0 < c_3 < 1$, such that*

$$c_1 \|x\|^2 \leq v(\tau, x, \alpha) \leq c_2 \|x\|^2,$$
$$v(\tau + 1, x(\tau + 1, \tau_0, x_0), \alpha) \leq c_3 v(\tau, x(\tau), \alpha)$$

for all $x \in R^n$ and $\tau \in \mathcal{N}_\tau^+$.

2.3.2.2 Vector approach Vector approach to the analysis of solution behavior of system (2.2.1) is based on the following assertion.

Theorem 2.3.9 *Assume that*

$$u(\tau + 1) \leq \phi(\tau, u(\tau)), \quad \tau \geq \tau_0,$$

where $u \in R^m$, $\phi\colon \mathcal{N}_\tau^+ \times R^m \to R^m$, $\tau \in \mathcal{N}_\tau^+$, *and function* $\phi(\tau, r)$ *is nondecreasing in* r. *Let* $r_n(\tau)$ *be the solution of the comparison system*

$$u(\tau + 1) = \phi(\tau, u(\tau))$$

with the initial condition $u(\tau_0) \leq r(\tau_0)$.

Then for all $\tau \geq \tau_0$ *the component-wise estimate*

$$u(\tau) \leq r_n(\tau).$$

holds.

For the proof of this assertion see Lakshmikantham and Trigiante [1].
For function (2.3.8) we consider the first difference

$$(2.3.13) \quad \Delta V(\tau, x(\tau))\big|_{(2.2.1)} = V(\tau + 1, x(\tau + 1)) - V(\tau, x(\tau))$$

and assume that the function $g\colon \mathcal{N}_\tau^+ \times R^m \to R^m$, exists so that

$$(2.3.14) \quad \Delta V(\tau, x(\tau))\big|_{(2.2.1)} \leq g(\tau, V(\tau, x(\tau))).$$

From (2.3.14) and (2.3.13) imply the estimate

$$(2.3.15) \quad V(\tau + 1, x(\tau + 1)) \leq V(\tau, x(\tau)) + g(\tau, V(\tau, x(\tau))).$$

Inequality (2.3.15) corresponds to the comparison system

$$(2.3.16) \quad u(\tau + 1) = u(\tau) + g(\tau, u(\tau)) = \phi(\tau, u(\tau)).$$

For system (2.2.1) the following assertion takes place.

Theorem 2.3.10 *Assume that for system of equations (2.2.1):*

(1) *there exist the vector-function (2.3.8), whose components are locally Lipschitzian and satisfy inequalities (2.3.9);*

(2) *there exists the function* $g\colon \mathcal{N}_\tau^+ \times R_+^m \to R^m$, $g(\tau, 0) = 0$ *for all* $\tau \in \mathcal{N}_\tau^+$ *satisfying inequality (2.3.14) such that the function* $\phi(\tau, u)$ *is nondecreasing in* u;

(3) *for all* $(\tau, x) \in \mathcal{N}_\tau^+ \times R^n$ *for the function (2.3.11) the estimates*

$$(2.3.17) \quad (\psi_1^{1/2}(\|x\|))^{\mathrm{T}} B_1 \psi_1^{1/2}(\|x\|) \leq v(\tau, x, \alpha) \leq (\psi_2^{1/2}(\|x\|))^{\mathrm{T}} B_2 \psi_2^{1/2}(\|x\|).$$

are fulfilled.

Then, if the zero solution of system (2.3.16) is

(a) *stable in the whole and matrix B_1 in the estimate (2.3.17) is positive definite, the state $x = 0$ of system (2.2.1) is stable in the whole;*
(b) *uniformly stable in the whole and matrices B_1 and B_2 in the estimate (2.3.17) positive definite, the state $x = 0$ of system (2.2.1) is uniformly stable in the whole;*
(c) *asymptotically stable in the whole and matrix B_1 in the estimate (2.3.17) is positive definite, the state $x = 0$ of system (2.2.1) is asymptotically stable in the whole;*
(d) *uniformly asymptotically stable in the whole and matrices B_1 and B_2 in the estimate (2.3.17) are positive definite, the equilibrium state $x = 0$ of system (2.2.1) is uniformly asymptotically stable in the whole;*

The proof of this Theorem is standard in the framework of the comparison principle and so is omitted here.

Further we consider the nonlinear system

$$x(\tau + 1) = f(\tau, x(\tau)), \tag{2.3.18}$$

where $x \in R^n$ is a state vector of the system, $\mathcal{N}_\tau^+ = \{t_0 + k,\ k \in \mathcal{N}^+\}$, $\mathcal{N}^+ = \{0, 1, 2, \ldots\}$, $t_0 \in R$. Assume that the function $f \colon \mathcal{N}_\tau^+ \times R^n \to R^n$ such that for all $\tau \in T_\tau$ there exists a unique solution of system (2.3.18) for any initial data $(\tau_0, x_0) \in \mathcal{N}_\tau^+ \times R^n$ and the state $x(\tau) \equiv 0$ is the only equilibrium state.

System (2.3.18) is represented as

$$x_i(\tau + 1) = f_i(\tau, x(\tau)), \qquad i = 1, 2, \ldots, s, \tag{2.3.19}$$

where $x_i \in R^{n_i}$, $x = (x_1^T, x_2^T, \ldots, x_s^T)^T$, $n = n_1 + n_2 + \cdots + n_s$, $f_i \colon \mathcal{N}_\tau^+ \times R^n \to R^{n_i}$. Besides, we consider also the decomposition of system (2.3.18)

$$x_i(\tau + 1) = p_i(\tau, x_i(\tau)) + g_i(\tau, x(\tau)), \qquad i = 1, 2, \ldots, s, \tag{2.3.20}$$

where $p_i \colon \mathcal{N}_\tau^+ \times R^{n_i} \to R^{n_i}$, $g_i \colon \mathcal{N}_\tau^+ \times R^n \to R^{n_i}$, $p_i(\tau, 0) = g_i(\tau, 0) = 0$ for all $\tau \in \mathcal{N}_\tau^+$.

Theorem 2.3.11 *For the system of (2.3.19) assume:*

(1) *there exist constants $F_i > 0$, such that $\|f_i(\tau, x)\| \leq F_i \|x_i\|$ for all $(\tau, x) \in \mathcal{N}_\tau^+ \times R^n$, $i \in \mathcal{N}_s$;*

(2) there exist functions u_i, $w_i \in K$ and the functions $v_i \colon \mathcal{N}_\tau^+ \times R^{n_i} \to [0, +\infty)$ such that for all $(\tau, x) \in \mathcal{N}_\tau^+ \times R^n$ and $i \in \mathcal{N}_s$,

$$u_i(\|x_i\|) \leq v_i(\tau, x_i) \leq w_i(\|x_i\|);$$

(3) for each $i \in \mathcal{N}_s$ there exist sequences of real numbers $\{h_i^j\}_{j=0}^\infty$, $\{H_i^j\}_{j=0}^\infty$ such that $[h_i^j, H_i^j] \cap [h_i^k, H_i^k] = \varnothing$ for $j \neq k$, $0 < h_i^j < H_i^j$, $\lim_{j \to +\infty} H_i^j = 0$ and $w_i(F_i h_i^j) \leq u_i(H_i^j)$ for all $j \in N^+$;

(4) if for any solution $x(\tau)$ of system (2.3.19) for some $\overline{\xi} \in \mathcal{N}_{\tau_0}^+$, $\overline{i} \in \mathcal{N}_s$, $\overline{j} \in \mathcal{N}_+$ the inequality $h_{\overline{i}}^{\overline{j}} \leq \|x_{\overline{i}}(\overline{\xi})\| \leq H_{\overline{i}}^{\overline{j}}$ is satisfied,

$$\Delta v_{\overline{i}}(\overline{\xi}, x_{\overline{i}}(\overline{\xi})) = v_{\overline{i}}(\overline{\xi}+1, f_{\overline{i}}(\overline{\xi}, x(\overline{\xi}))) - v_{\overline{i}}(\overline{\xi}, x_{\overline{i}}(\overline{\xi})) \leq 0.$$

is satisfied too.

Then the equilibrium state $x = 0$ of system (2.3.19) is uniformly stable.

Proof Take arbitrary $\varepsilon > 0$. By conditions (3) of Theorem 2.3.11 for $\varepsilon_1 = \varepsilon/\sqrt{s}$ for any $i \in \mathcal{N}_s$ exists $j_i \in \mathcal{N}^+$, such that $H_i^{j_i} \leq \varepsilon_1$. For the sake of convenience designate $H_i^{j_i} = H_i$, $h_i^{j_i} = h_i$ and set $\delta < \min\{h_i, F_i h_i\}$. Then, if $\|x_0\| \leq \delta$, for any $i \in N_s$ we get

$$v_i(\tau_0, x_{0i}) \leq w_i(\|x_{0i}\|) \leq w_i(\delta) \leq w_i(F_i h_i) \leq u_i(H_i).$$

If we assume that there exist $\tau_1 > \tau_0$ and $\alpha \in \mathcal{N}_s$ such that $v_\alpha(\tau_1 + 1, x_\alpha(\tau_1 + 1)) > u_\alpha(H_\alpha)$, then without loss of generality one may consider that $v_\alpha(\tau_1, x_\alpha(\tau_1)) \leq u_\alpha(H_\alpha)$. Then $\|x_\alpha(\tau_1)\| \leq H_\alpha$ and

(2.3.21) $\quad v_\alpha(\tau_1 + 1, x_\alpha(\tau_1 + 1)) - v_\alpha(\tau_1, x_\alpha(\tau_1)) > 0.$

Let now $\|x_\alpha(\tau_1)\| < h_\alpha$. Applying conditions (1)–(3) of Theorem 2.3.11, one gets

$$v_\alpha(\tau_1 + 1, x_\alpha(\tau_1 + 1)) = v_\alpha(\tau_1 + 1, f_\alpha(\tau_1, x(\tau_1)))$$
$$\leq w_\alpha(\|f_\alpha(\tau_1, x(\tau_1))\|) \leq w_\alpha(F_\alpha \|x_\alpha(\tau_1)\|) < w_\alpha(F_\alpha h_\alpha) \leq u_\alpha(H_\alpha),$$

which contradicts the choice of the constant τ_1. Hence $h_\alpha \leq \|x_\alpha(\tau_1)\| \leq H_\alpha$, which yields, by condition (4) of Theorem 2.3.11

$$v_\alpha(\tau_1 + 1, x_\alpha(\tau_1 + 1)) - v_\alpha(\tau_1, x_\alpha(\tau_1))$$
$$= v_\alpha(\tau_1 + 1, f_\alpha(\tau_1, x(\tau_1))) - v_\alpha(\tau_1, x_\alpha(\tau)) \leq 0.$$

2. DISCRETE-TIME SYSTEMS

This contradicts inequality from the condition (4) of the Theorem. Consequently for all $\tau > \tau_0$ and $i \in \mathcal{N}_s$ $v_i(\tau, x_i(\tau)) \leq u_i(H_i)$, hence $\|x_i(\tau)\| \leq H_i \leq \varepsilon_1$ and therefore $\|x(\tau)\| \leq \varepsilon_1 \sqrt{s} = \varepsilon$ for all $\tau \in \mathcal{N}_{\tau_0}^+$.

The Theorem 2.3.11 is proved.

Following Yoshizawa [1], and Hahn [1] we shall cite the assertion necessary for subsequent presentation.

Definition 2.3.2 Solutions of system (2.3.18) are *uniformly bounded*, if for all $x_0 \in R^n$ and $\eta > 0$ there exists $B = B(\eta) > 0$ such that the inequality $\|x_0\| \leq \eta$ implies inequality $\|x(\tau)\| \leq B$ for all $\tau \in \mathcal{N}_{\tau_0}^+$.

Theorem 2.3.12 *Assume that*

(1) *there exist constants $F_i > 0$, such that $\|f_i(\tau, x)\| \leq F_i \|x_i\|$ for all $(\tau, x) \in \mathcal{N}_\tau^+ \times R^n$, $i \in \mathcal{N}_s$;*

(2) *there exist functions $u_i, w_i \in KR$ and functions $v_i \colon \mathcal{N}_\tau^+ \times R^{n_i} \to [0, +\infty)$ such that for all $(\tau, x) \in \mathcal{N}_\tau^+ \times R^n$ and $i \in \mathcal{N}_s$,*

$$u_i(\|x_i\|) \leq v_i(\tau, x_i) \leq w_i(\|x_i\|);$$

(3) *for every $i \in \mathcal{N}_s$ there exist sequences of real numbers $\{h_i^j\}_{j=0}^\infty$, $\{H_i^j\}_{0=1}^\infty$ such that $[h_i^j, H_i^j] \cap [h_i^k, H_i^k] = \emptyset$ for $j \neq k$, $0 < h_i^j < H_i^j$, $\lim\limits_{j \to +\infty} h_i^j = +\infty$ and for all $j \in N^+$ the inequalities*

$$w_i(F_i h_i^j) \leq u_i(H_i^j)$$

are fulfilled;

(4) *if for any solution $x(\tau)$ of system (2.3.19) for some $\overline{\xi} \in \mathcal{N}_{\tau_0}^+$, $\overline{i} \in \mathcal{N}_s$, $\overline{j} \in \mathcal{N}_+$ the inequality $h_{\overline{i}}^{\overline{j}} \leq \|x_{\overline{i}}(\overline{\xi})\| \leq H_{\overline{i}}^{\overline{j}}$ is satisfied, the inequality*

$$\Delta v_{\overline{i}}(\overline{\xi}, x_{\overline{i}}(\overline{\xi})) = v_{\overline{i}}(\overline{\xi}+1, f_{\overline{i}}(\overline{\xi}, x(\overline{\xi}))) - v_{\overline{i}}(\overline{\xi}, x_{\overline{i}}(\overline{\xi})) \leq 0$$

is satisfied too.

Then the solutions of system (2.3.19) are uniformly bounded.

Proof We take arbitrary $\eta > 0$. By conditions (2) and (3) of Theorem 2.3.12 for every $i \in \mathcal{N}_s$ there exists $j_i \in \mathcal{N}^+$, such that $u_i(H_i^{j_i}) \geq w_i(\eta)$. For the sake of convenience designate $H_i^{j_i} = H_i$, and $h_i^{j_i} = h_i$. Then, under condition $\|x_0\| \leq \eta$, for every $i \in \mathcal{N}_s$ we get

$$v_i(\tau_0, x_{0i}) \leq w_i(\|x_{0i}\|) \leq w_i(\eta) \leq u_i(H_i).$$

Further, similarly to the proof of Theorem 2.3.11 we show that for all $\tau > \tau_0$ and $i \in \mathcal{N}_s$ $v_i(\tau, x_i(\tau)) \leq u_i(H_i)$, hence $\|x_i(\tau)\| \leq H_i$ and consequently $\|x(\tau)\| \leq (\sum_{i=1}^{s} H_i^2)^{1/2} = B$ for all $\tau \in \mathcal{N}_{\tau_0}^+$. Theorem 2.3.12 is proved.

Theorem 2.3.13 *Assume that*

(1) *there exist constants $P_i, G_i > 0$, such that $\|p_i(\tau, x_i)\| \leq P_i \|x_i\|$, $\|g_i(\tau, x)\| \leq G_i$ for all $(\tau, x) \in \mathcal{N}_\tau^+ \times R^n$, $i \in \mathcal{N}_s$;*

(2) *there exist functions $u_i, w_i \in KR$ and functions $v_i \colon \mathcal{N}_\tau^+ \times R^{n_i} \to [0, +\infty)$ such that for all $(\tau, x) \in \mathcal{T} \times R^n$ and $i \in \mathcal{N}_s$,*

$$u_i(\|x_i\|) \leq v_i(\tau, x_i) \leq w_i(\|x_i\|);$$

(3) *for every $i \in \mathcal{N}_s$ there exist sequences of real numbers $\{h_i^j\}_{j=0}^{\infty}$, $\{H_i^j\}_{j=1}^{\infty}$ such that $[h_i^j, H_i^j] \cap [h_i^k, H_i^k] = \emptyset$ for $j \neq k$, $0 < h_i^j < H_i^j$, $\lim\limits_{j \to +\infty} h_i^j = +\infty$ and for all $j \in N^+$ the inequality*

$$w_i(P_i h_i^j + G_i) \leq u_i(H_i^j)$$

is satisfied;

(4) *if for any solution $x(\tau)$ of system (2.3.20) for some $\overline{\xi} \in \mathcal{N}_{\tau_0}^+$, $\overline{i} \in \mathcal{N}_s$, $\overline{j} \in \mathcal{N}_+$ the inequality $h_{\overline{i}}^{\overline{j}} \leq \|x_{\overline{i}}(\overline{\xi})\| \leq H_{\overline{i}}^{\overline{j}}$ is satisfied, the inequality*

$$\Delta v_{\overline{i}}(\overline{\xi}, x_{\overline{i}}(\overline{\xi})) = v_{\overline{i}}(\overline{\xi}+1, p_{\overline{i}}(\overline{\xi}, x_{\overline{i}}(\overline{\xi})) + g_{\overline{i}}(\overline{\xi}, x(\overline{\xi}))) - v_{\overline{i}}(\overline{\xi}, x_{\overline{i}}(\overline{\xi})) \leq 0$$

is satisfied too.

Then the solutions of system (2.3.20) are uniformly bounded.

Proof We take arbitrary $\eta > 0$. By conditions (2) and (3) of Theorem 2.3.13 for every $i \in \mathcal{N}_s$ there exists $j_i \in \mathcal{N}^+$, such that $u_i(H_i^{j_i}) \geq w_i(\eta)$. For convenience we designate $H_i^{j_i} = H_i$, $h_i^{j_i} = h_i$. Then for $\|x_0\| \leq \eta$, for every $i \in \mathcal{N}_s$ we get

$$v_i(\tau_0, x_{0i}) \leq w_i(\|x_{0i}\|) \leq w_i(\eta) \leq u_i(H_i).$$

If one assumes that there exist $\tau_1 > \tau_0$ and $\alpha \in \mathcal{N}_s$ such that $v_\alpha(\tau_1+1, x_\alpha(\tau_1+1)) > u_\alpha(H_\alpha)$, then without loss of generality one may consider that

(2.3.22) $$v_\alpha(\tau_1, x_\alpha(\tau_1)) \leq u_\alpha(H_\alpha).$$

Then inequality (2.3.22) implies $\|x_\alpha(\tau_1)\| \le H_\alpha$ and
(2.3.23) $\quad v_\alpha(\tau_1+1, x_\alpha(\tau_1+1)) - v_\alpha(\tau_1, x_\alpha(\tau_1)) > 0.$

Now let $\|x_\alpha(\tau_1)\| < h_\alpha$. Applying conditions of Theorem 2.3.13 we obtain
$$v_\alpha(\tau_1+1, x_\alpha(\tau_1+1)) = v_\alpha(\tau_1+1, f_\alpha(\tau_1, x_\alpha(\tau_1)) + g_\alpha(\tau_1, x(\tau_1)))$$
$$\le w_\alpha(P_\alpha\|x_\alpha(\tau_1)\| + G_\alpha) \le w_\alpha(P_\alpha h_\alpha + G_\alpha)\|) \le u_\alpha(H_\alpha),$$
which contradicts the choice of the constant τ_1. So, $h_\alpha \le \|x_\alpha(\tau_1)\| \le H_\alpha$, and hence, by condition (4) of Theorem 2.3.13 we get the inequality
$$v_\alpha(\tau_1+1, x_\alpha(\tau_1+1)) - v_\alpha(\tau_1, x_\alpha(\tau_1))$$
$$= v_\alpha(\tau_1+1, f_\alpha(\tau_1, x(\tau_1))) - v_\alpha(\tau_1 x_\alpha(\tau)) \le 0,$$
which contradicts the inequality (2.3.23). Therefore for all $\tau > \tau_0$ and $i \in \mathcal{N}_s$ $v_i(\tau, x_i(\tau)) \le u_i(H_i)$, hence $\|x_i(\tau)\| \le H_i$ and consequently $\|x(\tau)\| \le (\sum_{i=1}^s H_i^2)^{1/2} = B$ for all $\tau \in \mathcal{N}_{\tau_0}^+$. The Theorem 2.3.13 is proved.

Example 2.3.1 We consider the discrete system
(2.3.24)
$$x_1(\tau+1) = \frac{5}{8}x_1(\tau) + \frac{5}{16}x_1(\tau)\left((-1)^{r(|x_1(\tau)|)} + \cos(x_1^2(\tau) + x_2^2(\tau))\right),$$
$$x_2(\tau+1) = \frac{5}{8}x_2(\tau) + \frac{5}{16}x_2(\tau)\left((-1)^{r(|x_2(\tau)|)} + \cos(x_1^2(\tau) + x_2^2(\tau))\right),$$
where $x_1, x_2 \in R$, $\tau \in \{1, 2, 3, \dots\}$,
$$r(t) = \begin{cases} 1, & t \in \left[2^{-j}, \frac{3}{2} \cdot 2^{-j}\right], \quad j \in \mathcal{N}^+, \\ 0, & \text{in the other cases.} \end{cases}$$

Set $f_i(x) = 5/8\, x_i + 5/16\, x_i(-1)^{h(|x_1|)} + 5/16\, x_i \cos(x_1^2 + x_2^2)$, $v_i(x_i) = |x_i|$, $u_i(t) = w_i(t) = t$ for $t \in [0, +\infty)$, $s = 2$ and we get the constants $F_i = 5/4$, $h_i^j = 2^{-j}$, $H_i^j = \frac{3}{2}2^{-j}$. Since $\frac{5}{4}2^{-j} \le \frac{3}{2}2^{-j}$, the condition (3) of Theorem 2.3.11 is satisfied. We shall verify now condition (4):
$$\Delta v_i = \left|\frac{5}{8}x_i + \frac{5}{16}x_i\left((-1)^{r(|x_i|)} + \cos(x_1^2 + x_2^2)\right)\right| - |x_i|$$
$$= \left|\frac{5}{8}x_i - \frac{5}{16}x_i + \frac{5}{16}x_1\cos(x_1^2+x_2^2)\right| - |x_i|$$
$$= \left|\frac{5}{16}x_i + \frac{5}{16}x_i\cos(x_1^2+x_2^2)\right| - |x_i| \le \frac{5}{8}|x_i| - |x_i| = -\frac{3}{8}|x_i| < 0$$
for $2^{-j} \le \|x_i(\xi)\| \le \frac{3}{2}2^{-j}$, $j \in \mathcal{N}^+$. All conditions of Theorem 2.3.11 are satisfied, therefore the equilibrium state $x = 0$ of system (2.3.24) is uniformly stable.

Example 2.3.2 Consider the discrete system

$$
\begin{aligned}
x_1(\tau+1) &= x_1(\tau) + a\,\frac{x_1(\tau)}{1+x_1^2(\tau)}(\sin|x_1(\tau)| - \sin^2|x_2(\tau)|), \\
x_2(\tau+1) &= x_2(\tau) + a\,\frac{x_2(\tau)}{1+x_2^2(\tau)}(\sin|x_2(\tau)| - \sin^2|x_1(\tau)|),
\end{aligned}
$$
(2.3.25)

where $x_1, x_2 \in R$, $0 < a < 2$ and we set: $p_i(\tau, x_i) = x_i$, $g_i(\tau, x) = a\,\frac{x_i}{1+x_i^2}(\sin|x_i| - \sin^2|x_j|)$, $v_i(x_i) = x_i^2$, $u_i(t) = w_i(t) = t^2$, for $t \in [0, +\infty)$, $i = 1, 2$, $j \neq i$. We get the constants $P_i = 1$, $G_i = a$.

We check condition (4) of the Theorem 2.3.13:

$$
\Delta v_i(x_i) = \left(x_i + a\,\frac{x_i}{1+x_i^2}(\sin|x_i| - \sin^2|x_j|)\right) - x_i^2
$$

$$
= a\,\frac{x_i^2}{1+x_i^2}\left(2 + a\,\frac{x_i}{1+x_i^2}(\sin|x_i| - \sin^2|x_j|)\right)(\sin|x_i| - \sin^2|x_j|) \leq 0
$$

for $\sin|x_i| \leq 0$ and $0 < a \leq 2$. Then, if $\pi + 2\pi j \leq \|x_i\| \leq 2\pi + 2\pi j$, $j \in \mathcal{N}^+$, the condition (4) of Theorem 2.3.13 is satisfied for $h_i^j = \pi + 2\pi j$, $H_i^j = 2\pi + 2\pi j$, $i = 1, 2$.

We check inequalities $w_i(P_i h_i^j + G_i) \leq u_i(H_i^j)$, which after substitution of all constants, become $(\pi + 2\pi j + a)^2 \leq (2\pi + 2\pi j)^2$, or $a \leq \pi$ is true.

All conditions of Theorem 2.3.13 are satisfied, and so the solution of system (2.3.25) is uniformly bounded.

2.4 Stability Analysis of Linear Systems

2.4.1 A new Liapunov's function for linear system We consider a linear system of difference equations of arbitrary order in the form

(2.4.1) $\quad x_i(\tau+1) = A_{ii} x_i(\tau) + \sum_{\substack{j=1 \\ j \neq i}}^{m} A_{ij}(\tau) x_j(\tau), \quad i = 1, 2, \ldots, m,$

where $x = (x_1^T, \ldots, x_m^T)^T$, $\tau \in \mathcal{N}_\tau^+ = \{\tau_0 + k,\ k = 0, 1, \ldots,\}$ $\tau_0 > 0$, $x_i \in R^{n_i}$, $x \in R^n$, $n = \sum_{i=1}^{m} n_i$, A_{ii}, $i = 1, \ldots, m$, are constant matrices of appropriate dimensions, $A_{ij}(\tau)$, $i, j = 1, \ldots, m$, $i \neq j$, are determined

2. DISCRETE-TIME SYSTEMS

on the set N_τ^+. The transformation is made by means of mathematical decomposition for the preassigned order of independent subsystems or in terms of some physical speculations formed in the description of a real physical system by a system of difference equations.

For system (2.4.1) we construct the matrix-valued function $U(\tau, x)$. The diagonal elements $v_{ii}(x_i)$, $i = 1, 2, \ldots, m$, are taken as the quadratic forms

$$v_{ii}(x_i) = x_i^T P_{ii} x_i, \quad i = 1, 2, \ldots, m, \quad (2.4.2)$$

where P_{ii} are symmetric positive definite matrices. We assume that at least one of the matrices A_{ij} or A_{ji} is not equal to constant and take the corresponding non-diagonal elements $v_{ij}(\tau, x_i, x_j)$ as the bilinear form

$$\begin{aligned} v_{ij}(\tau, x_i, x_j) = v_{ji}(\tau, x_i, x_j) = x_i^T P_{ij}(\tau) x_j, \\ i, j = 1, 2, \ldots, m, \quad i \neq j, \end{aligned} \quad (2.4.3)$$

where the matrix $P_{ij}(\tau)$ satisfies difference equation

$$\begin{aligned} P_{ij}(\tau+1) - P_{ij}(\tau) + A_{ii}^T P_{ij}(\tau+1) A_{jj} - P_{ij}(\tau+1) \\ = -\frac{\eta_i}{\eta_j} A_{ii} P_{ii} A_{ij}(\tau) - \frac{\eta_j}{\eta_i} A_{ji}^T(\tau) P_{jj} A_{jj}. \end{aligned} \quad (2.4.4)$$

Equation (2.4.4) can be solved in the explicit form. Consider two cases.

Case 1. Assume that the matrices A_{ii} and A_{jj} are such that

$$q = \max_{k,l} |\lambda_k(A_{ii}) \lambda_l(A_{jj})| < 1.$$

We consider the linear operators

$$F_{ij}: R^{n_i \times n_j} \to R^{n_i \times n_j}, \quad F_{ij} X = A_{ii}^T X A_{jj}.$$

and present equation (2.4.4) as

$$P_{ij}(\tau) = -F_{ij} P_{ij}(\tau+1) + \frac{\eta_i}{\eta_j} A_{ii} P_{ii} A_{ij}(\tau) + \frac{\eta_j}{\eta_i} A_{ji}^T(\tau) P_{jj} A_{jj}. \quad (2.4.5)$$

Using the method of mathematical induction it is easy to show that

$$\begin{aligned} P_{ij}(\tau) = F_{ij}^\nu P_{ij}(\tau+\nu) + \sum_{k=0}^{\nu} F_{ij}^k \bigg[\frac{\eta_i}{\eta_j} A_{ii} P_{ii} A_{ij}(\tau+k) \\ + \frac{\eta_j}{\eta_i} A_{ji}^T(\tau+k) P_{jj} A_{jj} \bigg] \end{aligned} \quad (2.4.6)$$

for any positive integer ν. It is shown (see Daletskii and Krene [1]) that the eigenvalues of the operators F_{ij} are $\lambda_k(A_{ii})\lambda_l(A_{jj})$, and therefore the norm of the operator F_{ij}^ν admits the estimate

$$\|F_{ij}^\nu\| = \left\|\frac{1}{2\pi i}\int_{|z|=\frac{1+q}{2}} z^\nu R_z(F_{ij})\,dz\right\| \leq \frac{c}{2\pi}\int_{|z|=\frac{1+q}{2}} |z|^\nu\,dl = c\left(\frac{1+q}{2}\right)^\nu,$$

where $c = \max\limits_{|z|=\frac{1+q}{2}} \|R_z(F_{ij})\|$, $R_z(F_{ij})$ is a resolvent of the operator F_{ij}.

Taking into account $\frac{1+q}{2} < 1$, we get $\|F_{ij}^\nu\| \to 0$ as $\nu \to \infty$.

Further we are interested only in bounded solutions of equation (2.4.4). Passing to the limit in (2.4.6) as $\nu \to \infty$, we get

$$(2.4.7) \qquad P_{ij}(\tau) = \sum_{k=\tau}^{\infty} F_{ij}^{-\tau+k}\left\{\frac{\eta_i}{\eta_j}A_{ii}P_{ii}A_{ij}(k) + \frac{\eta_j}{\eta_i}A_{ji}^{\mathrm{T}}(k)P_{jj}A_{jj}\right\}.$$

Further it is assumed that the series in the right-side part of (2.4.7) converges.

Case 2. Assume that

$$q = \max_{k,l}|\lambda_k(A_{ii})\lambda_l(A_{jj})| \geq 1.$$

It is easy to notice that the operator F_{ij} is non-degenerated. We present equation (2.4.4) as

$$(2.4.8)\ P_{ij}(\tau+1) = F_{ij}^{-1}P_{ij}(\tau) - F_{ij}^{-1}\left\{\frac{\eta_i}{\eta_j}A_{ii}P_{ii}A_{ij}(\tau) + \frac{\eta_j}{\eta_i}A_{ji}^{\mathrm{T}}(\tau)P_{jj}A_{jj}\right\}.$$

Using the method of mathematical induction it is easy to show that

$$P_{ij}(\tau) = F_{ij}^{-\tau+\tau_0}P_{ij}(\tau_0) - \sum_{k=0}^{\tau-\tau_0-1} F_{ij}^{-\tau+\tau_0+k}$$
$$\times \left[\frac{\eta_i}{\eta_j}A_{ii}P_{ii}A_{ij}(\tau_0+k) + \frac{\eta_j}{\eta_i}A_{ji}^{\mathrm{T}}(\tau_0+k)P_{jj}A_{jj}\right].$$

Setting $P_{ij}(\tau_0) = 0$ we find partial solution of equation (2.4.4) in the form

$$(2.4.9) \qquad P_{ij}(\tau) = -\sum_{k=0}^{\tau-\tau_0-1} F_{ij}^{-\tau+\tau_0+k}$$
$$\times \left[\frac{\eta_i}{\eta_j}A_{ii}P_{ii}A_{ij}(\tau_0+k) + \frac{\eta_j}{\eta_i}A_{ji}^{\mathrm{T}}(\tau_0+k)P_{jj}A_{jj}\right].$$

2. DISCRETE-TIME SYSTEMS

Assuming that the matrices $P_{ij}(\tau)$ are bounded for all $\tau \geq \tau^*$ we introduce designations

$$\bar{c}_{ii} = \lambda_M(P_{ii}), \quad \bar{c}_{ij} = \sup_{\tau \geq \tau^*} \|P_{ij}(\tau)\|,$$

$$\underline{c}_{ii} = \lambda_m(P_{ii}), \quad \underline{c}_{ij} = -\sup_{\tau \geq \tau^*} \|P_{ij}(\tau)\|.$$

In view of the results from Krasovskii [1] and Djordjevic [1] the estimates for the elements matrix-valued function $U(\tau, x)$ are

$$\underline{c}_{ii}\|x_i\|^2 \leq v_{ii}(x_i) \leq \bar{c}_{ii}\|x_i\|^2, \quad i = 1, 2, \ldots, m,$$

$$\underline{c}_{ij}\|x_i\|\,\|x_j\| \leq v_{ij}(\tau, x_i, x_j) \leq \bar{c}_{ij}\|x_i\|\,\|x_j\|, \quad i,j = 1, 2, \ldots, m, \quad i \neq j.$$

Therefore for scalar function $v(\tau, x, \eta) = \eta^{\mathrm{T}} U(\tau, x)\eta$, $\eta \in R_+^m$, $\eta > 0$, the bilateral inequality

$$(2.4.10) \qquad w^{\mathrm{T}} H^{\mathrm{T}} \underline{C} H w \leq v(\tau, x, \eta) \leq w^{\mathrm{T}} H^{\mathrm{T}} \overline{C} H w,$$

is satisfied, where $\overline{C} = [\bar{c}_{ij}]_{i,j=1}^m$, $\underline{C} = [\underline{c}_{ij}]_{i,j=1}^m$, $H = \mathrm{diag}\,(\eta_1, \ldots, \eta_m)$, $w = (\|x_1\|, \ldots, \|x_m\|)^{\mathrm{T}}$. For the first difference of function $v(\tau, x, \eta)$ along solutions of system (2.4.1) in view of (2.4.4) one can get the estimate

$$(2.4.11) \qquad \Delta v(\tau, x, \eta)\Big|_{(2.4.1)} \leq w^{\mathrm{T}} S(\tau) w,$$

where $w = (\|x_1\|, \ldots, \|x_m\|)^{\mathrm{T}}$, $S(\tau) = [\sigma_{ij}(\tau)]_{i,j=1}^m$. The elements of matrix $S(\tau)$ have the following structure

$$\sigma_{ii}(\tau) = -\lambda_m(G_{ii})\eta_i^2 + \sum_{\substack{j=1 \\ j \neq i}}^m \|A_{ji}\|^2 \|P_{jj}\|\eta_j^2$$

$$+ \sum_{\substack{k,j=1 \\ k \neq j}}^m \lambda_M(A_{ki}^{\mathrm{T}} P_{kj} A_{ji} + A_{ji}^{\mathrm{T}} P_{jk}^{\mathrm{T}} A_{ki})\eta_k \eta_j,$$

$$\sigma_{ij}(\tau) = \sum_{\substack{k=1 \\ k \neq j, k \neq i}}^m \eta_i^2 \|A_{kj}\|\,\|P_{kk}\|\,\|A_{ki}\|$$

$$+ \sum_{\substack{k,l=1 \\ k \neq i, k \neq j, l \neq j}}^m \|A_{ki}\|\,\|P_{kl}\|\,\|A_{lj}\|\eta_j\eta_i, \quad i \neq j,$$

where $G_{ii} = -(A_{ii}^T P_{ii} A_{ii} - P_{ii})$, $\|\cdot\|$ is a spectral norm of the corresponding matrix. Using the function $U(\tau, x)$, estimate (2.4.10) of the scalar function $v(\tau, x, \eta)$ and estimate (2.4.11) of the first difference of this function along solutions of system (2.4.1), we formulate sufficient conditions of stability and uniform asymptotic stability of the equilibrium state $x = 0$ of system (2.4.1).

Theorem 2.4.1 *Let system of equations (2.4.1) be such that*
(1) *matrices $\overline{C}$ and $\underline{C}$ in estimate (2.4.10) are positive definite;*
(2) *there exist negative semi-definite (definite) matrix $\overline{S}$, such that*

$$\frac{1}{2}[S(\tau) + S^T(\tau)] \leq \overline{S} \quad \text{for all} \quad \tau \in N_{\tau_0}^+.$$

Then the equilibrium state $x = 0$ is uniformly stable (uniformly asymptotically stable).

Proof Condition (2) of Theorem 2.4.1 ensures the existence of $\tau_1 \in N_{\tau_0}^+$ such that for all $\tau \geq \tau_1$ for matrix $S(\tau)$ the generalized Silvester conditions are satisfied. So, for function $v(\tau, x, \eta) = \eta^T U(\tau, x)\eta$ for all $\tau \geq \tilde{\tau} = \max\{\tau_1, \tau^*\}$ all conditions of Theorem 16.3 from Hahn [1] are satisfied. Thus, the equilibrium state $x = 0$ is stable (uniformly asymptotically stable) with respect to $N_{\tilde{\tau}}^+$. Taking into account continuity of solutions $x(\tau, \tau_0, x_0)$ of system (2.4.1) in x_0 and discreteness of the set $N_{\tau_0}^+$, one can confirm the stability (uniform asymptotic stability) of the equilibrium state of system (2.4.1).

2.4.2 Examples
Consider the system

(2.4.12)
$$x(\tau + 1) = \rho_1 x(\tau) + \alpha A(\omega, \tau) y(\tau),$$
$$y(\tau + 1) = \rho_2 y(\tau) + \beta A^T(\omega, \tau) x(\tau),$$

where $x, y \in R^2$, and $\alpha, \beta, \rho_1, \rho_2 \in R$, $\omega \in [0, 2\pi)$,

$$A(\omega, \tau) = \begin{pmatrix} \cos \omega\tau & \sin \omega\tau \\ -\sin \omega\tau & \cos \omega\tau \end{pmatrix}, \quad \tau \in N_0^+.$$

Moreover, we designate $q = \rho_1 \rho_2$. Applying the approach proposed in Section 2.4.1 for system (2.4.12) we construct an auxiliary function

(2.4.13) $$v(\tau, x, y) = x^T x + y^T y + 2x^T P(\tau) y,$$

where

$$P(\tau) = \begin{cases} \dfrac{\alpha\rho_1 + \beta\rho_2}{1 - 2q\cos\omega + q^2} A(\omega, \tau - 1)(A(\omega, 1) - qI), & \text{if } |q| \leq 1; \\ -\dfrac{\alpha\rho_1 + \beta\rho_2}{1 - 2q\cos\omega + q^2}\Big[qA(\omega, \tau - 1) - A(\omega, \tau) - \\ \qquad q^{-\tau+1}A^{\mathrm{T}}(1) + q^{-\tau}I\Big], & \text{if } |q| > 1, \end{cases}$$

and I is an identify matrix of dimension 2. Theorem 2.4.1 allows us to establish sufficient stability conditions of system (2.4.12) in the form of a system of inequalities

(2.4.14)
$$|\alpha\rho_1 + \beta\rho_2| < \sqrt{1 - 2q\cos\omega + q^2};$$
$$\sigma_{11} < 0, \quad \sigma_{11}\sigma_{22} - \sigma_{12}^2 > 0,$$

where

$$\sigma_{11} = \rho_1^2 - 1 - \frac{2\rho_1\beta(\alpha\rho_1 + \beta\rho_2)(q - \cos\omega)}{1 - 2q\cos\omega + q^2} + \beta^2,$$

$$\sigma_{22} = \rho_2^2 - 1 - \frac{2\rho_2\alpha(\alpha\rho_1 + \beta\rho_2)(q - \cos\omega)}{1 - 2q\cos\omega + q^2} + \alpha^2,$$

and

$$\sigma_{21} = \sigma_{12} = |\alpha\beta|\frac{|\alpha\rho_1 + \beta\rho_2|}{\sqrt{1 - 2q\cos\omega + q^2}}.$$

It this case the equilibrium state $x = y = 0$ of system (2.4.12) is uniformly asymptotically stable, and the constructed function (2.4.13) is the Liapunov function.

In order to compare the obtained stability conditions with the conditions obtained in terms of vector Liapunov function we employ the results from Michel and Miller [1]. Construct the vector function $V(x, y) = (v_1(x), v_2(y))^{\mathrm{T}}$ with the components $v_1(x) = x^{\mathrm{T}}x$ and $v_2(y) = y^{\mathrm{T}}y$. Applying Theorem 3.3.14 from Michel and Miller [1] we present sufficient conditions of uniform asymptotic stability of system (2.4.12) in the form of the system of inequalities

(2.4.15)
$$\rho_1^2 + \beta^2 - 1 < 0;$$
$$(\rho_1^2 + \beta^2 - 1)(\rho_2^2 + \alpha^2 - 1) - 4|\alpha\beta||\rho_1\rho_2| > 0.$$

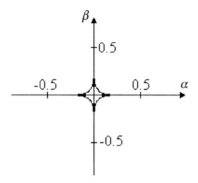

Figure 2.4.1. The domain of stability of (2.4.12) in the parameter space via Liapunov's vector function.

To compare conditions (2.4.15) and (2.4.14) obtained in terms of Theorem 2.4.1 we consider a system of difference equations

(2.4.16)
$$x(\tau + 1) = 0.95\,x + \alpha A\left(\frac{\pi}{3}, \tau\right) y,$$
$$y(\tau + 1) = -0.95\,y + \beta A^{\mathrm{T}}\left(\frac{\pi}{3}, \tau\right) x$$

and construct in the space of parameters (α, β) the domains of stability of the equilibrium space $x = y = 0$ of system (2.4.16). Figures 2.4.1 and 2.4.2 show that the domain constructed in terms of conditions (2.4.14) is wider than the domain constructed in terms of conditions (2.4.15).

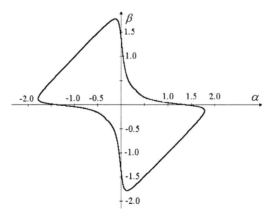

Figure 2.4.2. The domain of stability of (2.4.12) in the parameter space via Liapunov's matrix-valued function.

Note that for the system

(2.4.17)
$$x(\tau+1) = 1.2\,x + \alpha A\left(\frac{\pi}{3},\tau\right)y,$$
$$y(\tau+1) = -0.8\,y + \beta A^{\mathrm{T}}\left(\frac{\pi}{3},\tau\right)x$$

it is impossible to apply the vector function, because subsystem $x(\tau+1) = 1.2\,x$ is not exponentially stable. Nevertheless, conditions (2.4.14) allow us to construct for system (2.4.16) in the space (α,β) a domain of stability shown on Figure 2.4.3.

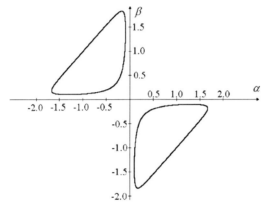

Figure 2.4.3. The domain of stability of (2.4.16) in the parameter space.

The system

(2.4.18)
$$x(\tau+1) = 1.05\,x + \alpha A\left(\frac{\pi}{3},\tau\right)y,$$
$$y(\tau+1) = -1.05\,y + \beta A^{\mathrm{T}}\left(\frac{\pi}{3},\tau\right)x$$

has exponentially unstable subsystem. However in this case conditions (2.4.14) also allow us to construct for system (2.4.18) a domain of stability in the space of parameters shown on Figure 2.4.4.

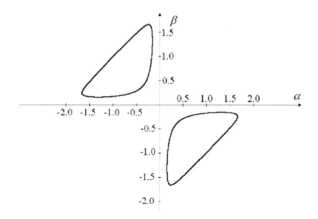

Figure 2.4.4. The domain of stability of (2.4.18) with exponentially unstable subsystem.

2.5 Some Applications of Hierarchical Liapunov's Functions

The hierarchical structure of the physical and/or other aspects of systems have suggested for the investigators an idea that the hierarchical Liapunov functions can be a natural tool for the analysis of such systems dynamics. In stability theory of continuous systems, two general classes of hierarchical dependence of subsystems are known. One is a regular hierarchy (see Ikeda and Šiljak [1], Martynyuk [1]), and the other is a mixed hierarchy of subsystems (see Martynyuk and Krapyvny [1], Martynyuk [2]). This section presents some results obtained in the framework of both approaches to stability analysis of discrete-time system.

2.5.1 Hierarchical decomposition and stability conditions

Assume that system (2.2.1) is decomposed into m interconnected subsystems

$$(2.5.1) \qquad x_i(\tau+1) = f_i(\tau, x_i(\tau)) + g_i(\tau, x(\tau)), \quad i = 1, 2, \ldots, m,$$

where $x_i \in R^{n_i}$, $x = (x_1^T, x_2^T, \ldots, x_m^T)^T$, $R^n = R^{n_1} \times R^{n_2} \times \cdots \times R^{n_m}$, $f_i \colon N_\tau^+ \times R^{n_i} \to R^{n_i}$, $g_i \colon N_\tau^+ \times R^n \to R^{n_i}$.

The equations

$$(2.5.2) \qquad x_i(\tau+1) = f_i(\tau, x_i(\tau)), \quad i = 1, 2, \ldots, m,$$

describe the dynamics of independent subsystems of (2.5.1), the functions g_i represent the interactions of (2.5.2) with the rest of the systems (2.5.1).

2. DISCRETE-TIME SYSTEMS

We suppose that $f_i(\tau, x_i)$ are continuous in x_i and solutions $x_i(\tau; \tau_0, x_{i0})$ of systems (2.5.2) are definite for all $\tau \in N_\tau^+$ and $x_{i0} \in R^{n_i}$. Besides assume that $f_i(\tau, x_i) = x_i$ for all $\tau \in N_\tau^+$ iff $x_i = 0$, i.e. systems (2.5.2) admit zero solution $x_i \equiv 0$ and they correspond to the unique equilibrium states of systems (2.5.2).

Further each subsystem (2.5.2) is decomposed into m_i interconnected components

$$(2.5.3) \quad \begin{aligned} x_{ij}(\tau+1) &= p_{ij}(\tau, x_{ij}(\tau)) + q_{ij}(\tau, x_i(\tau)), \\ i &= 1, 2, \ldots, m, \ j = 1, 2, \ldots, m_i, \end{aligned}$$

where $x_{ij} \in R^{n_{ij}}$, $x_i = (x_{i1}^T, x_{i2}^T, \ldots, x_{im_i}^T)^T$, $R^{n_i} = R^{n_{i1}} \times R^{n_{i2}} \times \cdots \times R^{n_{im_i}}$, $p_{ij}: N_\tau^+ \times R^{n_{ij}} \to R^{n_{ij}}$, $q_{ij} \times R^{n_i} \to R^{n_{ij}}$.

The equations

$$(2.5.4) \quad x_{ij}(\tau+1) = p_{ij}(\tau, x_{ij}(\tau))$$

describe the dynamics of independent components of (2.5.2), the functions q_{ij} represent the interactions of (2.5.4) with the rest of the subsystems (2.5.2). We suppose that $p_{ij}(\tau, x_{ij})$ are continuous in x_{ij} and solutions $x_{ij}(\tau; \tau_0, x_{ij0})$ of systems (2.5.4) are definite for all $\tau \in N_\tau^+$ and $x_{ij0} \in R^{n_{ij}}$. Besides assume that $p_{ij}(\tau, x_{ij}) = x_{ij}$ for all $\tau \in N_\tau^+$ iff $x_{ij} = 0$, i.e. systems (2.5.4) admit zero solution $x_{ij} = 0$ and they correspond to the unique equilibrium states of systems (2.5.4).

We apply a two-level construction of Liapunov functions (see Ikeda and Šiljak [1]) in the stability analysis of (2.2.1). Assume that for each component (2.5.4) there is a Liapunov function $v_{ij}(\tau, x_{ij})$ which establishes the asymptotic stability of the equilibrium $x_{ij} = 0$ of (2.5.4). For each subsystem (2.5.2) we construct the auxiliary function as

$$(2.5.5) \quad v_i(\tau, x_i) = \sum_{j=1}^{m_i} d_{ij} v_{ij}(\tau, x_{ij}),$$

where d_{ij} are positive constants. Similarly, for the overall system (2.2.1) we construct the function

$$(2.5.6) \quad v(\tau, x) = \sum_{i=1}^{m} d_i v_i(\tau, x_i),$$

where d_i are positive constants. Under appropriate assumptions the function $v(\tau, x)$, constructed by (2.5.5)–(2.5.6), is a hierarchical vector Liapunov function for the system (2.2.1).

In order to state sufficient stability conditions we require the following assumptions.

Assumption 2.5.1 There exist:
(1) open discrete-time invariant connected neighborhoods $\mathcal{N}_{ij} \subset R^{n_{ij}}$ of the states $x_{ij} = 0$, $i = 1, 2, \ldots, m$, $j = 1, 2, \ldots, m_i$;
(2) the comparison functions $\varphi_{ij}, \phi_{ij}, \psi_{ij} \in K$, $i = 1, 2, \ldots, m$, $j = 1, 2, \ldots, m_i$;
(3) the functions $v_{ij} \colon N_\tau^+ \times R^{n_{ij}} \to R_+$ which are continuous functions in x_{ij} and satisfy the estimates:

(a) $\alpha_{ij}\varphi_{ij}(\|x_{ij}\|) \le v_{ij}(\tau, x_{ij}) \le \beta_{ij}\phi_{ij}(\|x_{ij}\|)$,
   for all $(\tau, x_{ij}) \in N_\tau^+ \times \mathcal{N}_{ij}$,

(b) $\Delta v_{ij}(\tau, x_{ij})\big|_{(2.5.4)} \le -\pi_{ij}\psi_{ij}(\|x_{ij}\|)$, for all $(\tau, x_{ij}) \in N_\tau^+ \times \mathcal{N}_{ij}$,

(c) $\Delta v_{ij}(\tau, x_{ij}(\tau))\big|_{(2.5.3)} - \Delta v_{ij}(\tau, x_{ij}(\tau))\big|_{(2.5.4)} \le \sum_{k=1}^{m_i} \xi_{jk}^i \psi_{ik}(\|x_{ik}\|)$,
   for all $(\tau, x_{ij}) \in N_\tau^+ \times \mathcal{N}_{ij}$,

where $\alpha_{ij}, \beta_{ij} > 0$, $\pi_{ij} > 0$, $\xi_{ij}^i \ge 0$ are real constants, $\|x_{ij}\|$ is the norm of the vectors x_{ij}, $i = 1, 2, \ldots, m$, $j = 1, 2, \ldots, m_i$.

Assumption 2.5.2 We assume that:
(1) there exist discrete-time invariant open connected neighborhoods $\mathcal{N}_i \subset R^{n_i}$ of the states $x_i = 0$ of the subsystems (2.5.2), $i = 1, 2, \ldots, m$;
(2) there exist the comparison functions $\psi_i \in K$, $i = 1, 2, \ldots, m$;
(3) the functions $v_i \colon N_\tau^+ \times R^{n_{ij}} \to R_+$, which are constructed by (2.5.5) satisfy the inequalities:

   (a) $\Delta v_i(\tau, x_i)\big|_{(2.5.2)} \le -\pi_i \psi_i(\|x_i\|)$, for all $(\tau, x_i) \in N_\tau^+ \times \mathcal{N}_i$,

   (b) $\Delta v_i(\tau, x_i(\tau))\big|_{(2.5.1)} - \Delta v_i(\tau, x_i(\tau))\big|_{(2.5.2)} \le \sum_{j=1}^{m} \xi_{ij} \psi_j(\|x_j\|)$,
      for all $(\tau, x_i) \in N_\tau^+ \times \mathcal{N}_i$,
   where $\pi_i > 0$, $\xi_{ij} \ge 0$ are real constants, $i = 1, 2, \ldots, m$.

Now we define the matrices $W_i = (w_{jk}^i)$ and $W = (w_{jk})$ with elements

$$w_{jk}^i = \begin{cases} \pi_{ij} - \xi_{jj}^i, & \text{if } j = k, \\ -\xi_{jk}^i, & \text{if } j \ne k, \end{cases} \quad \text{and} \quad w_{jk} = \begin{cases} \pi_j - \xi_{jj}, & \text{if } j = k, \\ -\xi_{jk}, & \text{if } j \ne k, \end{cases}$$

respectively.

The following result establishes sufficient stability conditions for the system (2.2.1).

Theorem 2.5.1 *Assume that for the system (2.2.1) the decomposition (2.5.1)–(2.5.4) takes place and all conditions of Assumptions 2.5.1 and 2.5.2 are satisfied. If the matrices $W_1, W_2, \ldots, W_m$ and W are M-matrices, then the equilibrium $x = 0$ of the system (2.2.1) is asymptotically stable.*

If all conditions of Assumptions 2.5.1 and 2.5.2 are satisfied for $\mathcal{N}_{ij} = R^{n_{ij}}$, $\mathcal{N}_i = R^{n_i}$ and the functions $\varphi_{ij}, \phi_{ij} \in KR$, global asymptotic stability takes place.

Proof The conditions (1)–(3a) of Assumption 2.5.1 imply that there exist the functions $\varphi, \phi \in K$ such that

$$\varphi(\|x\|) \leq v(\tau, x) \leq \phi(\|x\|)$$

for all $(\tau, x) \in N_\tau^+ \times \mathcal{N}$.

From the conditions (3b), (3c) of Assumption 2.5.1 for the first forward difference $\Delta v_{ij}(\tau, x_{ij}(\tau))\big|_{(2.5.3)}$ we have

$$\Delta v_{ij}(\tau, x_{ij}(\tau))\big|_{(2.5.3)} \leq -\pi_{ij}\psi_{ij}(\|x_{ij}\|) + \sum_{k=1}^{m_i} \xi_{jk}^i \psi_{ik}(\|x_{ik}\|)$$

$$= -(\pi_{ij} - \xi_{jj}^i)\psi_{ij}(\|x_{ij}\|) + \sum_{k=1, k \neq j}^{m_i} \xi_{jk}^i(\|x_{ik}\|)$$

for all $(\tau, x_{ij}) \in N_\tau^+ \times \mathcal{N}_{ij}$. Then for the first forward difference $\Delta v_i(\tau, x(\tau))\big|_{(2.5.2)}$ the following estimate holds

$$\Delta v_i(\tau, x_i(\tau))\big|_{(2.5.2)} = \sum_{j=1}^{m_i} d_{ij} \Delta v_{ij}(\tau, x_{ij}(\tau))\big|_{\widetilde{C}_{ij}}$$

$$\leq \sum_{j=1}^{m_i} d_{ij}\left(-(\pi_{ij} - \xi_{jj}^i)\psi_{ij}(\|x_{ij}\|) + \sum_{\substack{k=1 \\ k \neq j}}^{m_i} \xi_{jk}^i \psi_{ik}(\|x_{ik}\|)\right)$$

$$= -a_i^\mathrm{T} W_i z_i,$$

for all $(\tau, x_i) \in N_\tau^+ \times \mathcal{N}_i$, where $a_i = (d_{i1}, d_{i2}, \ldots, d_{im_i})^\mathrm{T}$, $z_i = (\psi_{i1}, \psi_{i2}, \ldots, \psi_{im_i})^\mathrm{T}$.

Since the matrix W_i is an M-matrix, there exists a vector a_i with positive elements such that the vector $a_i^T W_i$ has positive elements (see Ikeda and Šiljak [1]). Then the first forward difference of the function $v_i(\tau, x_i)$ along solutions of the system (2.5.2) is negative definite. Hence, the function v_i is a Liapunov function, which establishes the asymptotic stability of the equilibrium $x_i = 0$ of S_i, $i = 1, 2, \ldots, m$.

Using conditions of Assumption 2.5.2, as above, for the first forward difference of the function $v(\tau, x)$ along solutions of the overall system S we have

$$\Delta v(\tau, x(\tau))\big|_{(2.5.1)} \leq -a^T W z$$

for all $(\tau, x) \in N_\tau^+ \times \mathcal{N}$, where $a = (d_1, d_2, \ldots, d_s)^T$, $z = (\psi_1, \psi_2, \ldots, \psi_s)^T$.

As the matrix W is an M-matrix, then there exists a vector a with positive elements such that the vector $a^T W$ has positive elements. Therefore, the first forward difference of the function $v(\tau, x)$ along solutions of the system (2.2.1) is negative definite for all $(\tau, x) \in N_\tau^+ \times \mathcal{N}$. All conditions of the asymptotic stability theorem of Liapunov for a discrete-time system are satisfied (see LaSalle [2], Hahn [2], etc.) Hence, the equilibrium $x = 0$ of the system (2.2.1) is asymptotically stable.

If $\mathcal{N}_{ij} = R^{ij}$, $\mathcal{N}_i = R^{n_i}$ and $\varphi_{ij}, \phi_{ij} \in KR$, the function $v(\tau, x)$ is a positive definite and radially unbounded function, its first forward difference along solutions of S is negative definite for all $(\tau, x) \in N_\tau^+ \times R^n$. From here, the global asymptotic stability of the equilibrium $x = 0$ of the system (2.2.1) follows from well-known theorem (see Hahn [2]).

The proof of Theorem 2.5.1 is complete.

2.5.2 Hierarchical connective stability We assume that for the system (2.2.1) the decomposition (2.5.1)–(2.5.4) takes place. It is known (see Šiljak [1]), the interconnection functions $g_i(\tau, x)$ have the form

$$g_i(\tau, x) = g_i(\tau, \bar{e}_{i1} x_1, \bar{e}_{i2} x_2, \ldots, \bar{e}_{im} x_m), \quad i = 1, 2, \ldots, m,$$

where the matrix $\bar{E} = (\bar{e}_{ij})$ is a fundamental interconnection matrix of the system (2.5.1) with the elements

$$\bar{e}_{ij} = \begin{cases} 1, & \text{if } x_j \text{ is contained in } g_i(\tau, x), \\ 0, & \text{if } x_j \text{ is not contained in } g_i(\tau, x). \end{cases}$$

Let the functions of a discrete argument $e_{ij} \colon N_\tau^+ \to [0, 1]$ satisfy the inequalities

(2.5.7) $$e_{ij}(\tau) \leq \bar{e}_{ij}$$

for all $\tau \in N_\tau^+$. The binary numbers $\bar{e}_{ij}$ define the nominal strength of coupling between the independent subsystems (2.5.1) of (2.2.1). The matrix $E(\tau) = (e_{ij}(\tau))$ describes structural perturbations of (2.2.1).

If $E(\tau) \equiv 0$, the system (2.2.1) is decomposed into m independent subsystems (2.5.2), each of them being an interconnection of the components (2.5.3). The interconnection functions between the components (2.5.4) have the form

$$q_{ij}(\tau, x_i) = q_{ij}(\tau, \bar{\ell}^i_{j1} x_{i1}, \bar{\ell}^i_{j2} x_{i2}, \ldots, \bar{\ell}^i_{jm_i} x_{im_i}),$$
$$i = 1, 2, \ldots, m, \quad j = 1, 2, \ldots, m_i,$$

where

$$\bar{\ell}^i_{jk} = \begin{cases} 1, & x_{ik} \text{ is contained in } q_{ij}(\tau, x_i), \\ 0, & x_{ik} \text{ is not contained in } q_{ij}(\tau, x_i). \end{cases}$$

Let $\ell^i_{jk} \colon N_\tau^+ \to [0, 1]$ and the inequalities

(2.5.8) $\qquad e^i_{jk}(\tau) \le \bar{\ell}^i_{jk}, \quad i = 1, 2, \ldots, m, \quad j, k = 1, 2, \ldots, m_i,$

be satisfied for all $\tau \in N_\tau^+$.

The matrix $\bar{L}_i = (\bar{\ell}^i_{jk})$ is the fundamental interconnection matrix for the subsystem (2.5.2) and defines the nominal strength of coupling between the independent components (2.5.4) of (2.5.2). The matrix $L_i(\tau) = (\ell^i_{jk}(\tau))$ describes structural perturbations of (2.5.2).

The notion of hierarchical connective stability of the discrete-time system (2.2.1) is defined similarly to the continuous case (cf. Šiljak [1]).

Definition 2.5.1 The discrete-time system (2.2.1) is *hierarchically connectively stable* if:

(i) for $E(\tau) \equiv 0$, the equilibriums $x_i = 0$ of (2.5.1) are global asymptotically stable for all structural matrix $L_i(\tau)$, $i = 1, 2, \ldots, m$;
(ii) for $L_i(\tau) \equiv \bar{L}_i$, the equilibrium $x = 0$ of (2.2.1) is global asymptotically stable for all structural matrix $E(\tau)$.

Assumption 2.5.3 We assume that:

(1) the requirements (1)–(3)(b) of Assumption 2.5.1 are satisfied for $\mathcal{N}_{ij} = R^{n_{ij}}$ and the functions $\varphi_{ij}, \phi_{ij} \in KR$, $i = 1, 2, \ldots, m$, $j = 1, 2, \ldots, m_i$;

(2) the first forward differences of the functions v_{ij} satisfy the inequalities

$$\Delta v_{ij}(\tau, x_{ij}(\tau))|_{(2.5.3)} - \Delta v_{ij}(\tau, x_{ij}(\tau))|_{(2.5.4)} \leq \sum_{k=1}^{m_i} \ell_{jk}^i(\tau) \xi_{jk}^i \psi_{ik}(\|x_{ik}\|)$$

for all $(\tau, x_{ij}) \in N_\tau^+ \times R^{n_{ij}}$, where $\xi_{jk}^i \geq 0$ are real constants, $i = 1, 2, \ldots, m$, $j = 1, 2, \ldots, m_i$.

Assumption 2.5.4 We assume that:
(1) the requirements (1)–(3)(a) of Assumption 2.5.2 are satisfied for $\mathcal{N}_i = R^{n_i}$, $i = 1, 2, \ldots, m$;
(2) the first forward differences of the functions v_i satisfy the inequalities

$$\Delta v_i(\tau, x_i(\tau))|_{(2.5.1)} - \Delta v_i(\tau, x_i(\tau))|_{(2.5.2)} \leq \sum_{j=1}^{m} e_{ij}(\tau) \xi_{ij} \psi_j(\|x_j\|)$$

for all $(\tau, x_i) \in N_\tau^+ \times R^{n_i}$, where $\xi_{ij} \geq 0$ are real constants, $i = 1, 2, \ldots, m$.

In this case the elements of the matrices $W_i(\tau) = (w_{jk}^i(\tau))$ and $W(\tau) = (w_{ij}(\tau))$ depend on discrete time, i.e.

$$w_{jk}^i(\tau) = \begin{cases} \pi_{ij} - \ell_{jj}^i(\tau)\xi_{jj}^i, & \text{if } j = k, \\ -\ell_{jk}^i(\tau)\xi_{jk}^i, & \text{if } j \neq k, \end{cases}$$

$$w_{ij}(\tau) = \begin{cases} \pi_j - e_{jj}(\tau)\xi_{jj}(\tau), & \text{if } j = i, \\ -e_{ij}(\tau)\xi_{ij}, & \text{if } j \neq i. \end{cases}$$

By $\overline{W}_1, \overline{W}_2, \ldots, \overline{W}_m$ and $\overline{W}$ we denote the matrices for the nominal interconnections specified by the fundamental interconnection matrices $\overline{L}_1, \overline{L}_2, \ldots, \overline{L}_m$ and $\overline{E}$. The conditions of the hierarchical connective stability of the system (2.2.1) are in the following theorem.

Theorem 2.5.2 *We assume that for the system (2.2.1) the decomposition (2.5.1)–(2.5.4) takes place and all conditions of Assumptions 2.5.3 and 2.5.4 are satisfied. If the matrices $\overline{W}_1, \overline{W}_2, \ldots, \overline{W}_m$ and $\overline{W}$ are M-matrices, then the equilibrium $x = 0$ of the system (2.2.1) is hierarchically connectively stable.*

2. DISCRETE-TIME SYSTEMS

Proof From the condition (1) of Assumption 2.5.3, we obtain that there exist the functions $\varphi_i, \phi_i \in KR$, $\varphi, \phi \in KR$ such that

(2.5.9)
$$\varphi_i(\|x_i\|) \leq v_i(\tau, x_i) \leq \phi_i(\|x_i\|) \quad \text{for all} \quad (\tau, x_i) \in N_\tau^+ \times R^{n_i},$$
$$\varphi(\|x\|) \leq v(\tau, x) \leq \phi(\|x\|) \quad \text{for all} \quad (\tau, x) \in N_\tau^+ \times R^n.$$

Using conditions of Assumption 2.5.3 we obtain for the forward difference $\Delta v_{ij}(\tau, x_{ij}(\tau))\big|_{(2.5.3)}$ the estimate

$$\Delta v_{ij}(\tau, x_{ij}(\tau))\big|_{(2.5.3)} \leq -\pi_{ij}\psi_{ij}(\|x_{ij}\|) + \sum_{k=1}^{m_i} \ell_{jk}^i(\tau)\xi_{jk}^i\psi_{ik}(\|x_{ik}\|)$$

$$= -(\pi_{ij} - \ell_{jj}^i(\tau)\xi_{jj}^i)\psi_{ij}(\|x_{ij}\|) + \sum_{k=1,k\neq j}^{m_i} \ell_{jk}^i(\tau)\xi_{jk}^i(\|x_{ik}\|)$$

for all $(\tau, x_{ij}) \in N_\tau^+ \times R^{n_{ij}}$. Then we get for the forward difference $\Delta v_i(\tau, x(\tau))\big|_{(2.5.2)}$ the inequality

$$\Delta v_i(\tau, x_i(\tau))\big|_{(2.5.2)} = \sum_{j=1}^{m_i} d_{ij}\Delta v_{ij}[\tau, x_{ij}(\tau)]\big|_{(2.5.3)}$$

$$\leq \sum_{j=1}^{m_i} d_{ij}(-(\pi_{ij} - \ell_{jj}^i(\tau)\xi_{jj}^i)\psi_{ij}(\|x_{ij}\|)$$

$$+ \sum_{k=1,k\neq j}^{m_i} \ell_{jk}^i(\tau)\xi_{jk}^i\psi_{ik}(\|x_{ik}\|)) = -a_i^\mathrm{T} W_i(\tau)z_i,$$

for all $(\tau, x_i) \in N_\tau^+ \times R^{n_i}$, where $a_i = (d_{i1}, d_{i2}, \ldots, d_{im_i})^\mathrm{T}$, $z_i = (\psi_{i1}, \psi_{i2}, \ldots, \psi_{im_i})^\mathrm{T}$. From (2.5.7) we obtain $W_i(\tau) \leq \overline{W}_i$. Then

(2.5.10) $\quad \Delta v_i(\tau, x_i(\tau))\big|_{(2.5.2)} \leq -a_i^\mathrm{T} \overline{W}_i z_i, \quad i = 1, 2, \ldots, s.$

As the matrix $\overline{W}_i$ is an M-matrix, then there exists a vector a_i with positive elements such that the vector $a_i^\mathrm{T} \overline{W}_i$ has positive elements. Therefore, the first forward differences of the functions $v_i(\tau, x_i)$ along solutions of the subsystems (2.5.2) are negative definite for all $(\tau, x_i) \in N_\tau^+ \times R^{n_i}$. From here we conclude that the equilibriums $x_i = 0$ of the subsystems (2.5.2) are globally asymptotically stable for $E(\tau) \equiv 0$ and for all structural matrices $E(\tau) \equiv 0$. The part (1) of Definition 2.1 is executed.

Let all the interconnections between the components (2.5.4) be fixed, i.e. $L_i(\tau) \equiv \overline{L}_i$, $i = 1, 2, \ldots, m$. As above, using the inequality (2.5.8) we get

$$\Delta v(\tau, x(\tau))\big|_{(2.2.1)} \leq -a^T \overline{W} z$$

for all $(\tau, x) \in N_\tau^+ \times R^n$ and all structural matrices $E(\tau)$. From the condition of Theorem 2.5.2, the matrix $\overline{W}$ is M-matrix. Then there exists a vector a with positive elements such that the vector $a^T \overline{W}$ has positive elements. Therefore, the forward difference of the function $v(\tau, x)$ along solutions of the system (2.2.1) is negative definite for all $(\tau, x) \in N_\tau^+ \times R^n$. From here and (2.5.9), we conclude that the equilibrium $x = 0$ of the system (2.2.1) is global asymptotically stable for $L(\tau) \equiv \overline{L}$ and for all interconnection matrices $E(\tau)$.

The proof of Theorem 2.5.2 is complete.

Example 2.5.1 Consider the system

(2.5.11) $\qquad x(\tau+1) = \begin{pmatrix} 0.2 & -0.02 & 0.5 \\ 0.4 & 0.5 & 0.5 \\ 0.05 & 0.05 & 0.9 \end{pmatrix} x(\tau),$

where $\tau \in N_\tau^+$, $x \in R^3$, is an interconnection of two subsystems

$$x_1(\tau+1) = \begin{pmatrix} 0.2 & 0.2 \\ 0.4 & 0.5 \end{pmatrix} x_1(\tau),$$

$$x_2(\tau+1) = 0.9\, x_2(\tau).$$

The fundamental interconnection matrix and the interconnection matrix

$$\overline{E} = \begin{pmatrix} 0 & 1 \\ 1 & 0 \end{pmatrix}, \qquad E(\tau) = \begin{pmatrix} 0 & e_{12}(\tau) \\ e_{21}(\tau) & 0 \end{pmatrix},$$

where $e_{12}, e_{21}: N_\tau^+ \to [0, 1]$, correspond to such decomposition.

Prove that all conditions of Theorem 2.5.2 are satisfied, and the equilibrium $x = 0$ of the system (2.5.11) is hierarchically connected stable.

Prove that the equilibrium $x = 0$ of the system (2.5.11) is not connectedly stable in the standard sense (see Šiljak [1]).

2.5.3 Application of hierarchical matrix-valued function

Consider the autonomous discrete-time system

(2.5.12) $\qquad x(\tau+1) = f(x(\tau)),$

2. DISCRETE-TIME SYSTEMS

where $x \in R^n$, $\tau \in \mathcal{N}_\tau^+$, $f\colon R^n \to R^n$, $f(0) = 0$, $n \geq 2$, and its decomposition into m subsystems

(2.5.13) $$x_i(\tau + 1) = f_i(x_i(\tau)) + g_i(x(\tau)),$$

where $x_i \in R^{n_i}$, $f_i\colon R^{n_i} \to R^{n_i}$, $g_i\colon R^n \to R^{n_i}$, $\sum_{i=1}^m n_i = n$, $f_i(0) = g_i(0) = 0$ for all $i = 1, 2, \ldots m$.

Introduce the designations $(x_i, x_j) = (0, \ldots, 0, x_i^T, \ldots, x_j^T, 0, \ldots, 0,)^T$ for all $(i \neq j) \in [1, m]$. Then

$$q_i = f_i(0, \ldots, 0, x_i^T, \ldots, x_j^T, 0, \ldots, 0,),$$
$$q_j = f_j(0, \ldots, 0, x_i^T, \ldots, x_j^T, 0, \ldots, 0,)$$

for all $(i \neq j) \in [1, m]$ and we consider the couples of subsystems

(2.5.14) $$\begin{aligned} x_i(\tau + 1) &= q_i(x_i(\tau), x_j(\tau)), \\ x_j(\tau + 1) &= q_j(x_j(\tau), x_i(\tau)), \quad i \neq j \in [1, m]. \end{aligned}$$

Here $x_i \in R^{n_i}$, $q_i\colon R^{n_i} \to R^{n_i}$, $q_j\colon R^{n_j} \to R^{n_i}$, $q_i(0,0) = q_j(0,0) = 0$ for all $i \neq j \in [1, m]$.

Together with couples (2.5.14) we consider the interconnected subsystems

(2.5.15) $$\begin{aligned} x_i(\tau + 1) &= q_i(x_i(\tau), x_j(\tau)) + g_i^*(x(\tau)), \\ x_j(\tau + 1) &= q_j(x_j(\tau), x_i(\tau)) + g_j^*(x(\tau)), \quad i \neq j \in [1, m]. \end{aligned}$$

where $g_i^*\colon R^n \to R^{n_i}$, $g_j^*\colon R^n \to R^{n_j}$.

The subsystems (2.5.13), and (2.5.15) form structures of the first and second level decomposition.

The properties of solutions to system (2.5.12) are analysed by means of the matrix-valued function

(2.5.16) $$U(x(\tau)) = [v_{ij}(x_i, x_j)], \quad i, j = 1, 2, \ldots, m,$$

where $v_{ii} = v_{ii}(x_i)$, $v_{ij} = v_{ij}(x_i, x_j)$, $i \neq j$. Besides $v_{ii}\colon R^{n_i} \to R_+$, $v_{ij}\colon R^{n_i} \times R^{n_j} \to R$.

Proposition 2.5.1 If there exist functions (2.5.16), constants $\underline{\gamma}_{ii}, \overline{\gamma}_{ii} > 0$, $\underline{\gamma}_{ij}, \overline{\gamma}_{ij}$, $i, j = 1, 2, \ldots m$, and comparison functions $\varphi_i, \psi_i \in K(KR)$, such that

(a) $\underline{\gamma}_{ii} \varphi_i^2(\|x_i\|) \leq v_{ii}(x_i) \leq \overline{\gamma}_{ii} \psi_i^2(\|x_i\|), \quad \forall\, (x_i \neq 0) \in \mathcal{N}_i;$

(b) $\underline{\gamma}_{ij} \varphi_i(\|x_i\|) \varphi_j(\|x_j\|) \leq v_{ij}(x_i, x_j) \leq \overline{\gamma}_{ij} \psi_i(\|x_i\|) \psi_j(\|x_j\|),$
$\forall\, (x_i \neq 0, x_j \neq 0) \in \mathcal{N}_i \times \mathcal{N}_j,$

then for the function

(2.5.17) $$v(x,\eta)) = \eta^T U(x)\eta, \quad \eta \in R_+^m, \quad \eta > 0,$$

the estimate

(2.5.18) $$\varphi^T(\|x\|)H^T\underline{G}H\varphi(\|x\|) \le v(x,\eta) \le \psi^T(\|x\|)H^T\overline{G}H\psi(\|x\|),$$

holds true for all $(x_i \ne 0) \in \mathcal{N}$, where $H = \text{diag}\,(\eta_1, \eta_2, \ldots, \eta_m)$, $\varphi(\|x\|) = (\varphi_1(\|x_1\|), \ldots, \varphi_m(\|x_m\|))^T$, $\psi(\|x\|) = (\psi_1(\|x_1\|), \ldots, \psi_m(\|x_m\|))^T$, $\underline{G} = [\underline{\gamma}_{ij}]$, $\overline{G} = [\overline{\gamma}_{ij}]$, $i,j = 1,2,\ldots m$.

From (2.5.17) is follows that

(2.5.19) $$\Delta v(x,\eta)) = \eta^T \Delta U(x)\eta,$$

where

(2.5.20) $$\Delta U(x) = U(f(x)) - U(x)$$

is computed component-wise.

For the components $v_{ij}(\cdot)$ along solutions of system (2.5.13) for $i = j$ and along solutions of the system (2.5.15) for $i \ne j$ we have

$$\Delta v_{ii}(x_i) = v_{ii}(f_i(x_i)) - v_{ii}(x_i) + v_{ii}(f_i(x_i)$$
$$+ g_i(x)) - f_i(x_i), \quad i = 1, 2, \ldots, m;$$
$$\Delta v_{ij}(x_i, x_j) = v_{ij}(q_i(x_i, x_j), q_j(x_j, x_i)) - v_{ij}(x_i, x_j) + v_{ij}(q_i(x_i, x_j)$$
$$+ g_i^*(x), q_j(x_j, x_i) + g_j^*(x)) - v_{ij}(q_i(x_i, x_j), q_j(x_j, x_i)),$$
$$i \ne j, \quad i,j = 1, 2, \ldots, m.$$

Further we need the following assumption.

Proposition 2.5.2 Assume that for system (2.5.12) with decomposition (2.5.13), (2.5.15) there exist functions $v_{ij}(\cdot)$, $i,j = 1, 2, \ldots, m$, comparison functions $\psi_i(\|x_i\|)$, $\psi_i \in K$, $i = 1, 2, \ldots, m$, and constants α_i, β_{ij}, ε_i, ξ_{ij}, θ_j, λ_{ij}, σ_{ij}, $i \ne j$, $i,j = 1, 2, \ldots, m$, such that for all $(x_i \ne 0) \in \mathcal{N}_i$, $(x \ne 0) \in \mathcal{N}$:

(a) $v_{ii}(f_i(x_i)) - v_{ii}(x_i) \le \alpha_i \psi_i^2(\|x_i\|)$;

(b) $v_{ii}(f_i(x_i) + g_i(x_i)) - v_{ii}(f(x)) \le \psi_i(\|x_i\|) \sum_{k=1}^{m} \beta_{ij}\psi_k(\|x_k\|)$;

(c) $v_{ij}(q_i(x_i, x_j), q_j(x_j, x_i)) - v_{ij}(x_i, x_j) \le \varepsilon_i \psi_i^2(\|x_i\|)\xi_{ij}\psi_i(\|x_i\|)$
$\times \psi_j(\|x_j\|) + \theta_j \psi_j^2(\|x_j\|)$;

(d) $v_{ij}(q_i(x_i, x_j) + g_i^*(x), q_j(x_j, x_i) + g_j^*(x)) - v_{ij}(x_i, x_j) \le$
$\le \psi_i(\|x_i\|) \sum_{k=1}^{m} \lambda_{ik}\psi_k(\|x_k\|) + \psi_j(\|x_j\|) + \sum_{k=1}^{m} \sigma_{jk}\psi_k(\|x_k\|)$.

2. DISCRETE-TIME SYSTEMS

Then for the function (2.5.19) the estimate

(2.5.21) $$\Delta v(x,\eta)\big|_{(2.5.12)} \leq \psi^{\mathrm{T}}(\|x\|)P\psi(\|x\|),$$

holds for all $x \in \mathcal{N}$, where $P = [P_{ij}]$, $\psi(\|x\|) = (\psi_1(\|x_1\|), \ldots, \psi_m(\|x_m\|))^{\mathrm{T}}$,

$$P_{ii} = \eta_i^2(\alpha_i + \beta_{ii}) + \eta_i(\varepsilon_i + \theta_i + \sigma_{ii} + \lambda_{ii})\left(\sum_{j=1}^{m}\eta_j - \eta_i\right);$$

$$P_{ij} = \frac{1}{2}\eta_i^2(\beta_{ij} + \beta_{ji}) + \frac{1}{2}\eta_i\eta_j(\xi_{ij} + \xi_{ji}) + \frac{1}{2}\eta_i(\lambda_{ij} + \lambda_{ji} + \sigma_{ij} + \sigma_{ji})$$

$$\times \left(\sum_{k=1}^{m}\eta_k - \eta_i\right), \quad i,j = 1,2,\ldots,m, \quad i \neq j.$$

Besides $P_{ij} = P_{ji}$ for all $i,j = 1,2,\ldots,m$.

Remark 2.5.1 If in the conditions (a)–(d) of Proposition 2.5.2 the inequality sign "$\leq$" is reversed by "$\geq$", then the estimate (2.5.21) is satisfied with the inequality sign "$\geq$" for all $x \in \mathcal{N}$.

It is easy to show that for system (2.5.12) the following results are true.

Theorem 2.5.3 *Assume that for the system (2.5.12)*

(1) *there exist matrix-valued function (2.5.16) and a vector $\eta \in R_+^m$, $\eta > 0$, such that conditions of Propositions 2.5.1 and 2.5.2 are satisfied;*

(2) *matrices $\underline{G}$, $\overline{G}$ in estimate (2.5.18) are positive definite;*

(3) *matrix P in estimate (2.5.21) is negative semi-definite (negative definite).*

Then the equilibrium state $x = 0$ of system (2.5.12) is uniformly stable (asymptotically uniformly stable).

If in Proposition 2.5.1 the functions $\varphi_i \in KR$ and $\mathcal{N}_i = R^{n_i}$, $\mathcal{N}_j = R^{n_j}$, then the equilibrium state $x = 0$ of system (2.5.12) is uniformly stable in the whole (uniformly asymptotically stable in the whole).

Theorem 2.5.4 *Assume that conditions (1), (2) of Theorem 2.5.3, be satisfied as well as Remark 2.5.1 and matrix P in estimate (2.5.21) be positive definite.*

Then the equilibrium state $x = 0$ of system (2.5.12) will be unstable.

Proof Assertion of Theorems 2.5.3 and 2.5.4 are direct corollaries of Theorem 2.3.4 and so are omitted here (see also Djordjevic [1], Martynyuk [3], Krapivny and Martynyuk [1] for the details in the direction).

2.6 On Polydynamics of Systems on Time Scale

In this Section we introduce a new direction in the theory of dynamical systems on time scales. This is a polydynamics of nonlinear systems.

In order to present the concept of polydynamics of systems on time scales, let us first introduce some auxiliary results from the calculus on time scales.

2.6.1 Auxiliary results Let T be a one-dimensional time scale determined as any closed subset R. We assume that T is bounded and introduce designations $a = \sup T$ and $b = \inf T$, where $a, b < +\infty$. *The forward-jump and backward-jump functions* are defined by

$$\sigma(t) = \inf\{s \in T \colon s > t\} \text{ and } \pi(t) = \sup\{s \in T \colon s < t\}.$$

By means of the operators $\sigma(t)$ and $\pi(t)$ the values $\{t\}$ on time scale T are classified as follows:

(a) if $\sigma(t) = t$ ($\pi(t) = t$), then a point $t \in T$ is called *right-dense* (*left-dense*);
(b) if $\sigma(t) > t$ ($\pi(t) < t$), then a point $t \in T$ is called *right-scattered* (*left-scattered*) respectively.

Graininess of the time scale T is determined by

$$\mu(t) = \sigma(t) - t \text{ and } \eta(t) = t - \pi(t)$$

at forward jumps (backward jumps).

Together with the set T we shall consider the sets $T^k = T$ if b is left-dense and $T^k = T \backslash \{b\}$ if b is left-scattered. In the same way we define the set $T_k = T$ if a is left-dense and $T_k = T \backslash \{a\}$ if a is right-scattered. We designate $T^k \cap T_k = T_k^k$ and claim that T is a *uniform scale* for all $t \in T$ if $\mu(t) = \eta(t)$ for all $t \in T$.

Definition 2.6.1 Mapping u determined on T is Δ-differentiable on T^k if for any $\varepsilon > 0$ there exist a W-neighbourhood $t \in T^k$ such that for some γ the inequality

$$|u(\sigma(t)) - u(s) - \gamma(\sigma(t) - s)| < \varepsilon |\sigma(t) - s|$$

holds for all $s \in W$. In this case we write $u^\Delta(t) = \gamma$.

Definition 2.6.2 Mapping u determined on T is ∇-differentiable on T_k if for any $\varepsilon > 0$ there exists a V-neighbourhood $t \in T_k$ such that for some θ the inequality

$$|u(\pi(t)) - u(s) - \theta(\pi(t) - s)| < \varepsilon |\pi(t) - s|$$

holds true for all $s \in V$. in this case we write $u^\nabla(t) = \theta$.

Definition 2.6.3 Mapping u determined on T has a diamond-α derivative for all $t \in T_k^k$ iff the mapping u is Δ- and ∇-differentiable and

$$u^{\diamond_\alpha}(t) = \alpha u^\Delta(t) + (1-\alpha) u^\nabla(t), \quad \alpha \in [0, 1].$$

We note that the operators Δ, ∇ and $\diamond_\alpha$ are not, in general case, commuting.

2.6.2 Problems of polydynamics Let the state of some nonlinear system (S) at a time $t \in T$ be determined by the vector $x(t) \in R^n$, $n \geq 1$.
We cover the time scale T by the following subsets (see Sheng [1]):

$$\mathcal{A} = \{t \in T: \ t \text{ is left-dense and right-scattered}\};$$
$$\mathcal{B} = \{t \in T: \ t \text{ is left-scattered and right-dense}\};$$
$$\mathcal{C} = \{t \in T: \ t \text{ is left-scattered and right-scattered}\};$$
$$\mathcal{D} = \{t \in T: \ t \text{ is left-dense and right-dense}\}.$$

Without loss of generality, we may assume that $a \in \mathcal{A} \cup \mathcal{D}$ and $b \in \mathcal{B} \cup \mathcal{D}$ and denote the Euler derivative of vector $x(t)$ in t by $\dot{x}(t)$ if it exists.

We shall assume that the Δ-*dynamics* of system (S) on time scale T is described by the system of equations

(2.6.1) $$x^\Delta(t) = f(t, x(t)), \quad x(t_0) = x_0,$$

where

$$x^\Delta(t) = \begin{cases} \dfrac{x(\sigma(t)) - x(t)}{\mu(t)} & \text{if } t \in \mathcal{A} \cup \mathcal{C}; \\ \dot{x}(t) & \text{at other points,} \end{cases}$$

and $x \in R^n$, $f \colon T^k \times R^n \to R^n$.

Assume that the ∇-*dynamics* of system (S) on time scale T is described by the system of equations

(2.6.2) $$x^\nabla(t) = g(t, x(t)), \quad x(t_0) = x_0,$$

where

$$x^\nabla(t) = \begin{cases} \dfrac{x(t) - x(\pi(t))}{\eta(t)} & \text{if } t \in \mathcal{B} \cup \mathcal{C}; \\ \dot{x}(t) & \text{at other points,} \end{cases}$$

and $x \in R^n$, $g\colon T_k \times R^n \to R^n$.

Polydynamics of system (S) on time scale T is described by the system of equations

(2.6.3) $\quad x^{\Diamond_\alpha}(t) = F(t, x(t), \alpha), \quad x(t_0) = x_0, \quad \alpha \in [0, 1],$

where $F(t, x(t), \alpha) = \alpha f(t, x) + (1 - \alpha) g(t, x)$ and $x^{\Diamond_\alpha}$ — diamond-α derivative of vector $x(t) \in R^n$ calculated according to Definition 2.6.3.

Taking into account that $x^{\Diamond_\alpha}(t) = \dot{x}(t) + O(\max\{\mu(t), \eta(t)\})$ for all $t \in T$, we can rewrite system (2.6.3) as

(2.6.4) $\quad \begin{aligned} \dot{x}(t) &= F(t, x(t), \alpha) - O(\max\{\mu(t), \eta(t)\}), \\ x(t_0) &= x_0, \quad \alpha \in [0, 1]. \end{aligned}$

Thus, the problem of system (S) polydynamics on time scale consists in the investigation of behaviour of solutions to system (2.6.3) or (2.6.4) under different assumptions on dynamical properties of systems (2.6.1) and (2.6.2) on the corresponding subsets of time scale T.

2.6.3 Polydynamics analysis in general We assume that for system (2.6.3) a matrix-valued function $U(t, x) \in C(T_k^k \times R^n, R^{m \times m})$, $m \le n$, is constructed together with a vector $\phi \in R_+^m$ for which the function

(2.6.5) $\quad v(t, x, \phi) = \phi^T U(t, x) \phi$

satisfies the estimates

(2.6.6) $\quad h^T(t, x) A h(t, x) \le v(t, x, \phi) \le h^T(t, x) B h(t, x)$

for all $t \in T_k^k$ and $x \in \mathcal{D} \subseteq R^n$.

In inequalities (2.6.6) A and B are $m \times m$ constant matrices and the vector measure $h(t, x) = (h_1(t, x_1), \ldots, h_m(t, x_m))^T$ possesses measures of class $h_i \in \Gamma$ as its components. Besides, classes of functions Γ and Γ_0 are specified as:

$$\Gamma = \{h_i \in C(T_k^k \times R^{n_i}, R_+)\colon \inf_{(t, x_i)} h_i(t, x_i) = 0\},$$

$$\Gamma_0 = \{h_i \in \Gamma\colon \inf_{x_i} h_i(t, x_i) = 0 \text{ for every } t \in T_k^k\},$$

$x_i \in R^{n_i}$, $\sum_{i=1}^{n} n_i = n$, where n is dimension of state vector x of system (2.6.3).

We compute Δ and ∇–derivatives of function (2.6.5) along solutions of systems (2.6.1) and (2.6.2) and obtain the function

(2.6.7) $\qquad v^{\Diamond_\alpha}(t,x,\phi) = \alpha v^\Delta(t,x,\phi) + (1-\alpha)x^\nabla(t,x,\phi),$

where

$$v^\Delta(t,x,\phi) = v_t^\Delta(t,x,\phi) + (v_x^\Delta(t,x,\phi))^\mathrm{T} F(t,x,\alpha);$$
$$v^\nabla(t,x,\phi) = v_t^\nabla(t,x,\phi) + (v_x^\nabla(t,x,\phi))^\mathrm{T} F(t,x,\alpha).$$

Function (2.6.7) is called *diamond-α derivative of Liapunov function* on time scale T iff function (2.6.5) is positive definite with respect to measure and $v^{\Diamond_\alpha}(t,x,\phi) \leq 0$ on the set $D \subseteq R^n$ for all $t \in T_k^k$.

Definition 2.6.4 Polydynamics of system (2.6.3) is:

(a) (h_0,h)-*stable* if for any $\varepsilon > 0$ and $t_0 \in T_k^k$ there exists a positive function $\delta(t_0,\varepsilon) > 0$ such that for any $\varepsilon > 0$ the condition $h_0(t_0,x_0) < \delta$ implies the estimate $h(t,x(t)) < \varepsilon$ for all $t \geq t_0$ for any solution $x(t,t_0,x_0)$ of system (2.6.3);

(b) (h_0,h)-*uniformly stable* if in Definition 2.6.4 (a) the value δ does not depend on $t_0 \in T_k^k$.

Theorem 2.6.1 *Assume that*

(1) *the state of system (2.6.3) is estimated by the measures $h_0(t_0,x_0) \in \Gamma_0$ and $h(t,x) \in \Gamma$ for all $t \in T_k^k$;*

(2) *there exist a matrix-valued function $U \in C(T_k^k \times R^n, R^{m \times m})$ and a vector $\phi \in R_+^m$ such that function (2.6.5) is locally Lipschitzian in x;*

(3) *for all $(t,x) \in S(H,h) = \{(t,x) \in T_k^k \times R^n: h(t,x) < H,\ H > 0\}$*

$$v^{\Diamond_\alpha}(t,x,\phi)|_{(2.6.3)} \leq 0 \quad \text{for all} \quad \alpha \in [0,1];$$

(4) *the measure h is continuous with respect to measure h_0 in the domain of their definition;*

(5) *in inequality (2.6.7) the matrices A and B are ositive definite.*

Then, the polydynamics of system (2.6.3) is uniformly (h_0, h)-stable.

Proof Inequality (2.6.6) and condition (5) of Theorem 2.6.1 imply

$$(2.6.8) \qquad \lambda_m(A) p_0(t, x) \leq v(t, x, \phi) \leq \lambda_M(B) p(t, x),$$

where $\lambda_m(A)$, and $\lambda_M(B)$ are minimal and maximal eigenvalues of matrices A and B respectively and $p(t, x)$ and $p_0(t, x) \in \Gamma$ and

$$p_0(t, x) \leq h^T(t, x) h(t, x) \leq p(t, x)$$

for all $(t, x) \in S(H, h)$.

Estimate (2.6.8) implies that for any $t_0 \in T_k^k$ and $x_0 \in R^n$ there exists a constant $\delta_0 > 0$ such that

$$(2.6.9) \qquad v(t_0, x_0, \phi) \leq \lambda_M(B) p_0(t_0, x_0),$$

whenever $p_0(t_0, x_0) < \delta_0$. Lower bound of function $v(t, x, \phi)$ in inequality (2.6.8) yields

$$(2.6.10) \qquad \lambda_m(A) p(t, x) \leq v(t, x, \phi)$$

whenever $p(t, x) \leq H_0$, where $H_0 \in (0, H)$.

By condition (4) of Theorem 2.6.1 there exist a constant $\delta_1 > 0$ and function $\varphi \in K$ such that

$$(2.6.11) \qquad p(t_0, x_0) \leq \varphi(p_0(t_0, x_0))$$

whenever $p_0(t_0, x_0) < \delta_1$. We take δ_1 so that $\varphi(\delta_1) < H_0$.

Let $t_0 \in T_k^k$ and $\varepsilon \in (0, H_0)$ be given. We take $\delta_2 > 0$ so that

$$\lambda_M(B) \delta_2 < \lambda_m(A) \varepsilon$$

and define $\delta = \min(\delta_0, \delta_1, \delta_2)$. In this case condition $p_0(t_0, x_0) < \delta$ implies that

$$\lambda_m(A) p(t_0, x_0) \leq v(t_0, x_0, \phi) \leq \lambda_M(B) p_0(t_0, x_0) < \lambda_m(A) \varepsilon.$$

Hence it follows that $p(t_0, x_0) < \varepsilon$. Let us show that for any solution $x(t, t_0, x_0)$ of system (2.6.3) with the initial conditions $p_0(t_0, x_0) < \delta$ we have $p(t, x(t)) < \varepsilon$ for all $t \geq t_0$.

2. DISCRETE-TIME SYSTEMS

If this is not true, there must exist a $t_1 > t_0 \in T_k^k$ such that

(2.6.12) $\qquad p(t_1, x(t_1)) \geq \varepsilon \quad \text{and} \quad p(t, x(t)) < \varepsilon$

for $t \in [t_0, t_1)$ for at least one solution $x(t) = x(t, t_0, x_0)$ of system (2.6.3). In view of Definition 2.6.3 we rewrite condition (3) of Theorem 2.6.1 as:

(2.6.13)
$$v^{\Diamond_\alpha}(t, x, \phi) = \begin{cases} v^\nabla(t, x, \phi) \leq 0 & \text{for } \alpha = 0; \\ \alpha v^\Delta(t, x, \phi) + (1-\alpha) v^\nabla(t, x, \phi) \leq 0 & \text{for } 0 < \alpha < 1; \\ \alpha v^\Delta(t, x, \phi) \leq 0 & \text{for } \alpha = 1. \end{cases}$$

Taking into account that

$$\int_{t_0}^{t} v(\tau, x, \phi) \Diamond_\alpha \tau = \alpha \int_{t_0}^{t} v(\tau, x, \phi) \Delta\tau + (1-\alpha) \int_{t_0}^{t} v(\tau, x, \phi) \nabla\tau$$

for all $t \in T$ and $0 \leq \alpha \leq 1$, according to (2.6.13) we get the estimate

(2.6.14) $\qquad v(t, x(t), \phi) \leq v(t_0, x_0, \phi)$

for all $t \in T$ and for all $\alpha \in [0, 1]$.

Further from (2.6.12) and (2.6.14) we have

$$\lambda_m(A)\varepsilon \leq \lambda_m(A) p(t_1, x(t_1)) \leq v(t, x(t_1), \phi) \leq v(t_0, x_0, \phi) < \lambda_m(A)\varepsilon.$$

The obtained inequality is not compatible with the assumption on existence of $t_1 \geq t_0$ for which estimate (2.6.12) holds. This proves Theorem 2.6.1

Corollary 2.6.1 If for system (2.6.1) conditions (1), (2), (4), (5) of Theorem 2.6.1 are satisfied and for all $(t, x) \in S(H, h)$

$$v^\Delta(t, x, \phi)|_{(1.6.1)} \leq 0,$$

then Δ-dynamics of system (2.6.1) is uniformly (h_0, h)-stable.

Corollary 2.6.2 If for system (2.6.2) conditions (1), (2), (4), (5) of Theorem 2.6.1 and for all $(t, x) \in S(H, h)$

$$v^\nabla(t, x, \phi)|_{(1.6.1)} \leq 0,$$

then ∇-dynamics of system (2.6.2) is uniformly (h_0, h)-stable.

2.7 Problems for Investigation

2.7.1 In terms of function (2.3.1) with the elements (2.4.2)–(2.4.3) to construct an algorithm for estimation of domains of stability, attraction and asymptotic stability of system (2.2.1) and/or (2.2.2) in the sense of the following definitions (see Bouyekhf and Gruyitch [1]).

The equilibrium state $x = 0$ of system (2.2.1) has

(a) a domain of stability D_S iff for any $\tau_0 \in N_\tau^+$ and arbitrary $\varepsilon \in (0, H)$, for $x_0 \in D_S(\tau_0, \varepsilon)$ the condition $\|x(\tau_k, \tau_0, x_0)\| < \varepsilon$ is satisfied for all $\tau_k > \tau_0$, $k = 1, 2, \ldots$. Besides, $D_S(\tau_0, \varepsilon)$ is a neighborhood $x = 0$ for all $\tau_0 \in N_\tau$ and $D_S(\tau_k) = \cup\{D_S(\tau_k, \varepsilon) \colon \varepsilon \in (0, H)\}$, $D_S(N_\tau) = \{(\tau_k, x) \colon \tau_k \in N_\tau^+ \text{ and } x \in D_S(\tau_k)\}$;

(b) a domain of attraction D_a iff for any $\tau_0 \in N_\tau^+$ and arbitrary $\rho \in (0, H)$ there exists a $\Delta(\tau_0, x_0, \rho) > 0$ such that the condition $x_0 \in D_a(\tau_0)$ implies the estimate $\|x(\tau_k, \tau_0, x_0)\| < \rho$ for all $\tau_k > \tau_0 + \Delta(\tau_0, x_0, \rho)$, $\tau_k \in N_\tau^+$. Moreover, $D_a(\tau_0)$ is a neighborhood of the state $x = 0$ for any $\tau_0 \in N_\tau^+$ and $D_a(N_\tau^+) = \{(\tau_k, x) \colon \tau_k \in N_\tau^+, x \in D_a(\tau_k)\}$;

(c) a domain of asymptotic stability D iff the domains D_S and D_a exist and $D = D_S \cap D_a$.

2.7.2 To establish conditions for various types of boundedness of solutions to system (2.2.1) in terms of function (2.3.1) in view of the results by Yoshizawa [1] obtained for continuous systems.

2.7.3 To establish conditions for global asymptotic stability of positive solutions to system of difference equations (cf. Graef and Qian [1])

$$(2.7.1) \qquad x(\tau + 1) = \alpha x(\tau) + f(x(\tau - k)), \quad \tau = 0, 1, \ldots,$$

where $\alpha = \mathrm{diag}(\alpha_1, \alpha_2, \ldots, \alpha_n)$, $\alpha_i \in (0, 1)$, $k \in \{0, 1, \ldots\}$, $f \in C^1(R^n, R^n)$ and $f'(x) \leq 0$ element-wise.

Hint. Apply the direct Liapunov method base on multi-component functions (vector and/or matrix-valued).

2.7.4 To study dichotomy and asymptotic equivalence of solutions to the quasilinear

$$(2.7.2) \qquad x(\tau + 1) = A(\tau)x(\tau) + g(\tau, x(\tau))$$

and the linear

(2.7.3) $$y(\tau + 1) = A(\tau)y(\tau)$$

systems of difference equations. It is assumed that the vector-function $g(\tau, x(\tau))$ is sufficiently small with respect to the norm and system (2.7.3) possesses dichotomy properties, $A(\tau)$ is $n \times n$-matrix defined for all $\tau \in N_\tau^+$ and $g \colon N_\tau^+ \times R^n \to R^n$.

Hint. Apply the matrix Liapunov function constructed in Section 2.4.

2.8 Comments and References

Sections 2.1 – 2.2 Discrete systems appear to be efficient mathematical models in the investigation of many real world processes and phenomena. Note that in the works by Euler and Lagrange the so-called recurrent series and some problems of probability theory were studied, being described by discrete (finite difference) equations. The active investigation of discrete systems for the last three decades is stipulated by new problems of technical progress. Discrete equations prove to be the most efficient model in description of the mechanical system with impulse perturbations as well as the systems comprising digital computing devices. Recently discrete systems have been applied in the modelling of processes in population dynamics, macroeconomy, chaotic dynamics of economic systems, modelling of recurrent neuron networks, chemical reactions, dynamics of discrete Markov processes, finite and probably automatic machines and computing processes (see, e.g. Agarwal [1], Anapol'skii [1], Bromberg [1], Furasov [1], Elaydi [1], Hahn [1], Karakostas, *et al.* [1], Krapivny and Martynyuk [1], La Salle [1, 2], Luca, *et al.* [1], Martynyuk [3], Michel and Miller [1], Pielou [1], Samarskii [1], Šiljak [1], Smirnov [1], etc.)

Section 2.3 Theorems 2.3.1 – 2.3.3 presented in this Section are new (cf. Martynyuk [4]). Corollaries 2.3.1 – 2.3.4 formulated and proved earlier as independent theorems are found in the works by Bromberg [1], La Salle [2], Hahn [1], Michel, *et al.* [1], etc. Theorems 2.3.12 and 2.3.13 are new. They are established in terms of the results of decomposition of discrete-time system (see Lukyanova [1]) and some ideas by Yuli, *et al.* [1].

Section 2.4 In this section some results by Martynyuk and Slyn'ko [1] are used. To show the efficiency of the proposed approach an example from Michel and Miller [1] is taken where the vector Liapunov function was applied.

Section 2.5 This section is based on some results by Lukyanova and Martynyuk [2]. Theorems 2.5.3 and 2.5.4 are new. Also some ideas are employed from Djordjevic [1], Ikeda and Šiljak [1], and monograph Šiljak [1].

Section 2.6 This section presents a new direction in the theory of systems on time scales. For a long time there has been an interest in studying and understanding the similarities and analogies between the theories of differential equations and forward difference equations with constant step size (see Fort [1], etc.).

The study of dynamical systems on time scales was initiated by Aulbach and Hilger [1], after Ph.D. of S.Hilger in 1988. The development of time scales theory and methods is still in its infancy, various types of dynamic equations have already been developed and their distinctive features investigated in many recent publications (see Sheng [1] and references therein). We used some results from the paper by Sheng [1] in this Section.

The dynamics and applications of discrete-time systems is in the focus of attention of many experts (see, for example, Agarwal [1], Bouyekhf and Gruyitch [1], Cruz-Hernandez, *et al.* [1, 2], Diamond [1], Medina and Pinto [1], Michel, Wang, *et al.* [1], Choi and Koo [1], Stipanovic and Šiljak [1], Zhang [1, 2], etc.).

References

Agarwal, R.P.
[1] *Difference Equations and Inequalities.* New York: Marcel Dekker, 1992.

Anapol'skii, L.Yu.
[1] Method of comparison in dynamics of discrete systems. . In *Vector Liapunov Functions and its Construction*, (Eds.: V.M.Matrosov and L.Yu.Anapol'skii). Novosibirsk: Nauka, 1980, P.92–128. [Russian]

Aulbach, B.
[1] Continuous and discrete dynamics near manifolds of equilibria. *Lecture Notes in Math.*, **1058**, Berlin: Springer-Verlag, 1984.

Bohner, M. and Peterson, A.
[1] *Dynamic Equations on Time Scales. An Introduction with applications.* Boston: Birkhauser, 2001.

Bouyekhf, R. and Gruyitch, L.T.
[1] Novel development of the Lyapunov stability theory for discrete-time systems I: Concepts and definitions. *Nonlin. Anal.* **42** (2000) 463–485.

Bromberg, B.V.
[1] *Matrix Methods in the Theory of Relay and Pulse Control.* Moscow: Nauka, 1967. [Russian]

Choi, S.K. and Koo, N.J.
[1] Variationally stable difference systems. *J. Math. Anal. Appl.* **256** (2001) 587–605.

Cruz-Hernandez, C., Alvarez-Gallegos, J. and Castro-Linares, R.
[1] Stability robustnes of linearizing controllers with state estimation for discrete-time nonlinear systems, *IMA Journal of Mathematical Control and Information* **18** (2001) 479-489.
[2] Stability of discrete nonlinear systems under nonvanishing perturbations: Application to a nonlinear model-matching problem, *IMA Journal Mathematical Control and Information* **16** (1999) 23-41.

Daletskii, Yu.L. and Krene, M.G.
[1] *Stability of Solutions of Differential Equations in Banach Space.* Moscow: Nauka, 1970. [Russian]

Diamond, P.
[1] Discrete Liapunov function with $V > 0$. *J. Austral. Math. Soc.* **20B** (1978) 280–284.

Djordjević, M.Z.
[1] *Zur Stabilitat nichtlinearer gekoppelter Systeme mit der Matrix-Ljapunov Methode.* Diss. ETH Nr. 7690, Zurich, 1984.

Elaydi, S.
[1] *An Introduction to Difference Equations.* New York: Springer-Verlag, 1996.

Eloe, P.W., Sheng, Q. and Henderson, J.
[1] Notes on crossed symmetry solutions of the two-point boundary value problems on time scales. *Journal of Difference Equations and Applications* **9**(1) (2003) 29–48.

Fort, T.
[1] *Finite Differences and Difference Equations in the Real Domain.* Oxford: Clarendon Press, 1948.

Furasov, V. D.
[1] *Stability and Stabilization of Discrete Processes.* Moscow: Nauka, 1982. [Russian]

Graef, J.R. and Qian, C.
[1] Global stability in a nonlinear difference equation, *J. of Differ. Eqns and Appl.* **5** (1999) 251–270.

Hahn, W.
[1] *Stability of Motion.* Berlin: Springer-Verlag, 1967.

Hilger, S.
[1] Analysis on measure chain – a unified approach to continuous and discrete calculus. *Results Math.* **18** (1990) 18-56.

Ikeda, M. and Šiljak, D.D.
[1] Hierarchical Liapunov functions. *J. Math. Anal. Appl.* **112**(1) (1985) 110–128.

Karakostas, G., Philos, Ch.G. and Sficas, Y.G.
[1] The dynamics of some discrete populatiom models, *Nonlin. Anal.* **17** (1991) 1069–1084.

Lakshmikanthan, V., Leela, S. and Martynyuk, A.A.
[1] *Stability Analysis of Nonlinear Systems.* New York: Marcel Dekker, 1989.

Lakshmikantham, V. and Trigiante, D.
[1] *Theory of Difference Equations: Numerical Analysis and Applications*, New York: Marcel Dekker, 2002.

LaSalle, J.P.
[1] Stability theory for difference equations. In: *Studies in Ordinary Differential Equations.* (Ed.: J.Hale). The Mathematical Association of America, 1977, P.1–31.
[2] *The Stability and Control of Discrete Processes.* Berlin: Springer-Verlag, 1986.

Luca, N. and Talpalaru, P.
[1] Stability and asymptotic behaviour of a class of discrete systems. *Ann. Mat. Pure Appl.* **112** (1977) 351–382.

Lukyanova, T.A.
[1] On a variant of stability conditions for discrete-time systems, (to appear).

Lukyanova, T.A. and Martynyuk, A.A.
[1] Robust stability: Three approaches for discrete-time systems. *Nonlinear Dynamics and Systems Theory* **2** (2002) 45–55.
[2] Hierarchical Lyapunov functions for stability analysis of discrete-time systems with applications to the neural networks. *Nonlinear Dynamics and Systems Theory* **4** (2004) 15–49.

Martynyuk, A.A.
[1] Hierarchical matrix Lyapunov function. *Diff. and Integ. Eqns* **2**(4) (1989) 411–417.
[2] *Stability by Liapunov's Matrix Function Method with Applications.* New York: Marcel Dekker, 1998.
[3] Stability analysis of discrete-time systems. *Int. Appl. Mech.* **36**(7) (2000) 3–35.
[4] *Qualitative Methods in Nonlinear Dynamics: Novel Approaches to Liapunov Matrix Functions.* New York: Marcel Dekker, 2002.

Martynyuk, A.A. and Krapivnyi, Yu.N.
[1] *Lyapunov Matrix Functions and Stability of Large-Scale Discrete Systems.* Preprint No. 11, Institute of Mathematics, Academy of Sciences of Ukr. SSR, Kiev (1988). [Russian]

Martynyuk, A.A. and Slyn'ko, V.I.
[1] An approach to constructing of matrix-valued Liapunov functions for a class of large scale discrete-time systems, (to appear).

Medina, R. and Pinto, M.
[1] Dichotomies and asymptotic equivalence of nonlinear difference systems, *J. of Differ. Eqns and Appl.* **5** (1999) 287–303.

Michel, A.N. and Miller, R.K.
[1] *Qualitative Analysis of Large Scale Dynamical Systems.* New York: Academic Press, 1977.

Michel, A.N., Wang, K. and Hu, B.
[1] *Qualitative Theory of Dynamical Systems. The Role of Stability Preserving Mappings.* New York: Marcel Dekker, 2001.

Ostrovskii, A.
[1] Uber die Derminanten mit uberwierender Hauptdiagonale. *Commentarii Mathematici Helvetici* **10** (1937) 69–76.

Pielou, E.C.
[1] *An Introduction to Mathematical Ecology.* New York: Wiley, 1969.

Samarskii. A.A.
[1] *Introduction to the Theory of Difference Scheme.* Moscow, Nauka, 1971. [Russian]

Sheng, Q.
[1] A view of dynamic derivatives on time scales from approximations. *Journal of Difference Equations and Allications* **11**(1) (2005) 63–81.

Sezer, M.E. and Šiljak, D.D.
[1] Robust stability of discrete systems. *International Journal of Control* **48** (1988) 2055–2063.

Šiljak, D.D.
[1] *Large-Scale Dynamic Systems: Stability and Structure.* New York: North Holland, 1978.

Smirnov, E.Ya.
[1] *Stabilization of Programmed Motions.* Amsterdam: Gordon and Breach Science Publishers, 2000.

Stipanovic, D.M. and Šiljak, D.D.
[1] Robust stability and stabilization of discrete-time non-linear systems: The LMI approach. *International Journal of Control* **74**(9) (2001) 873–879.

Yuli, F., Xiaoxin, L. and Qi, L.
[1] New criteria of stability and boundedness for discrete systems, *Nonlin. Anal.* **41** (2000) 779–785.

Zhang, S.
[1] Stability of infinite delay difference systems, *Nonlin. Anal.* **22** (1994) 1121–1129.
[2] Boundedness of infinite delay difference systems, *Nonlin. Anal.* **22** (1994) 1209–1219.

3

STABILITY IN FUNCTIONAL DIFFERENTIAL SYSTEMS

3.1 Introduction

Stability of continuous systems of ordinary differential equations has been studied by many authors for more than 100 years while stability investigation of solutions to functional differential equations has been undertaken quite recently. It is known (see Burton, Hale [1], Razumikhin [1], and Krasovskii [1], etc.) that there are two global approaches of qualitative investigation of the dynamical behavior of solutions to systems of this class. One is the method of Liapunov–Razumikhin functions and the other is the method of Liapunov–Krasovskii functionals. Each approach is being intensively developed and new interesting results are obtained in both directions (see References at the end of this Chapter).

The method of matrix-valued Liapunov functions worked out for the last years in qualitative stability theory of nonlinear systems (see Martynyuk [2, 3]) allows generalization in stability investigation of solutions to functional differential equations in the framework of both of the approaches mentioned above.

In this Chapter, stability analysis of large scale systems of functional differential equations with finite delay via the matrix-valued functionals of Liapunov–Krasovskii are considered. A new theorem on uniform asymptotic stability is established and some corollaries are obtained. These results generalize some theorems in the stability theory of functional differential equations via scalar Liapunov functionals obtained recently.

The results of this Chapter are arranged as follows.

Section 3.2 provides necessary information on functional differential equations and matrix-valued Liapunov–Krasovskii functionals.

Section 3.3 presents some theorems on stability of solutions to functional differential system. The general character of the results is shown by several corollaries, which are of independent importance.

Section 3.4 sets out a new approach to the construction of Liapunov–Krasovskii functionals for linear systems with finite delay. The advantages of the proposed approach are shown by stability investigation of zero solution for some systems with delay.

Section 3.5 proposes new forms of decomposition-aggregation of large-scale systems of functional differential equations via the application of matrix-valued functionals. This allows us to establish new sufficient stability conditions for the systems of equations under consideration in terms of special matrices property of fixed sign.

Section 3.6 deals with stability of autonomous systems with delay consisting of two interacting subsystems. Here the matrix-valued functional is applied which was constructed by the method mentioned in Section 3.4.

In Section 3.7, an oscillatory system is studied which is subject to the delay dependent force. Also, restrictions on parameters value are indicated for which oscillator motions are asymptotically stable.

In the final Section 3.8, statements of the problems are presented associated with further development of the method of matrix Liapunov functions (functionals) in stability theory of functional differential equations.

3.2 Preliminaries

3.2.1 Functional differential equations

For $x \in R^n$, $|\cdot|$ denotes the Euclidean norm of x. Let $\mathcal{C} = C([-\tau, 0], R^n)$ be the space of continuous functions which map $[-\tau, 0]$ into R^n and for $\varphi \in \mathcal{C}$, $\|\varphi\| = \sup_{-\tau \leq \theta \leq 0} |\varphi(\theta)|$, $\mathcal{C}_H$ is a set $\varphi \in \mathcal{C}$ for which $\|\varphi\| < H$, and x_t, as an element of $\mathcal{C}$, is defined by correlation $x_t(\theta) = x(t + \theta)$, $-\tau \leq \theta \leq 0$.

Throughout this Chapter we will use a seminorm $|\cdot|_\tau$ on $\mathcal{C}$ with $|\varphi|_\tau \leq \|\varphi\|$ for all $\varphi \in \mathcal{C}$.

We consider the large-scale systems described by functional differential equations

$$(3.2.1) \qquad \frac{dx}{dt} = f(t, x_t), \quad x_{t_0} = \varphi_0 \in \mathcal{C}, \quad t_0 \geq 0,$$

where $f \in C(R_+ \times \mathcal{C}, R^n)$, $x \in C([t_0 - \tau, \infty), R^n)$, and $x_t \in \mathcal{C}$. In (3.2.1), dx/dt denotes the righ-hand derivative of x at $t \in R_+$. We suppose that for every $\varphi \in \mathcal{C}$ and for every $t_0 \geq 0$, system (3.2.1) possesses a unique solution $x_t(t_0, \varphi)$ with $x_{t_0} = \varphi$ and we denote by $x(t) = x(t; t_0, \varphi)$ the value of $x_t(t_0, \varphi)$ at t (for details see Hale [1]).

Let the system (3.2.1) be decomposed into m interconnected subsystems described by the equations

$$\text{(3.2.2)} \qquad \frac{dx^i}{dt} = f_i(t, x_t^i) + g_i(t, x_t^1, \ldots, x_t^m),$$

where $i \in I_m = \{1, 2, \ldots, m\}$, $\sum_{i=1}^{m} n_i = n$, $f_i \in C(R_+ \times \mathcal{C}_{H_i}, R^{n_i})$, and $g_i \in C(R_+ \times \mathcal{C}_H, R^{n_i})$, $\mathcal{C}_H = \mathcal{C}_{H_1} \times \ldots \times \mathcal{C}_{H_m}$, i.e. $\mathcal{C}_{H_i}$ is a set of $\varphi_i \in \mathcal{C}$ for which $\|\varphi^i\| < H_i$, $i \in I_m$. When $g_i(t, x_t) \equiv 0$ for all $i \in I_m$, from (3.2.2) we obtain the isolated subsystems

$$\text{(3.2.3)} \qquad \frac{dx^i}{dt} = f_i(t, x_t^i), \quad x_{t_0}^i \in \varphi_0^i \in \mathcal{C}_{H_i}, \quad t_0 \geq 0.$$

We also assume that $f(t, 0, \ldots, 0) = 0$, $f_i(t, 0) = 0$, and $g_i(t, 0, \ldots, 0) = 0$ for all $i \in I_m$, so that $x = 0$ ($x^i = 0$) are the solutions of (3.2.1) or (3.2.2) respectively.

3.2.2 Auxiliary results
We will need some notions which we present now.

Definition 3.2.1 The functional $U(t, \varphi) = [v_{ij}(t, \cdot)]$, $i, j = 1, 2, \ldots, m$, continuous and defined on $R_+ \times \mathcal{C}_H$ which together with the upper right Dini derivative of $V(t, \varphi, \eta) = \eta^T U(t, \varphi) \eta$, $\eta \in R_+^m$, defined by

$$\text{(3.2.4)} \qquad D^+ V(t, \varphi, \eta)|_{(3.2.1)} = \limsup\{[V(t+h, x_{t+h}(t_0, \varphi), \eta) - V(t, x_t(t_0, \varphi), \eta)] h^{-1} \colon h \to 0^+\},$$

solves the problem of stability of solution $x = 0$, is called *Liapunov matrix-valued functional*.

Definition 3.2.2 The function $w \colon [0, \infty) \to [0, \infty)$, $w(0) = 0$, $w(r)$ strictly increasing with $w(0) = 0$ and $w(r) \to \infty$ for $r \to +\infty$, is called a *wedge function* (for short we will use $w \in W$).

Definition 3.2.3 The zero solution of (3.2.1) is

(i) *stable* if for each $\varepsilon > 0$ there exists $\delta = \delta(t_0, \varepsilon)$ such that $[(t_0, \varphi) \in R_+ \times \mathcal{C}_H, \|\varphi\| < \delta, t \geq t_0]$ implies $|x(t; t_0, \varphi)| < \varepsilon$ for all $t \geq t_0$;

(ii) *uniformly stable* if it is stable and if in the definition 3.2.3 (a) the value δ does not depend on t_0;

(iii) *asymptotically stable* (AS) if it is stable and for any $t_0 \geq 0$ there exists $\Delta > 0$ such that $[(t, \varphi) \in R_+ \times C_H, \|\varphi\| < \Delta, t \geq t_0]$ implies $|x(t; t_0, \varphi)| \to 0$ for $t \to +\infty$;

(iv) *uniformly asymptotically stable* (UAS) if it is uniformly stable and there exists $\delta^* > 0$ such that for each $\varepsilon > 0$ there exists $T > 0$ such that $[(t_0, \varphi) \in R_+ \times C_H, \|\varphi\| < \delta^*, t_0 \geq 0]$ implies $|x(t; t_0, \varphi)| < \varepsilon$ for all $t \geq t_0 + T$.

3.2.3 Some estimates for the Liapunov matrix-valued functional

For isolated subsystems (3.2.3) we assume that the matrix-valued functional

$$U(t, \varphi) = [v_{ij}(t, \cdot)], \quad i, j = 1, 2, \ldots, m, \quad m < n,$$

is constructed in some way with the elements satisfying the conditions:

(A) $v_{ii}(t, \varphi^i) \in C(R_+ \times C_{H_i}, R_+)$, $v_{ii}(t, 0) = 0$, $i = 1, 2, \ldots, m$, $v_{ii}(t, \varphi^i)$ are locally Lipschitzian in φ^i for all $i \in I_m$;

(B) $v_{ij}(t, \varphi^i, \varphi^j) \in C(R_+ \times C_{H_i} \times C_{H_j}, R_+)$, $v_{ij}(t, 0, 0) = 0$, $v_{ij}(t, \varphi^i, \varphi^j)$ locally Lipschitzian in (φ^i, φ^j) for all $(i \neq j) \in I_m$.

Then the functional

(3.2.5) $$V(t, \varphi, \eta) = \eta^T U(t, \varphi) \eta, \quad \eta \in R_+^m,$$

is applied in stability analysis of the state $x = 0$ of system (3.2.2). The upper right derivative of functional (3.2.5) is defined by

(3.2.6) $$D^+ V(t, \varphi, \eta)|_{(1.1)} = \eta^T D^+ U(t, \varphi) \eta,$$

where $D^+ U(t, \varphi) = [D^+ v_{ij}(t, \cdot)]$, $i, j = 1, 2, \ldots, m$, and $D^+ v_{ij}(t, \cdot)$ is specified according to (3.2.4) for all $(i, j) \in I_m$.

Assumption 3.2.1 *There exist wedge functions w_{ik}, $i = 1, 2, \ldots, m$, $k = 1, 2$ and constants $\underline{a}_{ij}$, $\overline{a}_{ij}$ such that for the elements $v_{ij}(t, \cdot)$, $i, j = 1, 2, \ldots, m$, of the matrix-valued functional $U(t, \varphi)$, $U(t, 0) = 0$ for all $t \in R_+$ the following estimates hold true*

(1) $\underline{a}_{ij} w_{i1}^2(|\varphi^i(0)|) \leq v_{ii}(t, x_t^i) \leq \overline{a}_{ij} w_{i2}(\|\varphi^i\|)$
   for all $(t, x^i) \in R_+ \times C_{H_i}$, $i = 1, 2, \ldots, m$;

(2) $\underline{a}_{ij} w_{i1}(|\varphi^i(0)|) w_{j1}(|\varphi^j(0)|) \leq v_{ij}(t, x_t^i, x_t^j) \leq \overline{a}_{ij} w_{i2}(\|\varphi^i\|) w_{j2}(\|\varphi^j\|)$
   for all $(t, x^i, x^j) \in R_+ \times C_{H_i} \times C_{H_j}$, $i \neq j$.

3. FUNCTIONAL DIFFERENTIAL SYSTEMS

Proposition 3.2.1 If all conditions of Assumption 3.2.1 are satisfied, then for the functional

(3.2.7) $\quad V(t, \varphi, \psi) = \psi^{\mathrm{T}} U(t, \varphi) \psi, \quad \psi \in R_+^m, \quad \psi > 0,$

the bilateral inequality

(3.2.8) $\quad w_1^{\mathrm{T}}(|\varphi(0)|) \Psi^{\mathrm{T}} A \Psi w_1(|\varphi(0)|) \leq V(t, \varphi, \psi) \leq w_2^{\mathrm{T}}(\|\varphi\|) \Psi^{\mathrm{T}} B \Psi w_2(\|\varphi\|)$

holds true for $(t, \varphi) \in R_+ \times C_H$, where

$$w_1^{\mathrm{T}}(|\varphi(0)|) = \bigl(w_{i1}(|\varphi^1(0)|), \ldots, w_{m1}(|\varphi^m(0)|)\bigr), \quad i = 1, 2, \ldots, m,$$
$$w_2^{\mathrm{T}}(\|\varphi\|) = \bigl(w_{i2}(\|\varphi^1\|), \ldots, w_{m2}(\|\varphi^m\|)\bigr), \quad i = 1, 2, \ldots, m,$$
$$A = [\underline{\alpha}_{ij}], \quad \underline{\alpha}_{ij} = \underline{\alpha}_{ji}, \quad B = [\overline{\alpha}_{ij}], \quad \overline{\alpha}_{ij} = \overline{\alpha}_{ji},$$
$$\Psi = \mathrm{diag}\,[\psi_1, \ldots, \psi_m].$$

Proof Proposition 3.2.1 is proved by the direct substitution by estimates (1), (2) from Assumption 3.2.1 in formula (3.2.5).

It is easy to see the positive definiteness and decrease conditions for functional (3.2.5) can be obtained from estimate (3.2.8).

Assumption 3.2.2 There exist wedge functions w_{ik}, $i = 1, 2, \ldots, m$, $k = 1, 2, 3$, functionals $v_{ii}(t, \varphi^i) \in C(R_+ \times C_{H_i}, R_+^{m_i})$, $1 \leq m_i \leq n_i$, $v_{ii}(t, 0) = 0$, $v_{ii}(t, \varphi^i)$ are locally Lipschitzian in φ^i, and $v_{ij}(t, \varphi^i, \varphi^j) \in C(R_+ \times C_{H_i} \times C_{H_j}, R_+^{m_i \times m_j})$, $v_{ij}(t, \varphi^i, \varphi^j)$ are locally Lipschitzian in φ^i and φ^j for all $(i \neq j) \in I_m$, and constants $\beta_{1i}, \beta_{2i}, \beta_{3i}, \beta_{kij}$, $k = 1, 2, 3$, $i, j = 1, 2, \ldots, m$ $(i \neq j)$, for which

(1) $\quad D^+ v_{ii}(t, \varphi^i)\big|_{(3.2.3)} \leq \beta_{1i} w_i^2(\|x_t^i\|) \quad$ for all $\quad (t, \varphi^i) \in R_+ \times C_{H_i};$

(2) $\quad D^+ v_{ii}(t, \varphi^i)\big|_{(3.2.2)} - D^+ v_{ii}(t, \varphi^i)\big|_{(3.2.3)} \leq \sum_{\substack{i=1 \\ i \neq j}}^m \beta_{1ij} w_i(\|x_t^i\|) w_j(\|x_t^j\|)$

for all $\quad (t, \varphi^i, \varphi^j) \in R_+ \times C_{H_i} \times C_{H_j}, \quad i \neq j;$

(3) $\quad D^+ v_{ij}(t, \varphi^i, \varphi^j)\big|_{(3.2.2)} \leq \beta_{2i} w_i^2(\|x_t^i\|) + \sum_{\substack{i=1 \\ i \neq j}}^m \beta_{2ij} w_i(\|x_t^i\|) w_j(\|x_t^j\|) +$

$+ \sum_{\substack{i=1 \\ i \neq j}}^m \beta_{3ij} w_i(\|x_t^i\|) w_j(\|x_t^j\|) + \beta_{3j} w_j^2(\|x_t^j\|)$

for all $\quad (t, \varphi^i, \varphi^j) \in R_+ \times C_{H_i} \times C_{H_j}, \quad i, j = 1, 2, \ldots, m, \quad i \neq j.$

Proposition 3.2.2 If all conditions of Assumption 3.2.2, are satisfied, then the change of the total derivative $D^+V(t,\varphi,\psi)\big|_{(3.2.2)}$ of functional $V(t,\varphi,\psi)$ along solutions of system (3.2.2) is estimated by

(3.2.9) $$D^+V(t,\varphi,\psi)\big|_{(3.2.2)} \leq w^{\mathrm{T}}(\|x_t\|)Gw(\|x_t\|)$$

for all $(t,\varphi) \in R_+ \times \mathcal{C}_H$, where

$$w(\|x_t\|) = \left(w_1(\|x_t^1\|), \ldots, w_m(\|x_t^m\|)\right)^{\mathrm{T}},$$
$$G = [\gamma_{ij}], \quad \gamma_{ij} = \gamma_{ji}, \quad i,j = 1,2,\ldots,m,$$
$$\gamma_{ii} = \psi_i^2 \beta_{1i} + 2\sum_{\substack{j=2\\i<j}}^{m} \psi_i \psi_j (\beta_{2i} + \beta_{3j}), \quad i = 1,2,\ldots,m,$$
$$\gamma_{ij} = \frac{1}{2}\psi_i^2[\beta_{1ij} + \beta_{1ji}] + \psi_i \psi_j \left[\sum_{\substack{i=1\\i\neq j}}^{m} \beta_{2ij} + \sum_{\substack{j=1\\j\neq i}}^{m} \beta_{3ij}\right],$$
$$(i \neq j) \in I_m.$$

Proof Inequality (3.2.9) is proved by the analysis of the expression of $D^+V(t,\varphi,\psi)$ under conditions (1)–(3) of Assumption 3.2.2.

3.3 Theorems on Stability

In this Section, the conditions of different types of stability of the equilibrium state $x = 0$ of system (3.2.1) are established in terms of the direct Liapunov method realized on the set of matrix-valued functionals. Note that only few steps have been made in this direction and actually the development of this branch of stability theory of functional differential equations is an area open for investigations.

Theorem 3.3.1 *Assume that system (3.2.1) is decomposed into subsystems (3.2.2), and for subsystems (3.2.3) the matrix-valued functional $U(t,\varphi)$ is constructed for which and for function (3.3.6) the conditions below are satisfied:*

(1) *Let there exist a wedge vector-function $\widetilde{w}_1$, a constant $K \in (0, H)$ and a constant $m \times m$ matrix A such that for any $(t,\varphi) \in R_+ \times \mathcal{C}_K$*

(i) $\widetilde{w}_1^T(|\varphi(0)|)Aw_1(|\varphi(0)|) \leq V(t,\varphi,\eta)$, $V(t,0,\eta) = 0$,

(ii) $D^+V(t,\varphi,\eta)|_{(3.2.1)} \leq 0$,

(iii) $\lambda_m(A) > 0$.

Then the equilibrium state $x = 0$ of system (3.2.1) is stable.

(2) Let there exist wedge vector-functions $\widetilde{w}_1, \widetilde{w}_2, \widetilde{w}_3$, constant $m \times m$-matrices A, B, C and a constant $r_0 > 0$ such that for any $(t,\varphi) \in R_+ \times C_K$

(i) $\widetilde{w}_1^T(|\varphi(0)|)A\widetilde{w}_1(|\varphi(0)|) \leq V(t,\varphi,\eta) \leq \widetilde{w}_1^T(|\varphi|_\tau)B\widetilde{w}_2(|\varphi|_\tau) + \widetilde{w}_3^T(\|\varphi\|)C\widetilde{w}_3(\|\varphi\|)$,

(ii) $D^+V(t,\varphi,\eta)|_{(3.2.1)} \leq 0$,

(iii) $\lambda_m(A)W_1(r) - \lambda_M(C)W_3(r) > 0$ for all $r \in (0, r_0)$, where $W_1, W_3 \in W$-class, so that $W_1(r) \leq \widetilde{w}_1^T(r)\widetilde{w}_1(r)$, $W_3 \geq \widetilde{w}_3^T(r)\widetilde{w}_3(r)$ for any $r \in (0, r_0)$,

(iv) $\lambda_m(A) > 0$, $\lambda_M(B) > 0$, $\lambda_M(C) > 0$.

Then the equilibrium state $x = 0$ of system (3.2.1) is uniformly stable.

(3) Let there exist wedge vector-functions $\widetilde{w}_1, \widetilde{w}_2, \widetilde{w}_3$, and $\widetilde{w}_4$, constant $m \times m$-matrices A, B and C, mentioned in Theorem 3.3.1(2) and a $m \times m$-matrix $D \in C(R_+, R^{m \times m})$, $D(t)$ is bounded on any finite interval $J \subset R_+$ and for any $(t,\varphi) \in R_+ \times C_K$

(i) condition (i) of Theorem 3.3.1(2) is satisfied,

(ii) $D^+V(t,\varphi,\eta)|_{(3.2.1)} \leq -\widetilde{w}_4^T(|\varphi|_\tau)D(t)\widetilde{w}_4(|\varphi|_\tau)$,

(iii) condition (iii) of Theorem 3.3.1(2) is satisfied,

(iv) condition (iv) of Theorem 3.3.1(2) is satisfied and $\lambda_M(D(t)) > 0$ for all $t \in R_+$.

Then the equilibrium state $x = 0$ of system (3.2.1) is uniformly asymptotically stable.

Proof The validity of assertion (1) of Theorem 3.3.1 follows by the following arguments. Let $\varepsilon > 0$ and $t_0 \in R_+$ be given. The construction of the functional $V(t,\varphi,\eta)$ implies that $V(t,\varphi,\eta) = 0$ whenever $\varphi = 0$. Hence it follows that, given (t_0, ε) there exists $\delta = \delta(t_0, \varepsilon) > 0$ such that $\sup V(t,\varphi,\eta) < \lambda_m(A)w_1(\varepsilon)$ for all $t \in R_+$, where $\lambda_m(A) > 0$, $|\varphi| < \delta$, $w_1(\varepsilon)$ is a wedge function. From conditions (i) and (ii) of Theorem 3.3.1(1) we find that

$$\lambda_m(A)w_1(|x(t;t_0,\varphi)|) \leq V(t,x_t(t),\eta) \leq V(t_0,\varphi,\eta) \leq \lambda_m(A)w_1(\varepsilon)$$

and consequently $|x(t;t_0,\varphi)| \leq \varepsilon$ for all $t \geq t_0$. This proves assertion (1) of Theorem 3.3.1.

Assertion 3.3.1(2) is proved as follows. Let value $\varepsilon > 0$ ($\varepsilon < H < +\infty$) be given. By condition (iii) of Theorem 3.3.1(2) the value $\delta > 0$ is taken so that
$$\lambda_M(B)w_2(\delta) + \lambda_M(C)w_3(\delta) < \lambda_m(A)w_1(\varepsilon).$$

Then for any $(t_0,\varphi) \in R_+ \times \mathcal{C}_K$ such that $\|\varphi\| < \delta$ by condition (ii) along solution $x(t) = x(t;t_0,\varphi)$ we get the estimate

$$\begin{aligned}\lambda_m(A)w_1(|x(t)|) &\leq V(t,x_t,\eta) \leq V(t_0,\varphi,\eta) \\ &\leq \lambda_M(B)w_2(|\varphi|_\tau) + \lambda_M(C)w_3(\|\varphi\|) \\ &\leq \lambda_M(B)w_2(\delta) + \lambda_M(C)w_3(\delta) < \lambda_m(A)w_1(\varepsilon).\end{aligned} \qquad (3.3.1)$$

In view of condition (iv) of Theorem 3.3.1(2) inequalities (3.3.1) yield $|x(t)| < \varepsilon$ for all $t \geq t_0$, whenever $\|\varphi\| < \delta$. This proves assertion (2) of Theorem 3.3.1.

Let us now prove assertion (3) of Theorem 3.3.1. It is clear that under conditions of this assertion all conditions of assertion (2) of this theorem are satisfied and the equilibrium state $x = 0$ of system (3.2.1) is uniformly stable. According to Definition 3.2.3(iv) it is necessary to verify that under conditions of assertion (3) of Theorem 3.3.1 for any $\varepsilon > 0$ there exist $T > 0$ and $\delta^* > 0$ such that the condition $[t_0 \geq 0, \|\varphi\| < \delta^*]$ implies $|x(t)| < \varepsilon$ for all $t \geq t_0 + T$. Assume that δ^* is taken from the condition $\delta^* = \min(K, r_0, 1)$, where $r_0 > 1$. Let solution $x(t) = x(t,t_0,\varphi)$ start for $\|\varphi\| < \delta^*$. Then $x_t(t_0,\varphi) \in \mathcal{C}_K$ and $|x_t(t,\varphi)| < 1 < r_0$ for all $t \geq t_0$. By conditions (ii) and (iii) of Theorem 3.3.1(3)

$$\begin{aligned}\lambda_m(A)w_1(|x(t)|) &\leq V(t,x_t,\eta) \\ &\leq \lambda_M(B)w_2(|x_t|_\tau) + \lambda_M(C)w_3(\|x_t\|),\end{aligned} \qquad (3.3.2)$$

$$D^+V(t,x_t,\eta)|_{(3.2.1)} \leq -\lambda_M(t)w_4(|x(t)|) \qquad (3.3.3)$$

for all $t \geq t_0$, where $\lambda_M(t) = \lambda_M(D(t)) > 0$ and $w_4(r) \geq \widetilde{w}_4^T(r)\widetilde{w}_4(r)$.

Then, given $\varepsilon > 0$ the constant $0 < \mu < 1$ is taken so that
$$\lambda_M(B)w_2(\mu) + \lambda_M(C)w_3(\mu) < \lambda_m(A)w_1(\varepsilon).$$

By condition (iii) of assertion (3) of Theorem 3.3.1 one can find $0 < \beta < \mu$ such that
$$\lambda_m(A)w_1(r) - \lambda_M(C)w_3(r) \geq \beta + \lambda_M(B)w_2(\beta)$$

for all $r \in [1, \mu]$. The fact that w_1 is a uniformly continuous function of r on the interval $[1, \beta]$, implies that a value $0 < \alpha < \mu - \beta$, can be indicated such that

(3.3.4) $\qquad \lambda_m(A)w_1(r) - \lambda_m(A)w_1(r-\alpha) < \beta \quad \text{for all} \quad r \in [1,\mu].$

Therefore

(3.3.5) $\qquad \begin{aligned}\lambda_m(A)w_1(r-\alpha) - \lambda_M(C)w_3(r) &> \lambda_m(A)w_1(r) - \beta \\ - \lambda_M(C)w_3(r) &> \lambda_M(B)w_2(\beta)\end{aligned}$

for any $r \in [1, \mu]$.

Condition (3.3.3) yields

(3.3.6) $\qquad \begin{aligned} V(t, x_t, \eta) &\leq V(t_0, \varphi, \eta) - \int_{t_0}^{T} \lambda_M(s) w_4(|x_s|_\tau)\,ds \\ &\leq \lambda_M(B) w_2(\delta) + \lambda_M(C) w_3(\delta) - \int_{t_0}^{T} \lambda_M(s) w_4(|x_s|_\tau)\,ds.\end{aligned}$

Let us show that $l > 0$ exists so that for any $t^* \geq t_0$ a $\Delta \in [t^*, t^* + l]$ can be found such that $|x_\Delta|_\tau < \beta$. Since $\lambda_M(t) > 0$ for all $t \geq t_0$, the value $l > 0$ is determined by the inequality

$$\lambda_M(B) w_2(\delta) + \lambda_M(C) w_3(\delta) < w_4(\beta) \int_\tau^{\tau+l} \lambda_M(s)\,ds.$$

Assume that $|x_s|_\tau \geq \beta$ for all $s \in [t^*, t^* + l]$. Let $t = t^* + l$, then (3.3.6) implies

(3.3.7) $\qquad \begin{aligned}V(t, x_t, \eta) &\leq V(t_0, \varphi, \eta) - \int_{t^*}^{T} \lambda_M(s) w_4(|x_s|_\tau)\,ds \\ &\leq \lambda_M(B) w_2(\delta) + \lambda_M(C) w_3(\delta) - w_4(\beta) \int_{t^*}^{t^*+l} \lambda_M(s)\,ds < 0.\end{aligned}$

Since $V(t, x_t, \eta) > 0$ by conditions (i) and (iv) of Theorem 4.1(3), the contradiction (3.3.7) proves the existence of $\Delta \in [t^*, t^* + l]$. Hence, there exists a sequence $\{t_n\}$ such that $t_{n-1} + \tau \leq t_n \leq t_{n-1} + \tau + l$, for which $|x_{t_n}|_\tau < \beta$ for $n = 1, 2, \ldots$. We designate $Q_j = [t_j - \tau, t_j]$ and for each value of j we shall consider two cases

(A) $|x_{t_j}| \leq \mu$ or
(B) $|x(\tilde{t})| > \mu$ for some $\tilde{t} \in Q_j$.

Let inequality (A) be satisfied, then for $t \geq t_j$

(3.3.8)
$$\begin{aligned}\lambda_m(A)w_1(|x(t)|) &\leq V(t, x_t, \eta) \leq V(t_j, x_{t_j}, \eta) \\ &\leq \lambda_M(B)w_2(|x_{t_j}|_\tau) + \lambda_M(C)w_3(|x_{t_j}|) \\ &\leq \lambda_M(B)w_2(\mu) + \lambda_M(C)w_3(\mu) < \lambda_m(A)w_1(\varepsilon).\end{aligned}$$

Hence, it follows that $|x(t, t_0, \varphi)| < \varepsilon$ for all $t \geq t_j$.

Then we consider case (B). Let $\mu_j = \|x_{t_j}\| > \mu$. Let us show that $|x(t)| < \mu_j - \alpha$ for all $t \geq t_j$. In fact, if there exists $\tau_j \geq t_j$ such that $|x(\tau_j)| = \mu_j - \alpha$, then

(3.3.9)
$$\begin{aligned}\lambda_m(A)w_1(\mu_j - \alpha) = \lambda_m(A)w_1(|x(\tau_j)|) &\leq V(\tau_j, x_{\tau_j}, \eta) \\ \leq V(t_j, x_{t_j}, \eta) &\leq \lambda_M(B)w_2(|x_{t_j}|_\tau) + \lambda_M(C)w_3(\|x_{t_j}\|) \\ &\leq \lambda_M(B)w_2(\beta) + \lambda_M(C)w_3(\mu_j).\end{aligned}$$

Inequality (3.3.9) contradicts condition (3.3.5) and therefore $|x(t, t_0, \varphi)| < \mu_j - \alpha$ for all $t \geq t_j$. Now we take N so that inequality $1 - N\alpha \leq \mu$ holds true. If case (B) takes place on Q_j for $j = 1, 2, \ldots, N$, then for $t \geq t_N$

$$|x(t, t_0, \varphi)| < \mu_N - \alpha < \mu_{N-1} - 2\alpha < \cdots < 1 - N\alpha \leq \mu.$$

Consequently, inequality (A) occurs on some Q_j, $j \leq N+1$, and then $|x(t, t_0, \varphi)| < \varepsilon$ for $t \geq t_{N+1} \geq t_j$. Since $t_{N+1} \leq t_0 + (N+1)(\tau+l) \equiv t_0 + T$, we have $|x(t, t_0, \varphi)| < \varepsilon$ for all $t \geq t_0 + T$ whenever $[\|\varphi\| < \delta^*, t_0 \geq 0]$. This proves assertion (3) of Theorem 3.3.1.

Theorem 3.3.1 has a series of corollaries set out below.

Corollary 3.3.1 For system (3.2.1) let functional

$$V(t, \varphi, \eta) = \eta^T U(t, \varphi)\eta$$

be constructed, $V: R_+ \times \mathcal{C}_H \times R_+^m \to R_+$ is continuous on $R_+ \times \mathcal{C}_H$ and

(i) $\tilde{w}_1^T(|\varphi(0)|)A\tilde{w}_1(|\varphi(0)|) \leq V(t, \varphi, \eta) \leq \tilde{w}_2^T(|\varphi|_\tau)B\tilde{w}_2(|\varphi|_\tau)$, $\tilde{w}_1, \tilde{w}_2 \in$ W-class;

(ii) $D^+V(t, \varphi, \eta)|_{(3.2.1)} \leq -\tilde{w}_4^T(|\varphi|_\tau)D(t)\tilde{w}_4(|\varphi|_\tau)$, $\tilde{w}_4 \in W$-class;

(iii) $\lambda_m(A) > 0$, $\lambda_M(B) > 0$, and $\lambda_M(t) = \lambda_M(D(t)) > 0$ for all $t \in R_+$.

Then the zero solution of system (3.2.1) is uniformly asymptotically stable.

Corollary 3.3.2 For system (3.2.1) let functional
$$V(t, \varphi, \eta) = \eta^{\mathrm{T}} U(t, \varphi) \eta$$
be constructed, $V \colon R_+ \times \mathcal{C}_H \times R_+^m \to R_+$ is continuous on $R_+ \times \mathcal{C}_H$ and

(i) $\widetilde{w}_1^{\mathrm{T}}(|\varphi(0)|) A \widetilde{w}_1(|\varphi(0)|) \leq V(t, \varphi, \eta) \leq \widetilde{w}_2^{\mathrm{T}}(\|\varphi\|) \widetilde{B} \widetilde{w}_2(\|\varphi\|)$, where $A, \widetilde{B}$ are constant matrices and $\widetilde{w}_1, \widetilde{w}_2 \in W$-class;

(ii) $D^+ V(t, \varphi, \eta)|_{(3.2.1)} \leq -\widetilde{w}_4^{\mathrm{T}}(\|\varphi\|) D(t) \widetilde{w}_4(\|\varphi\|)$, where $D(t)$ is continuous $m \times m$ matrix, $\widetilde{w}_4 \in W$-class;

(iii) $\lambda_m(A) > 0$, $\lambda_M(\widetilde{B}) > 0$, and $\lambda_M(t) = \lambda_M(D(t)) > 0$ for all $t \in R_+$.

Then the zero solution of system (3.2.1) is uniformly asymptotically stable.

Corollary 3.3.3 For system (3.2.1) let functional
$$V(t, \varphi, \eta) = \eta^{\mathrm{T}} U(t, \varphi) \eta$$
be constructed, $V \colon R_+ \times \mathcal{C}_H \times R_+^m \to R_+$ is continuous on $R_+ \times \mathcal{C}_H$ and

(i) $\widetilde{w}_1^{\mathrm{T}}(|\varphi(0)|) A \widetilde{w}_1(|\varphi(0)|) \leq V(t, \varphi, \eta) \leq \widetilde{w}_2^{\mathrm{T}}(\|\varphi\|) \widetilde{B} \widetilde{w}_2(\|\varphi\|)$, where $A, \widetilde{B}, \widetilde{w}_1, \widetilde{w}_2$ are the same as in Corollary 3.3.2;

(ii) $D^+ V(t, \varphi, \eta)|_{(3.2.1)} \leq -\widetilde{w}_4^{\mathrm{T}}(|\varphi(0)|) \widetilde{D}(t) \widetilde{w}_4(|\varphi(0)|)$, where $\widetilde{D}(t) \colon R_+ \to R^{m \times m}$ is continuous matrix on any finite interval $J \subset R_+$;

(iii) $\lambda_m(A) > 0$, $\lambda_M(\widetilde{B}) > 0$, and $\lambda_M(t) = \lambda_M(\widetilde{D}(t)) > 0$ for all $t \in R_+$;

(iv) for any $\gamma > 0$ there exists a constant $\Phi > 0$ such that $|f(t, \varphi)| \leq \Phi$ whenever $\|\varphi\| < \gamma$.

Then the zero solution of system (3.2.1) is uniformly asymptotically stable.

Corollary 3.3.4 For system (3.2.1) let continuous on $R_+ \times \mathcal{C}_H$ functional $V(t, \varphi, \eta) = \eta^{\mathrm{T}} U(t, \varphi) \eta$, $\eta \in R_+^m$, be constructed such that

(i) $\widetilde{w}_1^{\mathrm{T}}(|\varphi(0)|) A \widetilde{w}_1(|\varphi(0)|) \leq V(t, \varphi, \eta) \leq \widetilde{w}_2^{\mathrm{T}}(|\varphi(0)|) B \widetilde{w}_2(|\varphi(0)|) + \widetilde{w}_3^{\mathrm{T}}(|\varphi|_2) \widetilde{C} \widetilde{w}_3(|\varphi|_2)$, where $|\cdot|_2$ denotes L^2–norm on the space of functions $\mathcal{C}$;

(ii) $D^+V(t,\varphi,\eta)|_{(3.2.1)} \leq -\widetilde{w}_4^T(|\varphi(0)|)\widetilde{D}(t)\widetilde{w}_4(|\varphi(0)|)$, where $\widetilde{D}(t)$ — $m \times m$ matrix mentioned in Corollary 3.3.3;

(iii) $\lambda_m(A) > 0$, $\lambda_M(B) > 0$, $\lambda_M(\widetilde{C}) > 0$ and $\lambda_M(t) = \lambda_M(\widetilde{D}(t)) > 0$ for all $t \in R_+$.

Then the zero solution of system (3.2.1) is uniformly asymptotically stable.

Corollary 3.3.5 For system (3.2.1) let continuous on $R_+ \times \mathcal{C}$ functional $V(t,\varphi,\eta) = \eta^T U(t,\varphi)\eta$ be constructed and there exists a constant $M > 0$ such that

(i) $\widetilde{w}_1^T(|\varphi(0)|)A\widetilde{w}_1(|\varphi(0)|) \leq V(t,\varphi,\eta) \leq \widetilde{w}_2^T(\|\varphi\|)\widetilde{B}\widetilde{w}_2(\|\varphi\|)$;

(ii) $D^+V(t,\varphi,\eta)|_{(3.2.1)} \leq -\widetilde{w}_4^T(|\varphi(0)|)\widetilde{D}(t)\widetilde{w}_4(|\varphi(0)|) + M|f(t,\varphi)|$;

(iii) $\lambda_m(A) > 0$, $\lambda_M(\widetilde{B}) > 0$ and $\lambda_M(t) = \lambda_M(\widetilde{D}(t)) > 0$ for all $t \in R_+$.

Then the zero solution of system (3.2.1) is uniformly asymptotically stable.

Corollary 3.3.6 For system (3.2.1) let continuous on $R_+ \times \mathcal{C}$ functional $V(t,\varphi,\eta) = \eta^T U(t,\varphi)\eta$, $\eta \in R_+^m$, be constructed and there exists a constant $H > r_0 > 0$ such that

(i) $\widetilde{w}_1^T(|\varphi(0)|)A\widetilde{w}_1(|\varphi(0)|) \leq V(t,\varphi,\eta) \leq \widetilde{w}_2^T(|\varphi(0)|)B\widetilde{w}_2(|\varphi(0)|) + \widetilde{w}_3^T(\|\varphi\|)C\widetilde{w}_3(\|\varphi\|)$;

(ii) $D^+V(t,\varphi,\eta)|_{(3.2.1)} \leq -\widetilde{w}_4^T(|\varphi(0)|)\widetilde{D}(t)\widetilde{w}_4(|\varphi(0)|)$;

(iii) $\lambda_m(A)w_1(r) - \lambda_M(C)w_3(r) > 0$ for all $r \in (0, r_0)$;

(iv) $\lambda_m(A) > 0$, $\lambda_M(B) > 0$, $\lambda_M(C) > 0$ and $\lambda_M(t) = \lambda_M(\widetilde{D}(t)) > 0$ for all $t \in R_+$.

Then the zero solution of system (3.2.1) is uniformly asymptotically stable.

Corollary 3.3.7 For system (3.2.1) let continuous on $R_+ \times \mathcal{C}$ functional $V(t,\varphi,\eta) = \eta^T U(t,\varphi)\eta$, $\eta \in R_+^m$, be constructed and there exists a constant $H > r_0 > 0$ such that

(i) $\widetilde{w}_1^T(|\varphi(0)|)A\widetilde{w}_1(|\varphi(0)|) \leq V(t,\varphi,\eta) \leq \widetilde{w}_2^T(|\varphi(0)|+|\varphi|_2)\widehat{B}\widetilde{w}_2(|\varphi(0)|+|\varphi|_2) + \widetilde{w}_3^T(\|\varphi\|)C\widetilde{w}_3(\|\varphi\|)$;

(ii) $D^+V(t,\varphi,\eta)|_{(3.2.1)} \leq -\widetilde{w}_4^T(|\varphi(0)|)\widetilde{D}(t)\widetilde{w}_4(|\varphi(0)|)$;

(iii) $\lambda_m(A)w_1(r) - \lambda_M(C)w_3(r) > 0$ for all $r \in (0, r_0)$;

(iv) $\lambda_m(A) > 0$, $\lambda_M(\widehat{B}) > 0$, $\lambda_M(C) > 0$ and $\lambda_M(t) = \lambda_M(\widetilde{D}(t)) > 0$ for all $t \in R_+$.

Then the zero solution of system (3.2.1) is uniformly asymptotically stable.

3.4 Large-Scale System of Functional Differential Equations

It is known that the problem of constructing appropriate Liapunov–Krasovskii functionals which solve the stability problem of solution $x = 0$ of system (3.2.1) is a central one in the direct Liapunov method for functional differential equations (3.2.1). The application of matrix-valued functionals allows one to extend the class of admissible components for construction of the appropriate Liapunov–Krasovskii functional.

Further we shall investigate systems of connected equations (3.2.2) and independent subsystems (3.2.3) in the framework of general methodology of qualitative analysis of large scale systems motion.

3.4.1 Approach A_1
This approach is based on the system of conditions below.

Assumption 3.4.1 There exist functionals $v_{ii}(t, \varphi^i)\colon R_+ \times C_{H_i} \to R_+$ and $v_{ij}(t, \varphi^i, \varphi^j)\colon R_+ \times C_{H_i} \times C_{H_j} \to R$, wedge functions $\widetilde{w}_1, \widetilde{w}_2, \widetilde{w}_3 \in W$-class and constants $\widetilde{a}_{ij}$, $\widetilde{b}_{ij}$, $\widetilde{c}_{ij}$, $i, j = 1, 2, \ldots, m$, and $0 < K_i < H$ such that

(1) $\widetilde{a}_{ii} \widetilde{w}_{i1}^2(|\varphi^i(0)|) \leq v_{ii}(t, \varphi^i) \leq \widetilde{b}_{ii} \widetilde{w}_{i2}^2(|\varphi^i|_\tau) + \widetilde{c}_{ii} \widetilde{w}_{i3}^2(\|\varphi^i\|)$ for all $(t, \varphi^i) \in R_+ \times C_{K_i}$, $i = 1, 2, \ldots, m$;

(2) $\widetilde{a}_{ij} \widetilde{w}_{i1}(|\varphi^i(0)|) \widetilde{w}_{j1}(|\varphi^j(0)|) \leq v_{ij}(t, \varphi^i, \varphi^j) \leq \widetilde{b}_{ij} \widetilde{w}_{i2}(|\varphi^i|_\tau) \times \widetilde{w}_{j2}(|\varphi^j|_\tau) + \widetilde{c}_{ij} \widetilde{w}_{i3}(\|\varphi^i\|) \widetilde{w}_{j3}(\|\varphi^j\|)$ for all $(t, \varphi^i, \varphi^j) \in R_+ \times C_{K_i} \times C_{K_j}$, $i \neq j \in I_m$.

The following assertion holds true.

Proposition 3.4.1 If all conditions of Assumption 3.4.1 are satisfied, then in the domain of values $(t, \varphi) \in R_+ \times C_K$ for the functional

(3.4.1) $$V(t, \varphi, \eta) = \eta^\mathrm{T} U(t, \varphi) \eta, \quad \eta \in R_+^m, \quad \eta > 0,$$

the bilateral inequality

(3.4.2) $$\widetilde{w}_1^\mathrm{T}(|\varphi(0)|) H^\mathrm{T} \widetilde{A} H \widetilde{w}_1(|\varphi(0)|) \leq V(t, \varphi, \eta)$$
$$\leq \widetilde{w}_2^\mathrm{T}(|\varphi|_\tau) H^\mathrm{T} \widetilde{B} H \widetilde{w}_2(|\varphi|_\tau) + \widetilde{w}_3^\mathrm{T}(\|\varphi\|) H^\mathrm{T} \widetilde{C} H \widetilde{w}_3(\|\varphi\|)$$

is fulfilled for all $(t, \varphi) \in R_+ \times C_K$, where

$$\tilde{w}_1^{\mathrm{T}}(|\varphi(0)|) = (\tilde{w}_{11}(|\varphi^1(0)|), \ldots, \tilde{w}_{m1}(|\varphi^m(0)|)),$$
$$\tilde{w}_2^{\mathrm{T}}(|\varphi|_\tau) = (\tilde{w}_{12}(|\varphi^1|_\tau), \ldots, \tilde{w}_{m2}(|\varphi^m|_\tau)),$$
$$\tilde{w}_3^{\mathrm{T}}(\|\varphi\|) = (\tilde{w}_{13}(\|\varphi^1\|), \ldots, \tilde{w}_{m3}(\|\varphi^m\|)),$$
$$\tilde{A} = [\tilde{a}_{ij}], \quad \tilde{a}_{ij} = \tilde{a}_{ji}, \quad \tilde{B} = [\tilde{b}_{ij}], \quad \tilde{b}_{ij} = \tilde{b}_{ji},$$
$$\tilde{C} = [\tilde{c}_{ij}], \quad \tilde{c}_{ij} = \tilde{c}_{ji}, \quad H = \mathrm{diag}\,[\eta_1, \ldots, \eta_m],$$
$$C_K = C_{K_1} \times C_{K_2} \times \ldots \times C_{K_m}.$$

Proposition 3.4.1 is proved by direct substitution by estimates (1), (2) of Assumption 3.4.1 in the expression of functional $V(t, \varphi, \eta) = \sum_{i,j=1}^{m} \eta_i \eta_j v_{ij}(t, \cdot)$ with subsequent simple transformations.

Assumption 3.4.2 There exist functionals $v_{ii}(t, \varphi^i)$, $i = 1, 2, \ldots, m$, and $v_{ij}(t, \varphi^i, \varphi^j)$, $i \neq j \in I_m$, mentioned in Assumption 3.4.1 which are locally Lipschitzian in φ^i and (φ^i, φ^j) respectively, wedge functions $\tilde{w}_{i4}$, $i = 1, 2, \ldots, m$, and continuous positive functions $d_{ki}(t)$, $d_{kij}(t)$: $R_+ \to R_+$, $k = 1, 2, 3$, $i, j = 1, 2, \ldots, m$, $i \neq j$, such that

(1) $D^+ v_{ii}(t, \varphi^i)|_{(3.2.3)} \leq d_{1i} \tilde{w}_{i4}^2(|\varphi^i|_\tau)$ for all
$(t, \varphi^i) \in R_+ \times C_{K_i}$;

(2) $D^+ v_{ii}(t, \varphi^i)|_{(3.2.2)} - D^+ v_{ii}(t, \varphi^i)|_{(3.2.3)}$
$$\leq \sum_{\substack{i=1 \\ i \neq j}}^{m} d_{1ij} \tilde{w}_{i4}(|\varphi^i|_\tau) \tilde{w}_{j4}(|\varphi^j|_\tau) \quad \text{for all}$$
$(t, \varphi^i, \varphi^j) \in R_+ \times C_{K_i} \times C_{K_j}, \quad i \neq j;$

(3) $D^+ v_{ij}(t, \varphi^i, \varphi^j)|_{(3.2.2)} \leq d_{2i} \tilde{w}_{i4}^2(|\varphi^i|_\tau)$
$$\leq \sum_{\substack{i=1 \\ i \neq j}}^{m} d_{2ij} \tilde{w}_{i4}(|\varphi^i|_\tau) \tilde{w}_{j4}(|\varphi^j|_\tau)$$
$$+ \sum_{\substack{i=1 \\ i \neq j}}^{m} d_{3ij} \tilde{w}_{i4}(|\varphi^i|_\tau) \tilde{w}_{j4}(|\varphi^j|_\tau) + d_{3i} \tilde{w}_{j4}^2(|\varphi^j|_\tau) \quad \text{for all}$$
$(t, \varphi^i, \varphi^j) \in R_+ \times C_{K_i} \times C_{K_j}, \quad i \neq j.$

3. FUNCTIONAL DIFFERENTIAL SYSTEMS

Proposition 3.4.2 *If all conditions of Assumption 3.4.2 are satisfied, then*

(3.4.3) $$D^+V(t,\varphi,\eta)|_{(3.2.2)} \leq \tilde{w}_4^{\mathrm{T}}(|\varphi|_\tau)D(t)\tilde{w}_4(|\varphi|_\tau)$$

for all $(t,\varphi) \in R_+ \times C_K$, $C_K = C_{K_1} \times C_{K_2} \times \ldots \times C_{K_m}$, *where*

$$\tilde{w}_4(|\varphi|_\tau) = (\tilde{w}_{14}(|\varphi^1|_\tau), \ldots, \tilde{w}_{m4}(|\varphi^m|_\tau))^{\mathrm{T}},$$
$$D(t) = [d_{ij}(t)], \quad d_{ij} = d_{ji}, \quad i,j = 1,2,\ldots,m,$$
$$d_{ii}(t) = \eta_i^2 d_{1i} + 2\sum_{\substack{j=2\\j>i}}^m \eta_i\eta_j(d_{2i} + d_{3j}), \quad i = 1,2,\ldots,m,$$
$$d_{ij}(t) = \frac{1}{2}\eta_i^2(d_{1ij}(t) + d_{1ji}(t))$$
$$+ \eta_i\eta_j\left[\sum_{\substack{s=1\\s\neq j}}^m d_{2sj}(t) + \sum_{\substack{s=1\\s\neq j}}^m d_{3sj}(t)\right], \quad (i \neq j) \in I_m.$$

Proof Estimate (2.4.3) is obtained by direct substitution by estimates (1)–(3) from Assumption 3.4.2 in the expression

$$D^+V(t,\varphi,\eta) = \eta^{\mathrm{T}} D^+U(t,\varphi)\eta, \quad \eta \in R_+^m,$$

where $D^+U(t,\varphi)$ is computed component-wise along solutions of subsystems (3.2.3) and (3.2.2) respectively.

Estimates (3.4.2), (3.4.3) and Theorem 3.3.1 enable us to establish new stability conditions for the solution $x = 0$ of system (3.2.1) as follows.

Theorem 3.4.1 *Assume that for system (3.2.1) the functional $U(t,\varphi)$, is constructed for the components of which all estimates of Assumption 3.4.1 are fulfilled and for the upper right derivatives $D^+v_{ij}(t,\cdot)$ estimates (1)–(3) of Assumption 3.4.2 are fulfilled, and moreover*

(a) *in estimates (3.4.2) matrices $A = H^{\mathrm{T}}\tilde{A}H$, $B = H^{\mathrm{T}}\tilde{B}H$ and $C = H^{\mathrm{T}}\tilde{C}H$ are positive definite;*

(b) *there exist $r_0 \leq \min_i K_i$ such that $\lambda_m(A)w_1(r) - \lambda_M(C)w_3(r) > 0$ for all $r \in (0, r_0)$, where w_1, w_2 are the wedge functions such that $w_1(r) \leq \tilde{w}_1^{\mathrm{T}}(r)\tilde{w}_1(r)$, $w_3(r) \geq \tilde{w}_3^{\mathrm{T}}(r)\tilde{w}_3(r)$ for any $r \in (0, r_0)$.*

Then the zero solution of system (3.1.2) is

(a) *uniformly stable*, if the constant matrix $D_M \geq \frac{1}{2}(D^{\mathrm{T}}(t) + D(t))$ in inequality (3.4.3) is negative semi-definite;
(b) *uniformly asymptotically stable* if the matrix D_M mentioned in condition (a) is negative definite.

Proof Under condition (a) of Theorem 3.4.1 estimate (3.4.2) implies that the functional (3.4.1) is positive definite and decreasing and the upper right derivative of the functional $V(t, \varphi, \eta)$ satisfies the condition

$$D^+V(t,\varphi,\eta)|_{(3.2.1)} \leq 0 \quad \text{for all} \quad (t,\varphi) \in R_+ \times C_K.$$

Therefore all conditions of Theorem 3.3.1(2) are satisfied and the zero solution of system (3.2.1) is uniformly stable. Assertion (b) of this Theorem is proved in the same way in view of Theorem 3.3.1(3).

3.4.2 Approach A_2 System of conditions of Assumption 3.4.2 outlines a general approach to stability analysis of zero solution of system (3.2.1). The essence of this approach is that stability conditions of zero solution of system (3.2.1) are established based on the analysis of dynamical properties of independent subsystems (3.4.3) and qualitative estimates of interconnection functions between them. Below the other system of conditions is presented under which stability of system (3.2.1) can be studied in the framework of this approach.

Assumption 3.4.3 There exist functionals $v_{ii}(t, \varphi^i)$ and $v_{ij}(t, \varphi^i, \varphi^j)$, mentioned in Assumption 3.4.1 wedge functions $\widetilde{w}_{i4}$, $i = 1, 2, \ldots, m$, and constants d_{ki}, d_{ij}, $k = 1, 2$, $i, j = 1, 2, \ldots, m$, such that

(1) $D^+v_{ii}(t,\varphi^i)|_{(3.2.3)} \leq d_{1i}\widetilde{w}_{i4}^2(|\varphi^i|_\tau)$ for all $(t, \varphi^i) \in R_+ \times C_{K_i}$;

(2) $\displaystyle\sum_{i=1}^{m} \eta_i \Big\{ D^+v_{ii}(t,\varphi^i)|_{(3.2.2)} - D^+v_{ii}(t,\varphi^i)|_{(3.2.3)}$

$\displaystyle + \sum_{\substack{j=1 \\ j \neq i}}^{m} \eta_j (D^+v_{ij}(t,\varphi^i,\varphi^j)_{(3.2.2)}) \Big\}$

$\displaystyle \leq \sum_{i=1}^{m} d_{2i}\widetilde{w}_{i4}^2(|\varphi^i|_\tau) + \sum_{\substack{j=1 \\ j \neq i}}^{m} d_{ij}\widetilde{w}_{i4}(|\varphi^i|_\tau)\widetilde{w}_{j4}(|\varphi^j|_\tau)$

for all $(t, \varphi^i, \varphi^j) \in R_+ \times C_{K_i} \times C_{K_j}$.

Remark 3.4.1 In some cases condition (2) from Assumption 3.4.3 allows one to estimate more precisely the effect of the interconnection functions $g_i(t, x_t^1, \ldots, x_t^m)$, $i = 1, 2, \ldots, m$, on the whole system dynamics.

Proposition 3.4.3 *If all conditions of Assumption 3.4.3 are satisfied, then*

(3.4.4) $$D^+ V(t, \varphi, \eta)|_{(3.2.2)} \leq \tilde{w}_4^T(|\varphi|_\tau) E \tilde{w}_4(|\varphi|_\tau)$$

for all $(t, \varphi) \in R_+ \times C_K$, *where* $\tilde{w}_4(|\varphi|_\tau)$ *is determined as in Proposition 3.4.2*

$$E = [\varepsilon_{ij}], \quad \varepsilon_{ij} = \varepsilon_{ji}, \quad i, j = 1, 2, \ldots, m,$$
$$\varepsilon_{ii} = d_{1i} + d_{2i}, \quad i = 1, 2, \ldots, m,$$
$$\varepsilon_{ij} = \frac{1}{2} d_{ij}, \quad i \neq j, \quad (i, j) \in I_m.$$

This Proposition is proved in the same way as Proposition 3.4.2.

Theorem 3.4.2 *Assume that for system (3.2.2) the matrix-valued functional $U(t, \varphi)$ is constructed for which all estimates of Assumption 3.4.1 are fulfilled and for the upper right derivatives $D^+ v_{ij}(t, \cdot)$ estimates of Assumption 3.4.3 and conditions (a), (b) of Theorem 3.4.1 are satisfied. Then the zero solution of system (3.2.1) is*

(a) *uniformly stable, if the matrix E in estimate (3.4.4) is negative semi-definite;*
(b) *uniformly asymptotically stable, if the matrix E in estimate (3.4.4) is negative definite.*

The proof of Theorem 3.4.2 is omitted, since it is similar to the proof of Theorem 3.4.1.

3.4.3 Approach B_1
In distinction to the system of conditions from Assumptions 3.4.2 and 3.4.3, which is the basis for Theorem 2.4.2 we shall indicate the approach of establishing stability conditions for zero solution of system (3.2.1), in which the interacting subsystem is not divided into free subsystem and interconnection functions. Besides some restrictions are imposed simulteneously on all interacting subsystems (3.2.1).

Assumption 3.4.4 There exist functionals $v_{ii}(t,\varphi^i)$ and $v_{ij}(t,\varphi^i,\varphi^j)$, mentioned in Assumption 3.4.1, wedge functions $\tilde{w}_{i4}$ and bounded on any finite interval functions $\beta_{1i}(t)$, $\beta_{2i}(t)$, $\beta_{3i}(t)$, $\beta_{kij}(t)$, $k=1,2,3$, $(i\ne j)\in I_m$, such that

(1) $D^+v_{ii}(t,\varphi^i)|_{(3.2.2)} \le \beta_{1i}(t)\tilde{w}_{i4}^2(|\varphi^i|_\tau)$
$$+ \sum_{\substack{j=1\\j\ne i}}^m \beta_{1ij}(t)\tilde{w}_{i4}(|\varphi^i|_\tau)\tilde{w}_{j4}(|\varphi^j|_\tau) \quad \text{for all}$$
$(t,\varphi^j,\varphi^j)\in R_+\times C_{K_i}\times C_{K_j}, \quad i=1,2,\ldots,m;$

(2) $D^+v_{ij}(t,\varphi^i,\varphi^j)|_{(3.2.2)} \le \beta_{2i}(t)\tilde{w}_{i4}^2(|\varphi^i|_\tau)$
$$+\sum_{\substack{i=1\\i\ne j}}^m \beta_{2ij}(t)\tilde{w}_{i4}(|\varphi^i|_\tau)\tilde{w}_{j4}(|\varphi^j|_\tau)$$
$$+\sum_{\substack{i=1\\i\ne j}}^m \beta_{3ij}(t)\tilde{w}_{i4}(|\varphi^i|_\tau)\tilde{w}_{j4}(|\varphi^j|_\tau) + \beta_{3i}(t)\tilde{w}_{i4}^2(|\varphi^j|_\tau)$$
for all $(t,\varphi^i,\varphi^j)\in R_+\times C_{K_i}\times C_{K_j}$, $(i\ne j)\in I_m$.

Proposition 3.4.4 If all conditions of Assumption 3.4.4 are satisfied, then

(3.4.5) $$D^+V(t,\varphi,\eta)|_{(3.2.2)} \le \tilde{w}_4^T(|\varphi|_\tau)\theta(t)\tilde{w}_4(|\varphi|_\tau)$$

for all $(t,\varphi)\in R_+\times C_K$. Here the elements of matrix $\theta(t)=[\theta_{ij}(t)]$, $\theta_{ij}=\theta_{ji}$ for all $(i,j)\in I_m$, are determined as:

$$\theta_{ii}(t) = \eta_i^2\beta_{1i}(t) + 2\sum_{\substack{j=2\\j>i}}^m \eta_i\eta_j(\beta_{2i}(t)+\beta_{3j}(t)),$$
$i=1,2,\ldots,m;$

$$\theta_{ij}(t) = \frac{1}{2}\eta_i^2\beta_{1ij}(t) + \eta_i\eta_j\left(\sum_{\substack{i=1\\i\ne j}}^m \beta_{2ij}(t) + \sum_{\substack{j=1\\j\ne i}}^m \beta_{3ij}(t)\right),$$
$(i\ne j)\in I_m.$

The proof of this assertion is similar to those of Propositions 3.4.2 and 3.4.3 and so are omitted here.

3. FUNCTIONAL DIFFERENTIAL SYSTEMS

Theorem 3.4.3 *Assume that for system (3.2.1) the matrix-valued functional $U(t,\varphi)$, is constructed for components of which all conditions of Assumption 3.4.1 are fulfilled and for the upper derivatives $D^+v_{ij}(t,\cdot)$ estimates of Assumption 3.4.4 are satisfied as well as conditions (a) and (b) from Theorem 3.4.1.*

Then the zero solution of system (3.2.1):

(a) *uniformly stable, if the constant $m \times m$-matrix $\theta_M \geq \dfrac{1}{2}(\theta^{\mathrm{T}}(t)+\theta(t))$ in estimate (3.4.5) is negative semidefinite;*

(b) *uniformly asymptotically stable, if the matrix θ_M mentioned in condition (a) is negative definite.*

Proof of this Theorem is similar to that of Theorem 3.4.2.

3.4.4 Approach B_2

This approach is based on the following system of conditions.

Assumption 3.4.5 *There exist functionals $v_{ii}(t,\varphi^i)$ and $v_{ij}(t,\varphi^i,\varphi^j)$ for $(i \neq j) \in I_m$, mentioned in Assumption 3.4.1, wedge functions $\widetilde{w}_{i4}$ and constants β_{ii}, $i=1,2,\ldots,m$, β_{ij}, $(i \neq j) \in I_m$, such that*

$$\sum_{i=1}^{m}\eta_i^2(D^+v_{ii}(t,\varphi^i))|_{(3.2.2)} + 2\sum_{i=1}^{m-1}\sum_{\substack{j=2\\j>i}}^{m}\eta_i\eta_j(D^+v_{ij}(t,\varphi^i,\varphi^j))|_{(3.2.2)}$$
$$\leq \sum_{i=1}^{m}\beta_{ii}\widetilde{w}_{i4}^2(|\varphi^i|_\tau) + \sum_{i=1}^{m-1}\sum_{\substack{j=2\\j>i}}^{m}\beta_{ij}\widetilde{w}_{i4}(|\varphi^i|_\tau)\widetilde{w}_{j4}(|\varphi^j|_\tau)$$

for all $(t,\varphi^i,\varphi^j) \in R_+ \times C_{K_i} \times C_{K_j}$.

Similarly to the above proposition the following assertion takes place.

Proposition 3.4.5 *If all conditions of Assumption 3.4.5 are satisfied, then*

(3.4.6) $$D^+V(t,\varphi,\eta)|_{(3.2.2)} \leq \widetilde{w}_4^{\mathrm{T}}(|\varphi|_\tau)P\widetilde{w}_4(|\varphi|_\tau)$$

for all $(t,\varphi) \in R_+ \times C_K$, where the matrix $P = [\rho_{ij}]$, $\rho_{ij} = \rho_{ji}$ for all $(i,j) \in I_m$, has the elements

$$\rho_{ii} = \beta_{ii}, \quad i=1,2,\ldots,m; \quad \rho_{ij} = \dfrac{1}{2}\beta_{ij}, \quad (i \neq j) \in I_m.$$

The following result establishes stability conditions for zero solution of (3.2.1).

Theorem 3.4.4 *Assume that for interacting subsystems (3.2.2) the matrix-valued functional $U(t, \varphi)$, is constructed, for the components of which all conditions of Assumption 3.4.1 are satisfied and for the upper derivatives $D^+ v_{ij}(t, \cdot)$ the estimates of Assumption 3.4.5 are fulfilled as well as conditions (a) and (b) from Theorem 3.4.1.*

Then the zero solution of system (3.2.1) is

(a) *uniformly stable if the matrix P in the inequality (3.4.6) is negative semidefinite;*

(b) *uniformly asymptotically stable if the matrix P in the inequality (3.4.6) is negative definite.*

The proof of this theorem is similar to that of Theorem 3.3.1.

Remark 2.4.2 Theorems 3.4.1–3.4.4 given a series of Corollaries. Below a Corollary of Theorem 3.4.1, is presented being of importance in the next Section.

Corollary 3.4.1 Assume that for system (3.2.2) the matrix-valued functional $U(t, \varphi)$ be constructed such that the functional $V(t, \varphi, \eta)$ is continuous on $R_+ \times C_K$, and locally Lipschitzian in φ and

(i) $\widetilde{w}_1^T(|\varphi(0)|) A \widetilde{w}_1(|\varphi(0)|) \leq V(t, \varphi, \eta) \leq \widetilde{w}_2^T(\|\varphi\|) B \widetilde{w}_2(\|\varphi\|)$, where A, B are constant $m \times m$ matrices and $\widetilde{w}_1$, $\widetilde{w}_2$ are wedge functions;

(ii) $D^+ V(t, \varphi, \eta)|_{(3.2.2)} \leq -\widetilde{w}_3^T(|\varphi(0)|) \widetilde{D} \widetilde{w}_3(|\varphi(0)|)$ for all $(t, \varphi) \in R_+ \times C_K$, where $\widetilde{D}$ is a constant $m \times m$ matrix;

(iii) $\lambda_m(A) > 0$, $\lambda_M(B) > 0$, $\lambda_M(\widetilde{D}) > 0$.

Then the zero solution of system (3.2.2) is uniformly asymptotically stable.

3.4.5 Some applications Consider a linear delay system consisting of two interconnected subsystems. To analyse stability of its zero solution we shall apply Approach A_1. Let the system be of the form

(3.4.7)
$$\dot{x}_1 = A_1 x_1(t) + B_1 x_2(t) + C_1 x_1(t - \tau) + D_1 x_2(t - \tau),$$
$$\dot{x}_2 = A_2 x_1(t) + B_2 x_2(t) + C_2 x_1(t - \tau) + D_2 x_2(t - \tau),$$

where $\tau > 0$, $x_1 \in R^{m_1}$, $x_2 \in R^{m_2}$, $m_1 + m_2 = n$, A_i, B_i, C_i, D_i, $i = 1, 2$, are constant matrices of the corresponding dimensions.

Independent subsystems of system (3.4.7) are

(3.4.8)
$$\dot{x}_1 = A_1 x_1(t) + C_1 x_1(t - \tau),$$
$$\dot{x}_2 = B_2 x_2(t) + D_2 x_2(t - \tau).$$

3. FUNCTIONAL DIFFERENTIAL SYSTEMS

For system (3.4.8) we construct the matrix-valued functional $U(t, \varphi)$ with the elements

(3.4.9)
$$v_{11} = x_1^T(t)P_{11}x_1(t) + \int_{-\tau}^{0} x_1^T(t+s)P_{11}x_1(t+s)\,ds,$$
$$v_{22} = x_2^T(t)P_{22}x_2(t) + \int_{-\tau}^{0} x_2^T(t+s)P_{22}x_2(t+s)\,ds,$$
$$v_{12} = x_1^T(t)P_{12}x_2(t) + \int_{-\tau}^{0} x_1^T(t+s)P_{12}x_2(t+s)\,ds,$$

which satisfy estimates characteristic for the quadratic forms

(a) $\lambda_m(P_{11})|x_1(t)|^2 + \int_{-\tau}^{0} \lambda_m(P_{11})|x_1(t+s)|^2\,ds \leq v_{11}(t,\varphi_1)$

$\leq \lambda_M(P_{11})|x_1(t)|^2 + \int_{-\tau}^{0} \lambda_M(P_{11})|x_1(t+s)|^2\,ds,$

(b) $\lambda_m(P_{22})|x_2(t)|^2 + \int_{-\tau}^{0} \lambda_m(P_{22})|x_2(t+s)|^2, ds \leq v_{22}(t,\varphi_2)$ (3.4.10)

$\leq \lambda_M(P_{22})|x_2(t)|^2 + \int_{-\tau}^{0} \lambda_M(P_{22})|x_2(t+s)|^2\,ds,$

(c) $-\lambda_M^{1/2}(P_{12}P_{12}^T)|x_1(t)|\,|x_2(t)|$

$-\int_{-\tau}^{0} \lambda_M^{1/2}(P_{12}P_{12}^T)|x_1(t+s)|\,|x_2(t+s)|\,ds$

$\leq v_{12}(t,\varphi_1,\varphi_2) \leq \lambda_M(P_{12}P_{12}^T)|x_1(t)|\,|x_2(t)|$

$+\int_{-\tau}^{0} \lambda_M(P_{12}P_{12}^T)|x_1(t+s)|\,|x_2(t+s)|\,ds.$

Here P_{11}, P_{22} are symmetric positive definite matrices and P_{12} is a constant matrix.

It is easy to verify that for the functional

(3.4.11) $$V(t, x, \eta) = \eta^{\mathrm{T}} U(t, x(s))\eta, \quad \eta = (1, 1)^{\mathrm{T}},$$

the bilateral estimate

(3.4.12)
$$u_0^{\mathrm{T}}(t) H^{\mathrm{T}} A H u_0(t) + \int_{-\tau}^{0} u_0^{\mathrm{T}}(t+s) H^{\mathrm{T}} A H u_0(t+s)\, ds \leq V(t, x, \eta)$$
$$\leq u_0^{\mathrm{T}}(t) H^{\mathrm{T}} B H u_0(t) + \int_{-\tau}^{0} u_0^{\mathrm{T}}(t+s) H^{\mathrm{T}} B H u_0(t+s)\, ds,$$

where

$$u_0^{\mathrm{T}}(t) = (x_1^{\mathrm{T}}(t), x_2^{\mathrm{T}}(t)), \quad u_0^{\mathrm{T}}(t+s) = (x_1^{\mathrm{T}}(t+s), x_2^{\mathrm{T}}(t+s)),$$
$$H^{\mathrm{T}} = H = \operatorname{diag}(1, 1),$$

$$A = \begin{pmatrix} \lambda_m(P_{11}) & -\lambda_M^{1/2}(P_{12}P_{12}^{\mathrm{T}}) \\ -\lambda_M^{1/2}(P_{12}P_{12}^{\mathrm{T}}) & \lambda_m(P_{22}) \end{pmatrix},$$

$$B = \begin{pmatrix} \lambda_M(P_{11}) & \lambda_M^{1/2}(P_{12}P_{12}^{\mathrm{T}}) \\ \lambda_M^{1/2}(P_{12}P_{12}^{\mathrm{T}}) & \lambda_M(P_{22}) \end{pmatrix}.$$

It follows from estimate (3.4.12) that for functional (3.4.11) to be positive definite it is sufficient that the matrix $H^{\mathrm{T}} A H$ to be positive definite.

For the upper right derivative of functional (3.4.11) along solutions of system (3.4.7) one can easily obtain the estimate

(3.4.13) $$D^+ V(t, x(s))\big|_{(7)} \leq u^{\mathrm{T}} G u$$

where the following notations are used

$$u^{\mathrm{T}} = (|u_1(t, \cdot)|, |u_2(t, \cdot)|),$$

$$G = \begin{pmatrix} \lambda_M(C) + 2\lambda_M(F) & \lambda_M^{1/2}(KK^{\mathrm{T}}) \\ \lambda_M^{1/2}(KK^{\mathrm{T}}) & \lambda_M(D) + 2\lambda_M(L) \end{pmatrix},$$

where

$$C = \begin{pmatrix} \lambda_M(A_1^T P_{11} + P_{11}A_1 + P_{11}) & \lambda_M^{1/2}[(P_{11}C_1)(P_{11}C_1)^T] \\ \lambda_M^{1/2}[(P_{11}C_1)(P_{11}C_1)^T] & \lambda_M(-P_{11}) \end{pmatrix},$$

$$D = \begin{pmatrix} \lambda_M(B_2^T P_{22} + P_{22}B_2 + P_{22}) & \lambda_M^{1/2}[(P_{22}D_2)(P_{22}D_2)^T] \\ \lambda_M^{1/2}[(P_{22}D_2)(P_{22}D_2)^T] & \lambda_M(-P_{22}) \end{pmatrix},$$

$$F = \begin{pmatrix} \lambda_M(P_{12}A_2) & \frac{1}{2}\lambda_M^{1/2}((P_{12}C_2)(P_{12}C_2)^T) \\ \frac{1}{2}\lambda_M^{1/2}((P_{12}C_2)(P_{12}C_2)^T) & 0 \end{pmatrix},$$

$$K = \begin{pmatrix} \lambda_M^{1/2}(QQ^T) & \lambda_M^{1/2}(YY^T) \\ \lambda_M^{1/2}(WW^T) & 0 \end{pmatrix},$$

$$L = \begin{pmatrix} \lambda_M(B_1^T P_{12}) & \lambda_M^{1/2}((P_{12}^T D_1)(P_{12}^T D_1)^T) \\ \lambda_M^{1/2}((P_{12}^T D_1)(P_{12}^T D_1)^T) & 0 \end{pmatrix},$$

$$Q = P_{11}B_1 + A_2 P_{22}^T + A_1^T P_{12} + P_{12}B_2,$$
$$Y = P_{11}D_1 + P_{12}D_2,$$
$$W = C_2^T P_{22} + C_1^T P_{12}.$$

Estimates (3.4.12) and (3.4.13) yield stability criterion for the state $x = 0$ of system (3.4.7) formulated below.

Theorem 3.4.5 *Assume that for system (3.4.7) the matrix-valued functional $U(t, \varphi)$ is constructed with components (3.4.9). If in estimate (3.4.12) the matrices A and B are positive definite and in estimate (3.4.13) the matrix G is negative definite then the zero solution of system (3.4.7) is uniformly asymptotically stable.*

The proof of this theorem follows from Theorem 3.4.1.

Remark 3.4.3 It is easy to show that conditions of Theorem 3.4.5 are fulfilled provided that
(a) $\lambda_M(P_{11})\lambda_M(P_{22}) > \lambda_M(P_{12}P_{12}^T)$,
(b) $\lambda_M(C) + 2\lambda_M(F) < 0$,
(c) $\lambda_M(D) + 2\lambda_M(L) < 0$,
(d) $(\lambda_M(C) + 2\lambda_M(F))(\lambda_M(D) + 2\lambda_M(L)) > \lambda_M(KK^T)$.

Further, to study system (3.4.7) we apply Approach B_2 and the functional $U(\varphi)$ with the elements (3.4.9). Let $\eta \in R_+^2$, $\eta = (1,1)^T$. Then, in the framework of this approach one need to study sign-definiteness of the upper right derivative

$$D^+V(x,\eta)|_{(3.4.7)} = D^+v_{11}(x_1)|_{(3.4.7)} + D^+v_{22}(x_2)|_{(3.4.7)} \\ + 2D^+v_{12}(x_1,x_2)|_{(3.4.7)}.$$
(3.4.14)

Having accomplished simple transformations in the expression (3.4.14) we get the estimate

(3.4.15) $$D^+V(x,\eta)|_{(3.4.7)} \leq u^T S u,$$

where $u^T = (|u_1|, |u_2|)$ and

$$u_1^T = (|x_1(t)|, |x_1(t-\tau)|), \quad u_2^T = (|x_2(t)|, |x_2(t-\tau)|),$$

$$S = \begin{pmatrix} \lambda_M^{1/2}(C) & \lambda_M^{1/2}(KK^T) \\ \lambda_M^{1/2}(KK^T) & \lambda_M^{1/2}(D) \end{pmatrix},$$

$$C = \begin{pmatrix} c_{11} & c_{12} \\ c_{21} & c_{22} \end{pmatrix}, \quad D = \begin{pmatrix} d_{11} & d_{12} \\ d_{21} & d_{22} \end{pmatrix}, \quad K = \begin{pmatrix} k_{11} & k_{12} \\ k_{21} & k_{22} \end{pmatrix},$$

$$c_{11} = \lambda_M(A_1^T P_{11} + P_{11}A_1 + P_{12}A_2 + A_2^T P_{12} + P_{11}),$$

$$c_{21} = c_{12} = \lambda_M^{1/2}((P_{11}C_1 + P_{12}C_2)(P_{11}C_1 + P_{12}C_2)^T),$$

$$c_{22} = \lambda_M(-P_{11});$$

$$d_{11} = \lambda_M(B_2^T P_{22} + P_{22}B_2 + B_1 P_{12} + P_{12}^T B_1^T + P_{22}),$$

$$d_{21} = d_{12} = \lambda_M^{1/2}((P_{22}D_2 + P_{12}D_1)(P_{22}D_2 + P_{12}D_1)^T),$$

$$d_{22} = \lambda_M(-P_{22});$$

$$k_{11} = \lambda_M^{1/2}\left(\left(P_{11}B_1 + A_2^T P_{22} + A_1^T P_{12} + P_{12}B_2 + \frac{1}{2}P_{12}\right) \right. \\ \left. \times \left(P_{11}B_1 + A_2^T P_{22} + A_1^T P_{12} + P_{12}B_2 + \frac{1}{2}P_{12}\right)^T\right),$$

$$k_{12} = \lambda_M^{1/2}((P_{11}D_1 + P_{12}D_2)(P_{11}D_1 + P_{12}D_2)^T),$$

$$k_{21} = \lambda_M^{1/2}((P_{22}C_2 + P_{12}C_1)(P_{22}C_2 + P_{12}C_1)^T),$$

$$k_{22} = \frac{1}{2}\lambda_M^{1/2}(P_{12}P_{12}^T).$$

Estimates (3.4.6) and (3.4.15) provide the following result.

3. FUNCTIONAL DIFFERENTIAL SYSTEMS

Theorem 3.4.6 *Assume that for system (3.4.7) the matrix-valued functional $U(\varphi)$ is constructed with components (3.4.9). If in estimate (3.4.10) the matrices A and B are positive definite and in estimate (3.4.15) the matrix S is negative definite, then the zero solution of system (3.4.7) is uniformly asymptotically stable.*

Remark 3.4.4 It is easy to verify that conditions of Theorem 3.4.6 are satisfied provided that

(a) $\lambda_M(P_{11})\lambda_M(P_{22}) > \lambda_M(P_{12}P_{12}^T)$,

(b) $\lambda_M(C) < 0$,

(c) $\lambda_M(D) < 0$,

(d) $\lambda_M(C)\lambda_M(D) > \lambda_M(KK^T)$.

3.5 An Approach to Construction of Liapunov–Krasovskii Functionals

Consider the autonomous system with finite delay

$$(3.5.1) \qquad \frac{dx}{dt} = F(x, x_t), \quad x_{t_0} = \varphi_0 \in \mathcal{C}, \quad t_0 \geq 0,$$

where $x \in R^n$, $F \in C(R^n \times \mathcal{C}, R^n)$, which has the linear approximation

$$(3.5.2) \qquad \frac{dx}{dt} = Ax(t) + Bx(t-r) + f(x, x_t).$$

Here A, B are constant $n \times n$ matrices, $x(t)$ is n–dimensional vector, $r \geq 0$, $f \in C(R^n \times \mathcal{C}, R^n)$. The linear approximation of system (3.5.2)

$$(3.5.3) \qquad \frac{dx}{dt} = Ax(t) + Bx(t-r)$$

is decomposed into two subsystems

$$(3.5.4) \quad \begin{aligned} \frac{dx_1}{dt} &= A_{11}x_1(t) + A_{12}(t)x_2(t) + B_{11}x_1(t-r) + B_{12}(t)x_2(t-r), \\ \frac{dx_2}{dt} &= A_{21}(t)x_1(t) + A_{22}x_2(t) + B_{21}(t)x_1(t-r) + B_{22}x_2(t-r), \end{aligned}$$

where $x_i \in R^{n_i}$, $i = 1, 2$, $(x_1^T, x_2^T)^T = x$, A_{ij} and B_{ij} are matrices of corresponding dimensions for which the subsystems

(3.5.5)
$$\frac{dx_1}{dt} = A_{11}x_1(t) + B_{11}x_1(t - r)$$
$$\frac{dx_2}{dt} = A_{22}x_2(t) + B_{22}x_2(t - r).$$

are independent.

Now we consider autonomous linear system (3.5.3). Assume that for subsystems (3.5.5) the functionals

(3.5.6)
$$v_{11}(\varphi_1) = \varphi_1^T(0) P_{11} \varphi_1(0) + 2\varphi_1^T(0) \int_{-r}^{0} K_1(\theta) \varphi_1(\theta)\, d\theta$$
$$+ \int_{-r}^{0} \varphi_1^T(\theta) \Gamma_1(\theta) \varphi_1(\theta)\, d\theta + \int_{-r}^{0} \int_{-r}^{0} \varphi_1^T(\xi) \gamma_1(\xi, \eta) \varphi_1(\eta)\, d\xi\, d\eta,$$
$$v_{22}(\varphi_2) = \varphi_2^T(0) P_{22} \varphi_2(0) + 2\varphi_2^T(0) \int_{-r}^{0} K_2(\theta) \varphi_2(\theta)\, d\theta$$
$$+ \int_{-r}^{0} \varphi_2^T(\theta) \Gamma_2(\theta) \varphi_2(\theta)\, d\theta + \int_{-r}^{0} \int_{-r}^{0} \varphi_2^T(\xi) \gamma_2(\xi, \eta) \varphi_2(\eta)\, d\xi\, d\eta,$$

are constructed somehow, where P_{11}, P_{22} are constant symmetric positive definite matrices,

$$K_1, \Gamma_1 \in C([-r, 0], R^{n_1 \times n_1}), \quad K_2, \Gamma_2 \in C([-r, 0], R^{n_2 \times n_2}),$$
$$\gamma_1 \in C([-r, 0] \times [-r, 0], R^{n_1 \times n_1}), \quad \gamma_2 \in C([-r, 0] \times [-r, 0], R^{n_2 \times n_2}).$$

Further we employ the idea of approximation of system (3.5.4) by system of difference equations. With this in mind we divide the segment $[-r, 0]$ into N equal parts of length h, i.e. $Nh = r$; the derivatives $\frac{dx_i}{dt}$, $i = 1, 2$, are approximated by the differences $(x_i(t + h) - x_i(t))h^{-1}$. The system of

3. FUNCTIONAL DIFFERENTIAL SYSTEMS

difference equations corresponding to system (3.5.5) is (cf. Hale [1])*

(3.5.7)
$$\begin{aligned}
\widetilde{x}_{11}(\tau+1) &= (I_{n_1} + hA_{11})\widetilde{x}_{11}(\tau) + hB_{11}\widetilde{x}_{1N}(t) \\
&\quad + hA_{12}\widetilde{x}_{21}(t) + hB_{12}\widetilde{x}_{2N}(t) \\
\widetilde{x}_{12}(\tau+1) &= \widetilde{x}_{11}(\tau) \\
&\cdots\cdots\cdots\cdots\cdots\cdots\cdots\cdots\cdots\cdots \\
\widetilde{x}_{1N}(\tau+1) &= \widetilde{x}_{1N-1}(\tau) \\
\widetilde{x}_{21}(\tau+1) &= (I_{n_2} + hA_{22})\widetilde{x}_{21} + hB_{22}\widetilde{x}_{2N}(\tau) \\
&\quad + hA_{12}\widetilde{x}_{21}(\tau) + hB_{12}\widetilde{x}_{2N}(\tau) \\
\widetilde{x}_{22}(\tau+1) &= \widetilde{x}_{21}(\tau) \\
&\cdots\cdots\cdots\cdots\cdots\cdots\cdots\cdots\cdots\cdots \\
\widetilde{x}_{2N}(\tau+1) &= \widetilde{x}_{2N-1}(\tau),
\end{aligned}$$

where I_{n_1}, I_{n_2} are identity matrices of the corresponding dimensions.

The point $\widetilde{x}_i(0) = (\varphi_i(0), \varphi_i(-h), \ldots, \varphi_i(-Nh))^{\mathrm{T}}$ corresponds to the initial function specifying solution $\widetilde{x} = (\widetilde{x}_1^{\mathrm{T}}, \widetilde{x}_2^{\mathrm{T}})^{\mathrm{T}}$ of system of difference equations (3.5.7).

Further we present system (3.5.7) in matrix form

(3.5.8)
$$\begin{aligned}
\widetilde{x}_1(\tau+1) &= \widetilde{A}_{11}\widetilde{x}_1(\tau) + \widetilde{A}_{12}\widetilde{x}_2(\tau) \\
\widetilde{x}_2(\tau+1) &= \widetilde{A}_{21}\widetilde{x}_1(\tau) + \widetilde{A}_{22}\widetilde{x}_2(\tau),
\end{aligned}$$

where $\widetilde{x}_1 \in R^{n_1(N+1)}$, $\widetilde{x}_2 \in R^{n_2(N+1)}$ and

$$\widetilde{A}_{11} = \begin{pmatrix} I_{n_1} + hA_{11} & O_{n_1} & \cdots & O_{n_1} & hB_{11} \\ I_{n_1} & O_{n_1} & \cdots & O_{n_1} & O_{n_1} \\ \cdots & \cdots & \cdots & \cdots & \cdots \\ O_{n_1} & \cdots & \cdots & I_{n_1} & O_{n_1} \end{pmatrix},$$

$$\widetilde{A}_{22} = \begin{pmatrix} I_{n_2} + hA_{22} & O_{n_2} & \cdots & O_{n_2} & hB_{22} \\ I_{n_2} & O_{n_2} & \cdots & O_{n_2} & O_{n_2} \\ \cdots & \cdots & \cdots & \cdots & \cdots \\ O_{n_2} & \cdots & \cdots & I_{n_2} & O_{n_2} \end{pmatrix},$$

$$\widetilde{A}_{12} = \begin{pmatrix} hA_{12} & O_{n_1 \times n_2} & \cdots & hB_{12} \\ O_{n_1 \times n_2} & O_{n_1 \times n_2} & \cdots & O_{n_1 \times n_2} \\ \cdots & \cdots & \cdots & \cdots \\ O_{n_1 \times n_2} & \cdots & \cdots & O_{n_1 \times n_2} \end{pmatrix},$$

---

*It should be noted here that for stability analysis of the zero solution of system (3.5.1) with decomposition (3.5.4) a formal approach presented by Hale [1], p. 111–115, is employed.

$$\tilde{A}_{21} = \begin{pmatrix} hA_{21} & O_{n_2 \times n_1} & \cdots & hB_{21} \\ O_{n_2 \times n_1} & O_{n_2 \times n_1} & \cdots & O_{n_2 \times n_1} \\ \cdots & \cdots & \cdots & \cdots \\ O_{n_2 \times n_1} & \cdots & \cdots & O_{n_2 \times n_1} \end{pmatrix}.$$

Let k be arbitrary number, then vector $(\tilde{x}_1^T(kh), x_2^T(kh))^T$ is a phase vector for system (3.5.8) for any $t = kh$. For sufficiently small h vector $\tilde{x}_i(kh)$, $i = 1, 2$, is an exact enough approximation of solutions of system (3.5.5) at points kh, $k = 0, -1, \ldots, -N$.

Functionals $v_{11}(\varphi_1)$ and $v_{22}(\varphi_2)$ are approximated by the quadratic forms

(3.5.9) $\qquad \tilde{v}_{11}(\tilde{x}_1) = \tilde{x}_1^T \tilde{P}_{11} \tilde{x}_1, \quad \tilde{v}_{22}(\tilde{x}_2) = \tilde{x}_2^T \tilde{P}_{22} \tilde{x}_2,$

where

$$\tilde{P}_{11} = \begin{pmatrix} NP_{11} & k_{11}^T & k_{12}^T & \cdots & k_{1N}^T \\ k_{11} & \alpha_{11}^1 & \alpha_{12}^1 & \cdots & \alpha_{1N}^1 \\ k_{12} & \alpha_{12}^1 & \alpha_{22}^1 & \cdots & \alpha_{2N}^1 \\ \cdots & \cdots & \cdots & \cdots & \cdots \\ k_{1N} & \alpha_{1N}^1 & \alpha_{2N}^1 & \cdots & \alpha_{NN}^1 \end{pmatrix},$$

$$\tilde{P}_{22} = \begin{pmatrix} NP_{22} & k_{21}^T & k_{22}^T & \cdots & k_{2N}^T \\ k_{21} & \alpha_{11}^2 & \alpha_{12}^2 & \cdots & \alpha_{1N}^2 \\ k_{22} & \alpha_{12}^2 & \alpha_{22}^2 & \cdots & \alpha_{2N}^2 \\ \cdots & \cdots & \cdots & \cdots & \cdots \\ k_{2N}^2 & \alpha_{1N}^2 & \alpha_{2N}^2 & \cdots & \alpha_{NN}^2 \end{pmatrix}.$$

Here the constant matrices P_{11}, P_{22}, k_{ji}, α_{ij}^j, $i = 1, 2, \ldots, N$, $j = 1, 2$, of the corresponding dimensions are determined as

$$k_{ji} = K_j(-hi), \quad \alpha_{ii}^j = \Gamma_j(-ih), \quad i = 1, 2, \ldots, N, \quad j = 1, 2,$$
$$\alpha_{ij}^1 = \gamma_1(-hi, -hj), \quad \alpha_{ij}^2 = \gamma_2(-hi, -hj), \quad i, j = 1, 2, \ldots, N, \quad i \neq j.$$

We use equation (3.4) in order to construct the non-diagonal element $v_{12}(\tilde{x}_1, \tilde{x}_2)$ of the matrix-valued functional $U(\tilde{x}_1, \tilde{x}_2)$ in the bilinear form

(3.5.10) $\qquad v_{12}(\tilde{x}_1, \tilde{x}_2) = \tilde{x}_1^T \tilde{P}_{12} \tilde{x}_2,$

where matrix $\tilde{P}_{12}$ satisfies the equation

(3.5.11) $\qquad \tilde{A}_{11}^T \tilde{P}_{12} \tilde{A}_{22} - \tilde{P}_{12} = -\dfrac{\eta_1}{\eta_2} \tilde{A}_{11}^T \tilde{P}_{11} \tilde{A}_{12} - \dfrac{\eta_2}{\eta_1} \tilde{A}_{21}^T \tilde{P}_{22} \tilde{A}_{22}$

3. FUNCTIONAL DIFFERENTIAL SYSTEMS

and has the form

$$\widetilde{P}_{12} = \begin{pmatrix} NP_{12} & s_1^2 & s_2^2 & \ldots & s_N^2 \\ s_1^1 & q_{11} & q_{12} & \ldots & q_{1N} \\ s_2^1 & q_{21} & q_{22} & \ldots & q_{2N} \\ \cdots & \cdots & \cdots & \cdots & \cdots \\ s_N^1 & q_{N1} & q_{N2} & \ldots & q_{NN} \end{pmatrix}.$$

Here P_{12}, s_i^j, q_{ij} are matrices and η_1, η_2 are positive constants.

In terms of equation (3.5.11) we get

(3.5.12)
$$(NP_{12} + rA_{11}^T P_{12} + s_1^1)(I_{n_2} + hA_{22}) + s_1^2 + hA_{11}^T s_1^2 + q_{11} - NP_{12}$$
$$= -\frac{r\eta_1}{\eta_2} P_{11} A_{12} - hr\frac{\eta_1}{\eta_2} A_{11} P_{11} A_{12} - h\frac{\eta_1}{\eta_2} k_{11} A_{11} k_{1N}^T A_{12}$$
$$- r\frac{\eta_2}{\eta_1} A_{21}^T P_{22} - r\frac{\eta_1}{\eta_2} hA_{21}^T P_{22} A_{22} - \frac{\eta_2}{\eta_1} hA_{21}^T k_{2N} A_{21}^T k_{21}^T,$$

(3.4.13) $\quad s_i^2 - s_{i-1}^2 + hA_{11}^T s_i^2 + q_{1i} = -\frac{\eta_2}{\eta_1} hA_{21}^T k_{2i}, \quad i = 2, \ldots, N,$

(3.5.14) $\quad hs_i^1 B_{22} - q_{i-1,N} = -\frac{\eta_1}{\eta_2} hk_{1i}^T B_{12}, \quad i = 2, \ldots, N,$

(3.5.15) $\quad hB_{11}^T s_i^2 - q_{N,i-1} = -\frac{\eta_2}{\eta_1} hB_{21}^T k_{2i}, \quad i = 2, \ldots, N,$

(3.5.16)
$$rP_{12}B_{22} + hrA_{11}^T P_{12} B_{22} + hs_1^1 B_{22} - s_N^2 = -\frac{\eta_1}{\eta_2} rP_{11} B_{12}$$
$$- \frac{r\eta_1}{\eta_2} hA_{11}^T P_{11} B_{12} - \frac{\eta_1}{\eta_2} hk_{11} B_{12} - \frac{r\eta_2}{\eta_1} hA_{21}^T P_{22} B_{22},$$

(3.5.17)
$$rB_{11}^T P_{12} + hrB_{11}^T P_{12} A_{22} + hB_{11}^T s_1^2 - s_N^1 = -\frac{r\eta_1}{\eta_2} hB_{11}^T P_{11} A_{12}$$
$$- \frac{r\eta_2}{\eta_1} B_{21}^T P_{22} - \frac{r\eta_2}{\eta_1} hB_{21}^T P_{22} A_{22} - \frac{\eta_2}{\eta_1} hB_{21}^T k_{12}^T,$$

(3.5.18) $\quad s_i^1 + hs_i^1 A_{22} - s_{i-1}^1 + q_{i1} = -hk_{1i} A_{12}, \quad i = 2, \ldots, N,$

(3.5.19) $\quad q_{NN} = h\left(B_{11}^T B_{22} + \frac{\eta_1}{\eta_2} B_{11} B_{12} + \frac{\eta_2}{\eta_1} B_{21} B_{22}\right),$

(3.5.20) $\quad q_{ii} = \text{const}, \quad q_{ij} = q_{i-1, j-1}, \quad i, j = 2, \ldots, N.$

From equation (3.5.12), in view of (3.5.9) and $h \to 0$ we get*

$$(3.5.21) \quad A_{11}^{\mathrm{T}} P_{12} + P_{12} A_{22} = -\frac{\eta_1}{\eta_2} P_{11} A_{12} - \frac{\eta_2}{\eta_1} A_{21}^{\mathrm{T}} P_{22} - \frac{1}{r}(S_1(0) + S_2(0)).$$

Similarly in view of (3.5.20) we get from (3.5.14)

$$(3.5.22) \quad q_{1i} = q_{N+1-i,N} = h(s_{N+2-i}^1 B_{22} + k_{1,N+2-i}^{\mathrm{T}} B_{12}).$$

Then, in view of (3.5.22) equations (3.5.14) imply

$$(3.5.23) \quad \begin{aligned} (s_i^2 - s_{i-1}^2) h^{-1} + A_{11}^{\mathrm{T}} s_i^2 + s_{N+2-i}^1 B_{22} + \frac{\eta_1}{\eta_2} k_{1,N+2-i} B_{12} \\ = -\frac{\eta_2}{\eta_1} A_{21}^{\mathrm{T}} k_{2i}^{\mathrm{T}}, \quad i = 2, \ldots, N, \end{aligned}$$

and passing to the limit as $h \to 0$ we obtain

$$(3.5.24) \quad \begin{aligned} -\frac{dS_2}{d\theta} + A_{11}^{\mathrm{T}} S_2(\theta) + S_1(-r - \theta) B_{22} + \frac{\eta_1}{\eta_2} K_1(-r - \theta) B_{12} \\ = -\frac{\eta_2}{\eta_1} A_{21}^{\mathrm{T}} K_2^{\mathrm{T}}(\theta). \end{aligned}$$

Similarly to the above, in view of (3.5.19) we get from (3.5.24)

$$(3.5.25) \quad \begin{aligned} -\frac{dS_1}{d\theta} + S_1(\theta) A_{22} + B_{11}^{\mathrm{T}} S_2(-r - \theta) + \frac{\eta_2}{\eta_1} B_{21}^{\mathrm{T}} K_2^{\mathrm{T}}(-r - \theta) \\ = -\frac{\eta_1}{\eta_2} K_1(\theta) A_{12}. \end{aligned}$$

Taking into account (3.5.19) we find from (3.5.16) and (3.5.17) for $h \to 0$ the initial conditions

$$(3.5.26) \quad \begin{aligned} S_2(-1) &= r\left(P_{12} B_{22} + \frac{\eta_1}{\eta_2} P_{11} B_{12}\right), \\ S_1(-1) &= r\left(B_{11}^{\mathrm{T}} P_{12} + \frac{\eta_2}{\eta_1} B_{21}^{\mathrm{T}} P_{22}\right). \end{aligned}$$

---

*From here on in formulas (3.5.12)–(3.5.19) and (3.5.23) passage to the limit as $h \to 0$ is formal.

3. FUNCTIONAL DIFFERENTIAL SYSTEMS

In the expression of the bilinear form $\frac{1}{N}v_{12}(\tilde{x}_1,\tilde{x}_2)$ the formal limiting passage $(h \to 0)$ yields the expression for the functional
(3.5.27)

$$v_{12}(\varphi_1,\varphi_2) = \varphi_1^T(0)P_{12}\varphi_2(0) + \frac{1}{r}\varphi_1^T(0)\int_{-r}^{0} S_2(\theta)\varphi_2(\theta)\,d\theta$$

$$+ \frac{1}{r}\varphi_2^T(0)\int_{-r}^{0} S_1^T(\theta)\varphi_1(\theta)\,d\theta + \frac{1}{r}\int_{-r}^{0} d\xi\varphi_1^T(\xi)\int_{-r}^{\xi}\Big\{S_1(\xi-\eta-r)B_{22}$$

$$+ \frac{\eta_1}{\eta_2}K_1^T(\xi-\eta-r)B_{12}\Big\}\varphi_2(\eta)\,d\eta + \frac{1}{r}\int_{-r}^{0} d\xi\varphi_1^T(\xi)\int_{\xi}^{0}\Big\{B_{11}^T S_2(\eta-\xi-r)$$

$$+ \frac{\eta_1}{\eta_2}B_{21}^T K_2(\eta-\xi-r)\Big\}\varphi_2(\eta)\,d\eta.$$

In order to formulate stability conditions for system (3.5.4) in terms of the matrix-valued functional $U(\varphi_1,\varphi_2)$ with components (3.5.6) and (3.5.27) it is necessary to estimate its and their upper right derivative numbers along solutions of system (3.5.4). To this end we define concretely the choice of functionals (3.5.6) as

(3.5.28) $\quad v_{11}(\varphi_1) = \varphi_1^T(0)P_{11}\varphi_1(0) + \int_{-r}^{0} k(\theta)\varphi_1^T(\theta)D_1\varphi_1(\theta)\,d\theta,$

(3.5.29) $\quad v_{22}(\varphi_2) = \varphi_2^T(0)P_{22}\varphi_2(0) + \int_{-r}^{0} k(\theta)\varphi_2^T(\theta)D_2\varphi_2(\theta)\,d\theta,$

where P_{11}, P_{22}, D_1 and D_2 are positive definite matrices of the corresponding dimensions and $k(\theta) = 1 + \frac{1}{2r}\theta$.

Basing on the system of equations

(3.5.30) $\quad \dfrac{dS_2}{d\theta} = A_{11}^T S_2(\theta) + S_1(-r-\theta)B_{22},$

(3.5.31) $\quad \dfrac{dS_1}{d\theta} = S_1(\theta)A_{22} + B_{11}^T S_2(-r-\theta),$

(3.5.32) $\quad A_{11}^T P_{12} + P_{12}A_{22} = -\dfrac{\eta_1}{\eta_2}P_{11}A_{12} - \dfrac{\eta_2}{\eta_1}A_{21}^T P_{22} - \dfrac{1}{r}(S_1(0)+S_2(0))$

under initial conditions

(3.5.33)
$$S_2(-r) = r\left(P_{12}B_{22} + \frac{\eta_1}{\eta_2}P_{11}B_{12}\right),$$
$$S_1(-r) = r\left(B_{11}^T P_{12} + \frac{\eta_2}{\eta_1}B_{21}^T P_{22}\right),$$

where $P_{12} \in R^{n_1 \times n_2}$, $S_1, S_2 \in C^1([-r,0], R^{n_1 \times n_2})$, η_1, η_2 are positive constants we construct functional $v_{12}(\varphi_1, \varphi_2)$ in the form
(3.5.34)
$$v_{12}(\varphi_1,\varphi_2) = \varphi_1^T(0)P_{12}\varphi_2(0) + \frac{1}{r}\varphi_1^T(0)\int_{-r}^{0} S_2(\theta)\varphi_2(\theta)\,d\theta$$
$$+ \frac{1}{r}\varphi_2^T(0)\int_{-r}^{0} S_1^T(\theta)\varphi_1(\theta)\,d\theta + \frac{1}{r}\int_{-r}^{0} d\xi\,\varphi_1^T(\xi)\int_{-r}^{\xi} S_1(\xi-\eta-r)B_{22}\varphi_2(\eta)\,d\eta$$
$$+ \frac{1}{r}\int_{-r}^{0} d\xi\,\varphi_1^T(\xi)\int_{\xi}^{0} B_{11}^T S_2(\eta-\xi-r)\varphi_2(\eta)\,d\eta.$$

Since for the functionals $v_{ij}(\cdot)$, $i,j = 1,2$, the lower estimates
(3.5.35)
$$v_{11}(\varphi_1) \geq \lambda_m(P_{11})|\varphi_1(0)|^2 + \frac{1}{2}\lambda_m(D_1)\|\varphi_1\|_{L_2}^2$$
$$v_{22}(\varphi_2) \geq \lambda_m(P_{22})|\varphi_2(0)|^2 + \frac{1}{2}\lambda_m(D_2)\|\varphi_2\|_{L_2}^2$$
$$v_{12}(\varphi_1,\varphi_2) \geq -\|P_{12}\||\varphi_1(0)\||\varphi_2(0)| - \varkappa_2|\varphi_1(0)|\|\varphi_2\|_{L_2}$$
$$- \varkappa_1|\varphi_2(0)|\|\varphi_1\|_{L_2} - (\varkappa_{21}\|B_{11}\| + \varkappa_{12}\|B_{22}\|)\|\varphi_1\|_{L_2}\|\varphi_2\|_{L_2},$$

and the upper estimates
(3.5.36)
$$v_{11}(\varphi_1) \leq \lambda_M(P_{11})|\varphi_1(0)|^2 + \frac{1}{2}\lambda_M(D_1)\|\varphi_1\|_{L_2}^2$$
$$v_{22}(\varphi_2) \leq \lambda_M(P_{22})|\varphi_2(0)|^2 + \frac{1}{2}\lambda_M(D_2)\|\varphi_1\|_{L_2}^2$$
$$v_{12}(\varphi_1,\varphi_2) \leq \|P_{12}\||\varphi_1(0)\||\varphi_2(0)| + \varkappa_2|\varphi_1(0)|\|\varphi_2\|_{L_2}$$
$$+ \varkappa_{12}|\varphi_2(0)|\|\varphi_1\|_{L_2} + (\varkappa_{21}\|B_{11}\| + \varkappa_{12}\|B_{22}\|)\|\varphi_1\|_{L_2}\|\varphi_2\|_{L_2},$$

are satisfied, where

$$\varkappa_1 = \frac{1}{r}\left\{\int_{-r}^{0}\|S_1(\theta)\|^2 d\theta\right\}^{1/2}, \quad \varkappa_2 = \frac{1}{r}\left\{\int_{-r}^{0}\|S_2(\theta)\|^2 d\theta\right\}^{1/2},$$

$$\varkappa_{12} = \frac{1}{r}\left\{\int_{-r}^{0}\int_{-r}^{0}\|S_1(\xi-\eta-r)\|^2 d\xi d\eta\right\}^{1/2},$$

$$\varkappa_{21} = \frac{1}{r}\left\{\int_{-r}^{0}\int_{-r}^{0}\|S_2(\xi-\eta-r)\|^2 d\xi d\eta\right\}^{1/2},$$

then for the functional

$$v(\varphi_1,\varphi_2,\eta) = \eta^T U(\varphi_1,\varphi_2)\eta = \eta_1^2 v_{11}(\varphi_1) + 2\eta_1\eta_2 v_{12}(\varphi_1,\varphi_2) + \eta_2^2 v_{22}(\varphi_2)$$

the bilateral estimate

(3.5.37) $$u^T H^T \underline{C} H u \leq v(\varphi_1,\varphi_2,\eta) \leq u^T H^T \overline{C} H u,$$

is valid, where

$$u = (|\varphi_1(0)|, |\varphi_2(0)|, \|\varphi_1\|_{L_2}, \|\varphi_2\|_{L_2}),$$
$$H = \mathrm{diag}\,(\eta_1,\eta_2,\eta_1,\eta_2), \quad \zeta = \varkappa_{21}\|B_{11}\| + \varkappa_{12}\|B_{22}\|,$$

$$\overline{C} = \begin{pmatrix} \lambda_M(P_{11}) & \|P_{12}\| & 0 & \varkappa_2 \\ \|P_{12}\| & \lambda_M(P_{22}) & \varkappa_1 & 0 \\ 0 & \varkappa_1 & \lambda_M(D_1) & \zeta \\ \varkappa_2 & 0 & \zeta & \lambda_M(D_2) \end{pmatrix},$$

$$\underline{C} = \begin{pmatrix} \lambda_m(P_{11}) & -\|P_{12}\| & 0 & -\varkappa_2 \\ -\|P_{12}\| & \lambda_m(P_{22}) & -\varkappa_1 & 0 \\ 0 & -\varkappa_1 & \frac{1}{2}\lambda_m(D_1) & -\zeta \\ -\varkappa_2 & 0 & -\zeta & \frac{1}{2}\lambda_m(D_2) \end{pmatrix}.$$

Further together with functionals (3.5.28), (3.5.29) and (3.5.34) we use the upper right derivative numbers $D^+ v_{ij}(\cdot)|_{(3.5.4)}$, $i,j = 1,2$:

$$D^+ v_{11}(\varphi_1)|_{(3.5.4)} = \varphi_1^{\mathrm{T}}(0)(A_{11}^{\mathrm{T}} P_{11} + P_{11} A_{11} + D_1)\varphi_1(0)$$
(3.5.38)
$$- \frac{1}{2}\varphi_1^{\mathrm{T}}(-r) D_1 \varphi_1(-r) + \varphi_1^{\mathrm{T}}(0) P_{11} B_{11} \varphi_1(-r) + \varphi_1^{\mathrm{T}}(0) P_{11} A_{12} \varphi_2(0)$$
$$+ \varphi_1^{\mathrm{T}}(0) P_{11} B_{12} \varphi_2(-r) - \frac{1}{2r} \int_{-r}^{0} \varphi_1^{\mathrm{T}}(\theta) D_1 \varphi_1(\theta)\, d\theta,$$

$$D^+ v_{22}(\varphi_2)|_{(3.5.4)} = \varphi_2^{\mathrm{T}}(0)(A_{22}^{\mathrm{T}} P_{22} + P_{22} A_{22} + D_2)\varphi_2(0)$$
(3.5.39)
$$- \frac{1}{2}\varphi_2^{\mathrm{T}}(-r) D_1 \varphi_2(-r) + \varphi_2^{\mathrm{T}}(0) P_{22} B_{22} \varphi_2(-r) + \varphi_2^{\mathrm{T}}(0) P_{22} A_{21} \varphi_1(0)$$
$$+ \varphi_2^{\mathrm{T}}(0) P_{22} B_{21} \varphi_1(-r) - \frac{1}{2r} \int_{-r}^{0} \varphi_2^{\mathrm{T}}(\theta) D_2 \varphi_2(\theta)\, d\theta,$$

$$D^+ v_{12}(\varphi_1, \varphi_2)|_{(3.5.4)} = \varphi_1^{\mathrm{T}}(0)\left(A_{11}^{\mathrm{T}} P_{12} + P_{12} A_{22} + \frac{1}{r}(S_1(0) + S_2(0))\right)\varphi_2(0)$$
$$+ \frac{1}{2}\varphi_1^{\mathrm{T}}(0)(P_{12} A_{21} + A_{21}^{\mathrm{T}} P_{12}^{\mathrm{T}})\varphi_1(0) + \frac{1}{2}\varphi_2^{\mathrm{T}}(0)(A_{12}^{\mathrm{T}} P_{12} + P_{12}^{\mathrm{T}} A_{12})\varphi_2(0)$$
$$+ \varphi_1^{\mathrm{T}}(0) P_{12} B_{21} \varphi_1(-r) + \varphi_2^{\mathrm{T}}(-r) B_{12}^{\mathrm{T}} P_{12} \varphi_2(0)$$
(3.5.40)
$$+ \frac{1}{r}\varphi_2^{\mathrm{T}}(-r) B_{12}^{\mathrm{T}} \int_{-r}^{0} S_2(\theta)\varphi_2(\theta)\, d\theta + \frac{1}{r}\varphi_2^{\mathrm{T}}(0) A_{12}^{\mathrm{T}} \int_{-r}^{0} S_2(\theta)\varphi_2(\theta)\, d\theta$$
$$+ \frac{1}{r}\varphi_1^{\mathrm{T}}(0) A_{21}^{\mathrm{T}} \int_{-r}^{0} S_1^{\mathrm{T}}(\theta)\varphi_1(\theta)\, d\theta + \frac{1}{r}\varphi_1^{\mathrm{T}}(-r) B_{21}^{\mathrm{T}} \int_{-r}^{0} S_1^{\mathrm{T}}(\theta)\varphi_1(\theta)\, d\theta$$
$$- \frac{\eta_1}{\eta_2}\varphi_1^{\mathrm{T}}(0) P_{11} B_{12} \varphi_2(-r) - \frac{\eta_2}{\eta_1}\varphi_1^{\mathrm{T}}(-r) B_{21}^{\mathrm{T}} P_{22} \varphi_2(0).$$

In view of expressions (3.5.38)–(3.5.40) for the upper right derivative number of functional $v(\varphi_1, \varphi_2, \eta)$ in the domain of values $R^n \times C^n$ we have the estimate

(3.5.41) $\qquad D^+ v(\varphi_1, \varphi_2, \eta)|_{(3.5.4)} \leq u_1^{\mathrm{T}} \Sigma_1 u_1 + u_2^{\mathrm{T}} \Sigma_2 u_2,$

where
$$u_1 = (|\varphi_1(0)|, |\varphi_1(-r)|, \|\varphi_1\|_{L_2})^{\mathrm{T}},$$
$$u_2 = (|\varphi_2(0)|, |\varphi_2(-r)|, \|\varphi_2\|_{L_2})^{\mathrm{T}}$$

and $\Sigma_1 = [\sigma^1_{ij}]^3_{i,j=1}$, $\Sigma_2 = [\sigma^2_{ij}]^3_{i,j=1}$ are constant matrices with the elements

$$\sigma^1_{11} = \lambda_M(A^T_{11}P_{11} + P_{11}A_{11} + D_1)\eta^2_1 + \eta_1\eta_2\lambda_M(P_{12}A_{21} + A^T_{21}P^T_{12}),$$

$$\sigma^1_{22} = -\frac{1}{2}\lambda_m(D_1)\eta^2_1, \quad \sigma^1_{33} = -\frac{1}{2r}\lambda_m(D_1)\eta^2_1,$$

$$\sigma^1_{12} = \|P_{11}\|\|B_{11}\|\eta^2_1 + \|P_{12}\|\|B_{21}\|\eta_1\eta_2,$$

$$\sigma^1_{23} = \varkappa_1\|B_{21}\|\eta_1\eta_2, \quad \sigma^1_{13} = \varkappa_1\|A_{21}\|\eta_1\eta_2,$$

$$\sigma^2_{11} = \lambda_M(A^T_{22}P_{22} + P_{22}A_{22} + D_2)\eta^2_2 + \eta_1\eta_2\lambda_M(P_{21}A_{12} + A^T_{12}P^T_{21}),$$

$$\sigma^2_{22} = -\frac{1}{2}\lambda_m(D_2)\eta^2_2, \quad \sigma^2_{33} = -\frac{1}{2r}\lambda_m(D_2)\eta^2_2,$$

$$\sigma^2_{12} = \|P_{22}\|\|B_{22}\|\eta^2_2 + \|P_{12}\|\|B_{12}\|\eta_1\eta_2,$$

$$\sigma^2_{23} = \varkappa_2\|B_{12}\|\eta_1\eta_2, \quad \sigma^2_{13} = \varkappa_2\|A_{12}\|\eta_1\eta_2,$$

$$\sigma^1_{ij} = \sigma^1_{ji}, \quad \sigma^2_{ij} = \sigma^2_{ji}, \quad i,j = 1,2,3, \quad i \neq j.$$

In the partial case when $B_{11} = 0$ and $B_{22} = 0$ system (3.5.4) becomes

(3.5.42)
$$\frac{dx_1}{dt} = A_{11}x_1(t) + A_{12}x_2(t) + B_{12}x_2(t-r),$$
$$\frac{dx_2}{dt} = A_{21}x_1(t) + A_{22}x_2(t) + B_{21}x_1(t-r).$$

Besides, system of equations (3.5.30)–(3.5.32) becomes

(3.5.43)
$$\frac{dS_2}{d\theta} = A^T_{11}S_2(\theta), \quad \frac{dS_1}{d\theta} = S_1(\theta)A_{22}$$
$$A^T_{11}P_{12} + P_{12}A_{22} = -\frac{\eta_1}{\eta_2}P_{11}A_{12} - \frac{\eta_2}{\eta_1}A^T_{21}P_{22} - \frac{1}{r}(S_1(0) + S_2(0))$$

under the initial conditions

(3.5.44)
$$S_2(-r) = \frac{r\eta_1}{\eta_2}P_{11}B_{12}, \quad S_1(-r) = \frac{r\eta_2}{\eta_1}B^T_{21}P_{22}.$$

The first group of equations (3.5.43) can be integrated in the explicit form

(3.5.45)
$$S_1(\theta) = \frac{r\eta_2}{\eta_1}B^T_{21}P_{22}\exp\{A_{22}(\theta+r)\},$$
$$S_2(\theta) = \frac{r\eta_1}{\eta_2}\exp\{A^T_{11}(\theta+r)\}P_{11}B_{12}.$$

Letting $\theta = 0$ we find

$$S_1(0) = \frac{r\eta_2}{\eta_1} B_{21}^T P_{22} \exp\{A_{22}r\},$$

$$S_2(0) = \frac{r\eta_1}{\eta_2} \exp\{A_{11}^T r\} P_{11} B_{12}.$$

Therefore equation (3.5.32) becomes

(3.5.46)
$$A_{11}^T P_{12} + P_{12} A_{22} = -\frac{\eta_1}{\eta_2}(P_{11} A_{12} + \exp\{A_{11}^T r\} P_{11} B_{12})$$
$$- \frac{\eta_2}{\eta_1}(A_{21}^T P_{22} + B_{21}^T P_{22} \exp\{A_{22}r\}).$$

Necessary and sufficient existence conditions for unique solution of equation (3.5.46) follow from Lankaster [1].

Diagonal elements of the matrix-valued functional $U(\varphi_1, \varphi_2)$ are taken in the form (3.5.28), (3.5.29) for $w_1(\varphi_1) = v_{11}(\varphi_1)$ and $w_2(\varphi_2) = v_{22}(\varphi_2)$, and non-diagonal element $w_{12}(\varphi_1, \varphi_2)$ is represented as

(3.5.47)
$$w_{12}(\varphi_1, \varphi_2) = \varphi_1^T(0) P_{12} \varphi_2(0)$$
$$+ \frac{\eta_1}{\eta_2} \varphi_1^T(0) \int_{-r}^{0} \exp\{A_{11}^T(\theta + r)\} P_{11} B_{12} \varphi_2(\theta)\, d\theta$$
$$+ \frac{\eta_2}{\eta_1} \varphi_2^T(0) \int_{-r}^{0} B_{21}^T P_{22} \exp\{A_{22}(\theta + r)\} \varphi_1(\theta)\, d\theta.$$

For estimation of functional (3.5.47) we shall formulate one auxiliary result.

Lemma 3.5.1 *Let A be a constant $n \times n$-matrix, then estimate*

$$\|\exp At\| \leq e^{\Delta t} \sum_{k=0}^{n-1} \frac{1}{k!}(2t\|A\|)^k, \quad t \geq 0,$$

is valid, where $\Delta = \max\{\operatorname{Re}\lambda\colon \lambda \in \sigma(A)\}$, $\sigma(A)$ is a spectrum of matrix A.

Using this result one can estimate $\|\exp At\|$ as follows. Let $\varepsilon > 0$ be a sufficiently small positive number. Consider function

$$f(t) = e^{-\varepsilon t} \sum_{k=0}^{n-1} \frac{1}{k!}(2t\|A\|)^k, \quad t \geq 0.$$

3. FUNCTIONAL DIFFERENTIAL SYSTEMS

In view that $f(t) \to 0$ as $t \to \infty$ we conclude that there exists $M_\varepsilon = \max_{t \geq 0} f(t)$ and find the estimate

(3.5.48) $$\|\exp\{At\}\| \leq M_\varepsilon e^{(\Delta+\varepsilon)t} \quad \text{for} \quad t \geq 0.$$

Applying estimate (3.5.48) it is easy to find

$$\varkappa_1 = \frac{1}{r}\left\{\int_{-r}^{0} \lambda_M(S_1(\theta)S_1^T(\theta))\right\}^{1/2}$$

$$\leq \frac{\eta_2}{\eta_1}\|B_{21}\|\,\|P_{22}\|M_{\varepsilon 1}\left[\frac{e^{2(\Delta_1+\varepsilon)r}-1}{2(\Delta_1+\varepsilon)}\right]^{1/2} = \xi_1,$$

$$\varkappa_2 = \frac{1}{r}\left\{\int_{-r}^{0} \lambda_M(S_2(\theta)S_2^T(\theta))\right\}^{1/2}$$

$$\leq \frac{\eta_1}{\eta_2}\|B_{12}\|\,\|P_{11}\|M_{\varepsilon 2}\left[\frac{e^{2(\Delta_2+\varepsilon)r}-1}{2(\Delta_2+\varepsilon)}\right]^{1/2} = \xi_2,$$

where $M_{\varepsilon 1}$ and $M_{\varepsilon 1}$ are the corresponding constants, and Δ_1 and Δ_2 are maximal real values of spectra of matrices A_{11} and A_{22} respectively.

Thus, for the scalar functional

(3.5.49) $$w(\varphi_1, \varphi_2, \eta) = \eta^T U(\varphi_1, \varphi_2)\eta$$

the estimate

(3.5.50) $$u^T H^T \underline{C} H u \leq w(\varphi_1, \varphi_2, \eta) \leq u^T H^T \overline{C} H u,$$

is valid, where

$$u = (|\varphi_1(0)|, |\varphi_2(0)|, \|\varphi_1\|_{L_2}, \|\varphi_2\|_{L_2})^T, \quad H = \text{diag}\,(\eta_1, \eta_2, \eta_1, \eta_2),$$

$$\overline{C} = \begin{pmatrix} \lambda_M(P_{11}) & \|P_{12}\| & 0 & \xi_2 \\ \|P_{12}\| & \lambda_M(P_{22}) & \xi_1 & 0 \\ 0 & \xi_1 & \lambda_M(D_1) & 0 \\ \xi_2 & 0 & 0 & \lambda_M(D_2) \end{pmatrix},$$

$$\underline{C} = \begin{pmatrix} \lambda_m(P_{11}) & -\|P_{12}\| & 0 & -\xi_2 \\ -\|P_{12}\| & \lambda_m(P_{22}) & -\xi_1 & 0 \\ 0 & -\xi_1 & \frac{1}{2}\lambda_m(D_1) & 0 \\ -\xi_2 & 0 & 0 & \frac{1}{2}\lambda_m(D_2) \end{pmatrix}.$$

In view of estimates (3.5.41) and (3.5.48) in the region of values $R^n \times C^n$ it is easy to find the estimate of the upper right derivative number of functional (3.5.49) along solutions of system (3.5.42)

(3.5.51) $$D^+ v(\varphi_1, \varphi_2, \eta)\Big|_{(3.5.42)} \leq u_1^{\mathrm{T}} \Omega_1 w_1 + u_2^{\mathrm{T}} \Omega_2 w_2,$$

and $\Omega_1 = [w_{ij}^1]_{i,j=1}^3$, $\Omega_2 = [w_{ij}^2]_{i,j=1}^3$ are constant matrices with the elements

$$w_{11}^1 = \lambda_M(A_{11}^{\mathrm{T}} P_{11} + P_{11} A_{11} + D_1)\eta_1^2 + \eta_1 \eta_2 \lambda_M(P_{12} A_{21} + A_{21}^{\mathrm{T}} P_{12}^{\mathrm{T}}),$$

$$w_{22}^1 = -\frac{1}{2}\lambda_m(D_1)\eta_1^2, \quad w_{33}^1 = -\frac{1}{2r}\lambda_m(D_1)\eta_1^2,$$

$$w_{12}^1 = \|P_{12}\| \|B_{21}\| \eta_1 \eta_2, \quad w_{23}^1 = \xi_1 \|B_{21}\| \eta_1 \eta_2, \quad w_{13}^1 = \xi_1 \|A_{21}\| \eta_1 \eta_1,$$

$$w_{11}^2 = \lambda_M(A_{22}^{\mathrm{T}} P_{22} + P_{22} A_{22} + D_2)\eta_2^2 + \eta_1 \eta_2 \lambda_M(P_{21} A_{12} + A_{12}^{\mathrm{T}} P_{21}^{\mathrm{T}}),$$

$$w_{22}^2 = -\frac{1}{2}\lambda_m(D_2)\eta_2^2, \quad w_{33}^2 = -\frac{1}{2r}\lambda_m(D_2)\eta_2^2,$$

$$w_{12}^2 = \|P_{12}\| \|B_{12}\| \eta_1 \eta_2, \quad w_{23}^2 = \xi_2 \|B_{12}\| \eta_1 \eta_2, \quad w_{13}^2 = \xi_2 \|A_{12}\| \eta_1 \eta_2,$$

$$w_{ij}^1 = w_{ji}^1, \quad w_{ij}^2 = w_{ji}^2, \quad i, j = 1, 2, 3, \quad i \neq j.$$

Under some restrictions on the sign-definiteness of matrices $\overline{C}$, $\underline{C}$, and Σ_1, Σ_2 the constructed functional is the Liapunov–Krasovskii functional and applying this functional in Section 3.6 we shall establish new sufficient conditions for asymptotic stability of the equilibrium state $x = 0$ of quasilinear system. For system (3.5.42) the proposed method of constructing matrix-valued functional is more efficient, since system of equations (3.5.43)–(3.5.44) is integrable in the explicit form. By means of functional $v(\varphi_1, \varphi_2)$ in Section 3.6 we shall establish sufficient stability conditions for the equilibrium state of system (3.5.42).

3.6 Stability Analysis of Quasilinear Delay Systems

We consider an autonomous quasilinear delays system (3.5.2) with decomposition

(3.6.1)
$$\frac{dx_1}{dt} = A_{11} x_1(t) + A_{12} x_2(t) + B_{11} x_1(t-r) + B_{12} x_2(t-r) + f_1(x, x_t),$$

$$\frac{dx_2}{dt} = A_{21} x_1(t) + A_{22} x_2(t) + B_{21} x_1(t-r) + B_{22} x_2(t-r) + f_2(x, x_t),$$

where $x_i \in R^{n_i}$, $i = 1, 2$, $x = (x_1^{\mathrm{T}}, x_2^{\mathrm{T}})^{\mathrm{T}}$, A_{ij} and B_{ij} are constant matrices of appropriate dimensions.

We make the following assumptions on the functions $f_i(x, x_t)$, $i = 1, 2$.

3. FUNCTIONAL DIFFERENTIAL SYSTEMS

Assumption 3.6.1 The functions $f_i(x, x_t)$, $i = 1, 2$, satisfy the following conditions

(1) the functions $f_i \in C(R^n \times C^n, R^n)$ for $i = 1, 2$;
(2) the functions $f_i(0, 0) = 0$ iff $x = x_t = 0$;
(3) there exist constants $c_{ij}, l_{ij} > 0$, $i, j = 1, 2$, such that

$$|f_i(x, x_t)| \leq c_{i1}|x_1(0)| + c_{i2}|x_2(0)| + l_{i1}\|x_1\|_{L_2} + l_{i2}\|x_2\|_{L_2},$$

where $|\cdot|$ is Euclidean norm in R^{n_i}, $\|\cdot\|_{L_2}$ is the L_2-norm.

For the linear system

$$(3.6.2) \quad \begin{aligned} \frac{dx_1}{dt} &= A_{11}x_1(t) + A_{12}x_2(t) + B_{11}x_1(t-r) + B_{12}x_2(t-r), \\ \frac{dx_2}{dt} &= A_{21}x_1(t) + A_{22}x_2(t) + B_{21}x_1(t-r) + B_{22}x_2(t-r), \end{aligned}$$

using the results of Section 3.5 we construct the matrix-valued functional

$$U: R^n \times C^n \to R^{2 \times 2}$$

with the elements (3.5.28), (3.5.29) and (3.5.34).

Applying functional $U(\varphi_1, \varphi_2)$ one can establish sufficient stability conditions for solution $x = 0$ of system (3.6.1). First, we introduce the designations

$$\Delta\Sigma_1 = \begin{pmatrix} 2c_{11}\|P_{11}\| + c_{21}\|P_{12}\| & 0 & \|P_{11}\|l_{11} + \frac{1}{2}\|P_{12}\|l_{21} \\ 0 & 0 & 0 \\ \|P_{11}\|l_{11} + \frac{1}{2}\|P_{12}\|l_{21} & 0 & 0 \end{pmatrix},$$

$$\Delta\Sigma_2 = \begin{pmatrix} 2c_{22}\|P_{22}\| + c_{12}\|P_{12}\| & 0 & \|P_{22}\|l_{22} + \frac{1}{2}\|P_{12}\|l_{12} \\ 0 & 0 & 0 \\ \|P_{22}\|l_{22} + \frac{1}{2}\|P_{12}\|l_{12} & 0 & 0 \end{pmatrix},$$

$$\Delta\Sigma_{12} = \begin{pmatrix} 2c_{12}\|P_{11}\| + 2c_{21}\|P_{22}\| + \\ c_{22}\|P_{12}\| + c_{11}\|P_{12}\| & 0 & 2\|P_{22}\|l_{21} + \|P_{12}\|l_{12} \\ 0 & 0 & 0 \\ 2\|P_{11}\|l_{12} + \|P_{12}\|l_{22} & 0 & 0 \end{pmatrix}.$$

Theorem 3.6.1 *Let system of equations (3.6.1) be such that*

(1) *there exist solutions of equations (3.5.30) – (3.5.32) under initial conditions (3.5.33) for some $\eta \in R_+^2$, $\eta > 0$;*
(2) *matrices $\underline{C}$ and $\overline{C}$ in estimate (3.5.37) are positive definite;*
(3) *matrices $\Sigma_1 + \Delta\Sigma_1$ and $\Sigma_2 + \Delta\Sigma_2$ are negative definite;*
(4) *inequality*

$$\|\Sigma_{12}\| < \lambda_M(\Sigma_1 + \Delta\Sigma_1)\lambda_M(\Sigma_2 + \Delta\Sigma_2)$$

holds true.
Then the solution $x = 0$ of system (3.6.1) is uniformly asymptotically stable.

Proof Condition (2) of Theorem 3.6.1 ensures the possibility of constructing the "scalar" functional $v \colon R^n \times C^n \times R_+^2 \to R_+$, $v(\varphi_1, \varphi_2, \eta) = \eta^T U(\varphi_1, \varphi_2)\eta$, satisfying the conditions of definite positiveness and decrease. The upper right derivative number of the functional $v(\varphi_1, \varphi_2, \eta)$ admits the estimate

$$D^+ v(\varphi_1, \varphi_2, \eta)\Big|_{(3.6.1)} \leq u_1^T(\Sigma_1 + \Delta\Sigma_1)u_1 + 2u_1^T\Sigma_{12}u_2 + u_2^T(\Sigma_2 + \Delta\Sigma_2)u_2,$$

where

$$u_1 = (|\varphi_1(0)|, |\varphi_1(-r)|, \|\varphi_1\|_{L_2})^T, \quad u_2 = (|\varphi_2(0)|, |\varphi_2(-r)|, \|\varphi_2\|_{L_2})^T.$$

Conditions (3) and (4) ensure definite negativeness of $D^+v(\varphi_1, \varphi_2, \eta)|_{(3.6.1)}$. Thus, the solution $x = 0$ of system (3.6.1) is uniformly asymptotically stable and the constructed functional $v(\varphi_1, \varphi_2, \eta)$ is the matrix-valued Liapunov–Krasovskii functional.

Corollary 3.6.1 *Let system of equations (3.6.2) be such that*

(i) *there exist solutions of equations (3.5.30) – (3.5.32) under initial conditions (3.5.33) for some $\eta \in R_+^2$, $\eta > 0$;*
(ii) *matrices $\underline{C}$ and $\overline{C}$ in estimate (3.5.37) are positive definite;*
(iii) *matrices Σ_1 and Σ_2 are negative definite.*

Then solution $x = 0$ of system (3.6.2) is uniformly asymptotically stable.

In the partial case, when $B_{11} = 0$ and $B_{22} = 0$ sufficient conditions of of uniform asymptotic stability of solution $x = 0$ are formulated in terms of estimates (3.5.50) and (3.5.51) for matrix-valued functional $U(\varphi_1, \varphi_2)$ with the elements

$$w_{11}(\varphi_1) = \varphi_1^{\mathrm{T}}(0) P_{11} \varphi_1(0) + \int_{-r}^{0} k(\theta) \varphi_1^{\mathrm{T}}(\theta) D_1 \varphi_1(\theta)\, d\theta,$$

$$w_{22}(\varphi_2) = \varphi_2^{\mathrm{T}}(0) P_{22} \varphi_2(0) + \int_{-r}^{0} k(\theta) \varphi_2^{\mathrm{T}}(\theta) D_2 \varphi_2(\theta)\, d\theta,$$

$$w_{12}(\varphi_1, \varphi_2) = \varphi_1^{\mathrm{T}}(0) P_{12} \varphi_2(0)$$

$$+ \frac{\eta_1}{\eta_2} \varphi_1^{\mathrm{T}}(0) \int_{-r}^{0} \exp\{A_{11}^{\mathrm{T}}(\theta + r)\} P_{11} B_{12} \varphi_2(\theta)\, d\theta$$

$$+ \frac{\eta_2}{\eta_1} \varphi_2^{\mathrm{T}}(0) \int_{-r}^{0} B_{21}^{\mathrm{T}} P_{22} \exp\{A_{22}(\theta + r)\} \varphi_2(\theta)\, d\theta.$$

Theorem 3.6.2 *Let a system of equations (3.6.1) be such that*
(1) $B_{11} = 0$, $B_{22} = 0$;
(2) *matrices $\underline{C}$ and $\overline{C}$ in estimates (3.5.50) are positive definite;*
(3) *matrices $\Omega_1 + \Delta\Sigma_1$ and $\Omega_2 + \Delta\Sigma_2$ from estimate (3.5.51) are negative definite;*
(4) *inequality*

$$\|\Sigma_{12}\| < \lambda_M(\Omega_1 + \Delta\Sigma_1) \lambda_M(\Omega_2 + \Delta\Sigma_2)$$

holds true.

Then solution $x = 0$ of system (3.6.1) is uniformly asymptotically stable.

The proof is similar to the proof of Theorem 3.6.1.

Corollary 3.6.2 *Let system of equations (3.6.2) be such that*
(i) *matrices $B_{11} = 0$ and $B_{22} = 0$;*
(ii) *matrices $\underline{C}$ and $\overline{C}$ in estimate (3.5.50) are positive definite;*
(iii) *matrices Ω_1 and Ω_2 in estimate (3.5.51) are negative definite.*
Then solution $x = 0$ of system (3.6.2) is uniformly asymptotically stable.

3.6.1 Example As applications of results of Sections 3.5 and 3.6 we consider oscillations of a harmonic oscillator with delay force action. The perturbed motion equation of the oscillator is

$$\text{(3.6.3)} \qquad \frac{d^2x}{dt^2} + \mu \frac{dx}{dt} + \omega^2 x(t) + cx(t-r) = 0,$$

where x is a state variable, $\omega, c, \mu > 0$ are constants. Introduce an auxiliary variable $y = \dfrac{dx}{dt}$ and present equation (3.6.3) as a system

$$\text{(3.6.4)} \qquad \begin{aligned} \frac{dx}{dt} &= y, \\ \frac{dy}{dt} &= -\omega^2 x(t) - \mu y(t) - cx(t-r). \end{aligned}$$

Applying the proposed technique of construction of the Liapunov functionals for system (3.6.4) we construct a scalar functional $w(\varphi_1, \varphi_2)$ as

$$\text{(3.6.5)} \qquad w(\varphi_1, \varphi_2) = v_{11}(\varphi_1) + 2v_{12}(\varphi_1, \varphi_2) + v_{22}(\varphi_2),$$

where

$$v_{11}(\varphi_1) = \gamma^2 \varphi_1^2(0) + \gamma^2 d_1 \int_{-r}^{0} \left(1 + \frac{\theta}{2r}\right) \varphi_1^2(\theta)\, d\theta,$$

$$v_{22}(\varphi_2) = \varphi_2^2(0) d_2 \int_{-r}^{0} \left(1 + \frac{\theta}{2r}\right) \varphi_2^2(\theta)\, d\theta,$$

$$v_{12}(\varphi_1, \varphi_2) = 2\frac{\gamma^2 - \omega^2 - ce^{-\mu r}}{\mu} \varphi_1(0)\varphi_2(0) - 2ce^{-\mu r}\varphi_2(0) \int_{-r}^{0} e^{-\mu \theta} \varphi_1(\theta)\, d\theta,$$

and γ, d_1 and d_2 are indefinite positive constants.

Functional (3.6.5) can be estimated from below by means of the Cauchy–Bunyakovsky inequality

$$w(\varphi_1, \varphi_2) \geq \gamma^2 \varphi_1^2(0) + \varphi_2^2(0) - 2\frac{\gamma^2 - \omega^2 - ce^{-\mu r}}{\mu} |\varphi_1(0)|\,|\varphi_2(0)|$$

$$+ \frac{\gamma^2 d_1}{2} \|\varphi_1(\theta)\|_{L_2}^2 + \frac{d_2}{2} \|\varphi_2(\theta)\|_{L_2}^2 - 2|c|e^{-\mu r} \sqrt{\frac{e^{2\mu r} - 1}{2\mu}} |\varphi_2(0)|\,\|\varphi_1(\theta)\|_{L_2}.$$

The derivative of functional (3.6.5) along solutions of system (3.6.4) is
(3.6.6)
$$D^+w(\varphi_1,\varphi_2)\Big|_{(3.6.4)} = \left(d_1\gamma^2 - \frac{2\omega^2(\gamma^2 - \omega^2 - ce^{-\mu r})}{\mu}\right)\varphi_1^2(0)$$

$$-\frac{d_1\gamma^2}{2}\varphi_1^2(-r) - \frac{d_1\gamma^2}{2r}\|\varphi_1(\theta)\|_{L_2}^2 - \frac{2c(\gamma^2 - \omega^2 - ce^{-\mu r})}{\mu}\varphi_1(0)\varphi_1(-r)$$

$$+ 2ce^{-\mu r}\omega^2\varphi_1(0)\int_{-r}^{0} e^{-\mu\theta}\varphi_1(\theta)\,d\theta + 2c^2 e^{-\mu r}\varphi_1(-r)\int_{-r}^{0} e^{-\mu\theta}\varphi_1(\theta)\,d\theta$$

$$+\left(-2\mu + d_2 + \frac{2(\gamma^2 - \omega^2 - ce^{-\mu r})}{\mu}\right)\varphi_2^2(0) - \frac{d_2}{2}\varphi_2^2(-r) - \frac{d_2}{2r}\|\varphi_2(\theta)\|_{L_2}^2.$$

The analysis of (3.6.6) shows that it is reasonable to take constants $\gamma^2 = \omega^2 + ce^{-\mu r} + \frac{\mu^2}{2}$, $d_1 = \frac{\mu\omega^2}{2\gamma}$ and $d_2 = \frac{\mu}{2}$. Applying the Cauchy-Bunyakovsky inequality once again we estimate derivative (3.6.6) of functional (3.6.5)

$$D^+w(\varphi_1,\varphi_2)\Big|_{(3.6.3)} \leq \left(d_1\gamma^2 - \frac{2\omega^2(\gamma^2 - \omega^2 - ce^{-\mu r})}{\mu}\right)\varphi_1^2(0)$$

$$-\frac{d_1\gamma^2}{2}\varphi_1^2(-r) - \frac{d_1\gamma^2}{2r}\|\varphi_1(\theta)\|_{L_2}^2 + \frac{2|c|(\gamma^2 - \omega^2 - ce^{-\mu r})}{\mu}|\varphi_1(0)||\varphi_1(-r)|$$

$$+ 2|c|e^{-\mu r}\omega^2\sqrt{\frac{e^{2\mu r}-1}{2\mu}}|\varphi_1(0)|\,\|\varphi_1(\theta)\|_{L_2}$$

$$+ 2c^2 e^{-\mu r}\sqrt{\frac{e^{2\mu r}-1}{2\mu}}|\varphi_1(-r)|\,\|\varphi_1(\theta)\|_{L_2}^2$$

$$+\left(-2\mu + d_2 + \frac{2(\gamma^2 - \omega^2 - ce^{-\mu r})}{\mu}\right)\varphi_2^2(0) - \frac{d_2}{2}\varphi_2^2(-r) - \frac{d_2}{2r}\|\varphi_2(\theta)\|_{L_2}^2.$$

Conditions of positive definiteness of functional $w(\varphi_1,\varphi_2)$ and negative definiteness of functional $D^+w(\varphi_1,\varphi_2)\Big|_{(3.6.4)}$ yield a new condition of asymptotic stability of zero solution of equation (3.6.4) in the form of the system of inequalities

$$|c| < \frac{\mu}{2}\sqrt{\frac{\mu}{r(1-e^{-2\mu r})}}, \qquad \omega^2 > |c|\sqrt{\frac{24r(1-e^{-2\mu r}) + 2\mu^3}{\mu^3 - 4c^2 r(1-e^{-2\mu r})}},$$

$$\left(2\omega^2 + 2ce^{-\mu r} + \mu^2\right)\left(\mu^2\omega^2 - 2c^2(1-e^{-2\mu r})\right) \geq \frac{\mu^4\omega^2}{2}.$$

3.7 The Problems for Investigation

3.7.1 To construct the matrix-valued function $U(t,x)$ for system (3.2.1) and formulate a unified approach to the analysis of various types of stability in terms of comparison principle.

Hint: Use some results from the papers Martynyuk and Sun Zhenqi [1], Lakshmikantham, Leela, and Sivasundaram [1], and Taniguchi [1].

3.7.2 For the autonomous system

$$\text{(3.7.1)} \qquad \frac{dx}{dt} = f(x_t)$$

and the periodic system

$$\text{(3.7.2)} \qquad \frac{dx}{dt} = p(t, x_t)$$

of functional differential equations, where $f \in C(G, R^n)$, $p \in C(R_+ \times G, R^n)$, $G \subset \mathcal{C} = C([-\tau, 0], R^n)$, $0 \in G$, to construct the matrix-valued Liapunov–Razumikhin function and establishe instability conditions for the state $x = 0$.

3.7.3 To give a description of the most wide class of elements of the matrix-valued functional $U(t,x)$ for which the condition

$$\text{(3.7.3)} \qquad w_1(|\varphi(0)|) \leq \eta^T U(t, \varphi)\eta \leq w_2(\|\varphi\|)$$

can be weakened maximally in theorems on uniform asymptotic stability of the state $x = 0$ of system (3.2.1).

3.7.4 To indicate classes of matrix-valued functionals for which the total derivative due to system with aftereffect is efficiently computed: (a) in terms of the upper derivative number, (b) in terms of the invariant derivative (see Kim [1]), (c) in terms of the Frechet (Gateaux) derivative.

3.7.5 To compare efficiency of the results of stability analysis in parameter space of linear system with finite delay obtained by application of algorithm proposed in Section 3.4 and quadratic functional proposed by Infante and Castelan [1].

3.7.6 To generalize the method of matrix-valued Liapunov–Krasovskii functionals for the systems with infinite delay and establish conditions of uniform boundedness and uniform asymptotic stability for system with infinite delay.

3.7.7 To apply the method of matrix-valued functionals for solution of the problem on processes stability in neural network with finite and infinite delay

$$\text{(3.7.4)} \qquad \frac{dx}{dt} = -Dx(t) + T_0 g(x(t)) + T_1 g(x(t-\tau)) + I,$$

where $D = \operatorname{diag}(d_1, \ldots, d_n)$ is a matrix with positive diagonal elements, function $g\colon R^n \to R^n$ satisfies the conditions

$$0 \leq \frac{g_i(\alpha) - g_i(\beta)}{\alpha - \beta} \leq \gamma_i < +\infty$$

for all $\alpha \neq \beta \in R$, $i = 1, 2, \ldots, n$, T_0 and T_1 are constant matrices.

Hint. Compare the result with the one obtained by Joy [1].

3.8 Notes and References

Section 3.1 The intensive investigation of functional differential equations is motivated by many problem from mechanics, biology, elasticity, replicator dynamics, viscoelasticity, electricity, reactor dynamics, heat flow, chemical oscillations, and neural networks. The books of Andronov, Vitt and Khaikin [1], Azbelev and Simonov [1], Bellman and Cooke [1], Braun [1], Burton [1], Davis [1], Hale [1], McDonald [1], Haag [1], Kolmanovskii and Nosov [1], and papers by Ohta and Šiljak [1], So Joseph, *et al.* [1], etc. are wonderful sourse of the problems in the direction.

Section 3.2 This section is auxiliary. It is based on some known results by Krasovskii [1], Yoshizawa [1], Hale [1], as well as Burton [1], etc.

Section 3.3 This Section is based on the results by Martynyuk [1]. Moreover, it encorporates some other results: for Corollary 3.3.1 see Zhang Bo [1], for Corollaries 3.3.2 and 3.3.3 see Krasovskii [1], and Zhang Bo [1], for Corollaries 3.3.4 and 3.3.5 see Burton [1], and Zhang Bo [1], for Corollaries 3.3.6 and 3.3.7 see Hering [1], and Zhang Bo [1]. See also Zhang Bo [3], and Martynyuk and Stavroulakis [1].

Section 3.4 This section is based on some results by Martynyuk [1] and Martynyuk and Rizayev [1, 2].

Section 3.5 In this section some results by Martynyuk and Slyn'ko [1] are used as well as an idea from Hale [1] and some estimates from Martynyuk, Lakshmikantham and Leela [1]. Many authors try to find effective approaches to construction of Liapunov functionals (see Kim [1], Repin [1], Infante and Castelan [1], Zhang Wei [1], Gaishun and Knyazhishche [1], He Xue-zhong [1], Xu Bugong [1], Zhang Bo [2], etc.). However the problem is, in general, open.

Sections 3.6 In this Section some results from the paper by Martynyuk and Slyn'ko [1] are used as well as some estimates from Chapter 1 of the book.

The approaches presented in this Section are aimed at solving the problem of stability of the zero solution of systems functional differential equations. These approaches have a considerable potential for further development and applications.

First, we note that insignificant modification of the conditions of Theorem 3.3.1 allows us to establish a new boundedness conditions for motions in functional differential systems with finite delay by further development of results by Haddock and Zhao [1], and Kobayashi and Tsuruta [1]. Also note that the proposed approach enables us to determine the stabilizing (destabilizing) role of delay, since it admits the existence of both stable and unstable subsystems in the initial system (cf. Zuoxin Gan [1]).

Second, the method of constructing the auxiliary functions (functionals) developed in Chapter 1 and this chapter make it possible to apply efficiently some general results (see Burton and Hatvani [1], Hale [1], Hatvani [1], Krasovskii [1], Zang Bo [1, 3], etc.) on stability of solutions to functional differential equations which contain functionals satisfying estimates of the type

(i) $$w_1(|\varphi(0)|) \leq v(t,\varphi) \leq w_2(|\varphi|)$$

or its generalizations, for example, in the form

(ii) $$w_1(|\varphi(0)|) \leq v(t,\varphi) \leq w_2(|\varphi|) + w_3(|\varphi\|_{L_2}\|),$$

where $\|\cdot\|_{L_2}$ is the norm in space L_2, and $w_i(\cdot)$ is the comparison function of class $K(KR)$. We weaken conditions (i) or (ii) by expansion of the set of components v_{ij}, appropriate for construction of suitable functional. Some possibilities of the proposed technique of stability analysis are illustrated by the examples of quasilinear equations which remain an urgent object of investigations (see Sakata [1], Tsuruta [1], etc), including estimates of stability domains in parameter space (see Hara and Sugie [1]). We note that the proposed approach can be extended for nonautonomous systems and systems with several delays.

References

Andronov, A.A., Vitt, A.A. and Khaikin, S.E.
[1] *Theory of Oscillations.* Moscow: Nauka, 1981. [Russian]

Azbelev, N.V. and Simonov, P.M.
[1] *Stability of Differential Equations with Aftereffect.* London: Taylor and Francis, 2002.

Bellman, R., and Cooke, K.L.
[1] *Differential-Difference Equations.* New York: Academic Press, 1963.

Braun, M.
[1] *Differential Equations and their Applications.* New York: Springer-Verlag, 1975.

Burton, T.A.
[1] *Stability and Periodic Solutions of Ordinary and Functional Diferential Equations.* Orlando: Academic Press, 1985.

Burton, T.A. and Hatvani, L.
[1] Stability theorems for nonautonomous functional differential equations by Liapunov functionals. *Tohoku Math. J.* **41** (1989) 65–104.

Dao Yi Xu
[1] Uniform asymptotic stability in terms of two measures for functional differential equations. *Nonlin. Anal.* **27** (1996) 413–427.

Davis, H.T.
[1] *Introduction to Nonlinear Differential and Integral Equations.* New York: Dover, 1962.

Driver, R.D.
[1] Existence and stability of solutions of a delay-differential systems. *Archs. Ration. Mech. Anal.* **10** (1962) 401–426.

El'sgol'ts, L.E. and Norkin, S.B.
[1] *Introduction to the Theory and Application of Differential Equations with Deviating Arguments.* New York: Academic Press, 1973.

Infante, E.F. and Castelan, W.B.
[1] A Liapunov functional for a matrix difference-differential equation. *J. of Diff. Eqns* **29** (1978) 439–451.

Joy, M.
[1] On the global convergence of a class of functional diffrential equations with applications in neural network theory. *J. Math. Anal. Appl.* **232** (1999) 61–81.

Gaishun, V.I. and Knyazhishche, L.B.
[1] Nonmonotone Lyapunov functionals. Stability conditions for delay differential equations. *Differ. Uravn.* **30**(8) (1994) 1291–1298.

Godoy, S.M.S. and Reis, J.G.
[1] Stability and existence of periodic solutions of a functional differential equations. *J. Math. Anal. Appl.* **198** (1996) 381–398.

Haddock, J.R. and Zhao, J.
[1] Instability for autonomous and periodic functional differential equations with finite delay. *Funkcialaj Ekvacioj* **39** (1996) 553–570.

Haag, J.
[1] *Oscillatory Motions.* Belmont: Wadsworth, 1962.

Hale, J.K.
[1] *Theory of Functional Differential Equations.* Berlin: Springer-Verlag, 1977.
[2] Sufficient conditions for stability and instability of autonomous functional differential equations. *J. Diff. Eqns* **1** (1965) 452–482.

Hatvani, L.
[1] On Lyapunov's direct method for nonautonomous functional differential equations. In: *Dynamic Systems and Applications.* (Ed. G.S. Ladde). Dynamic Publishers, Atlanta, 2001, P.297–304.

He, Xue-Zhong
[1] The Lyapunov functionals for delay Lotka-Volterra-type models. *SIAM J. Appl. Math.* **58**(4) (1998) 1222–1236.

Hering, R.H.
[1] Uniform asymptotic stability in infinite delay systems *J. Math. Anal. Appl.* **180** (1993) 160–173.
[2] Boundednes and periodic solutions in infinite delay systems. *J. Math. Anal. Appl.* **163** (1992) 521–535.

Kim, A.V.
[1] *Direct Liapunov Method in the Stability Theory of the Systems with Aftereffect.* Ekaterinburg: Izd. Ural University, 1992. [Russian]

Kobayashi, K. and Tsuruta, K.
[1] Uniform boundedness and uniform asymptotic stability in functional differential equations with constant delay. *Funkcial. Ekvac.* **40** (1997) 79–92.

Kolmanovskii, V.B. and Nosov, V.R.
[1] *Stability of Functional Differential Equations.* London: Academic Press, 1986.

Krasovskii, N.N.
[1] *Stability of Motion.* Stanford: Stanford Univ. Press, 1963.

Ladas, G., Sficas, Y.G. and Stavroulakis, I.P.
[1] Asymptotic behavior of solutions of retarded differential equations. *Proc. Amer. Math. Soc.* **88** (1983) 247–253.

Lakshmikantham, V.
[1] Functional-differential systems and extension of Lyapunov's methods. *J. Math. Anal. Appl.* **8** (1964) 392–405.

Lakshmikantham, V. and Leela, S.
[1] A unified approach to stability theory for differential equations with infinite delay. *J. of Integral Eqns* **10** (1985) 147–156.

Lakshmikantham, V. and Martynyuk, A.A.
[1] Development of direct Liapunov method for the systems with delay. *Prikl. Mekh.* **29**(2) (1993) 3–16. [Russian]

Lakshmikantham, V., Leela, S. and Sivasundaram, S.
[1] Lyapunov functions on product spaces and stability theory of delay differential equations. *J. Math. Anal. Appl.* **154** (1991) 391–402.

Martynyuk, A.A.
[1] Matrix-valued functional approach for stability analysis of functional differential equations. *Nonlin. Anal.* **56** (2004) 793–802.
[2] *Stability by Liapunov's Matrix Function Method with Applications.* New York: Marcel Dekker, 1998.
[3] *Qualitative Methods in Nonlinear Dynamics: Novel Approaches to Liapunov's Matrix Functions.* New York: Marcel Dekker, 2002.
[4] Stability analysis of large scale functional differential equations, (to appear).

Martynyuk, A.A. and Slyn'ko, V.I.
[1] A new approach for construction of the matrix-valued Liapunov's functionals for functional differential equations, (to appear).

Martynyuk, A.A. and Rizaev, R.
[1] On the applications of matrix-valued functionals in studying stability of system with delay. *Dopov. Nats. Acad. Nauk Ukr.* (2) (2002) 15–19. [Russian]
[2] Sufficient stability conditions for systems with delay. *Int. Appl. Mech.* **37**(12) (2001) 1612–1617.

Martynyuk, A.A. and Sun Zhenqi
[1] The matrix-valued Liapunov functional and stability of systems with delay. *Dokl. Math.* **57**(2) (1998) 207–209.

Martynyuk, A.A. and Stavroulakis, I.P.
[1] A theorem on uniform asymptotic stability of solutions of the system with finite delay *Dokl. Acad. Nauk* **392**(4) (2003) 445–447. [Russian]

McDonald, N.
[1] *Time Lags in Biological Models.* Lect. Notes in Biomath., No 27, Berlin: Springer-Verlag, 1978.

Ohta, Y. and Šiljak, D.D.
[1] An inclusion principle for hereditary systems. *J. Math. Anal. Appl.* **98**(2) (1984) 581–598.

Razumikhin, B.S.
[1] The application of Lyapunov's method to problems in the stability of system with delay, Automat. *Remote Control* **21** (1960) 515–520.

Repin, I.M.
[1] Quadratic Liapunov functionals for systems with delays. *Prikl. Mat. Mech.* **29** (1965) 564–566.

So, Joseph W.-H., Wu, Jianhong and Zou, Xingfu
[1] Structured population on two patches: modeling dispersal and delay. *J. Math. Biol.* **43**(1) (2001) 37–51.

Sakata, S.
[1] Asymptotic stability for a linear system of differential-difference equations. *Funcial. Ekvac.* **41** (1998) 435–449.

Soo, R. Lee and Jamshidi, M.
[1] Stability analysis for large scale time delay systems via the matrix Lyapunov function. *Kybernetika* **28** (1992) 271-283.

Taniguchi, T.
[1] Asymptotic behavior theorems for non-autonomous functional differential equations via Lyapunov–Razumikhin method. *J. Math. Anal. Appl.* **189** (1995) 715–730.

Tsuruta, K.
[1] Asymptotic stability in functional differential equations with finite delay. *Nonlin. Anal.* **23** (1994) 999–1011.

Xu, Bugong
[1] Stability of retarded dynamical systems: A Lyapunov function approach. *J. Math. Anal. Appl.* **253** (2001) 590–615.

Zhang, Bo
[1] A stability theorem in functional differential equations. *Diff. and Integr. Eqns* **9** (1996) 199–208.
[2] Formulas of Liapunov functions for systems of linear ordinary and delay differential equations. *Funkcial. Ekvac.* **44** (2001) 253–278.
[3] Asymptotic stability in functional differential equations by Lyapunov functionals. *Trans. Am. Math. Soc.* **347**(4) (1995) 1375–1382.

Wei, Zhang
[1] Absolute stability criteria for retarded Lur'e type systems. *J. Syst. Sci. Math. Sci.* **18**(2) (1998) 129–132. [Chinese]

Gan, Zuoxin
[1] Lyapunov functional for multiple delay general Lur'e systems with multiple non-linearities. *J. Math. Anal. Appl.* **259** (2001) 596–608.

4

STABILITY ANALYSIS OF IMPULSIVE SYSTEMS

4.1 Introduction

In many cases the processes and phenomena described by ordinary differential equations and/or partial equations are subject to the effect of short-time perturbations. Sometimes these perturbations are ignored and in some cases their action is "smoothed" and presented by the functions of constant perturbations. However there exist processes for which the above approach proves to be crude and to not allow for peculiarities of the physical system under consideration. Among many examples of such systems we note the problems of clocks dynamics, population dynamics, electron technologies, pharmacokinetics and so on.

This chapter sets out some new approaches to the problems of qualitative analysis of like-stability properties of solutions for linear and non-linear impulsive systems. Major attention is focused on the development of the direct Liapunov method via multicomponent auxiliary functions.

The plan of the Chapter is as follows. Sections 4.2 – 4.4 are of the nature of reviews.

In Section 4.2 a system of ordinary differential equations is presented which describes a real system subject to short-term (impulsive) perturbations.

In Section 4.3 we impose the existence conditions solutions to the impulsive system and discuss the so-called "beating" problem characteristic of impulsive systems.

In Section 4.4 some results of stability analysis of solutions for the first approximation impulsive system are presented.

Section 4.5 deals with basic theorems of the direct Liapunov method via matrix-valued auxiliary functions. We consider two methods of matrix function application: scalar and vector ones.

In Section 4.6 new sufficient stability conditions are established for large-scale impulsive system. These conditions are formulated in terms of the property of definite sign of special matrices.

Section 4.7 deals with a statement of the problem and some results of stability analysis of an impulsive system with uncertain parameter values. Besides, the impulsive system is considered as a hybrid one, consisting of continuous and discrete components. It is assumed that both components have a similar effect on the dynamical properties of the whole system. A linear system is considered as an example and robust stability conditions are established for its solutions.

4.2 What is a Model of Impulsive Systems?

Mathematical modelling is a subject that is difficult in practice. In every case in the mathematical modelling of physical systems, one is invariably confronted with a dilemma: to use a more accurate model which is harder to manage, or to work with a simpler model which is easier to manipulate but with less confidence. As a result of many considerations of behavior of systems subject to short-time perturbations, a model was chosen which we present below. On the one hand this model is a natural generalization of the continuous description of the process, and on the other hand it allows one to take into account real impulsive perturbations.

Let some evolutionary process be described by

(a) the system of differential equations

$$(4.2.1) \qquad \frac{dx}{dt} = f(t,x),$$

where $f\colon R_+ \times \Omega \to R^n$, $\Omega \subset R^n$ is an open set in R^n;
(b) some sets $M(t)$, $N(t) \subset \Omega$ for every $t \in R_+$;
(c) the operator $A(t)\colon M(t) \to N(t)$ for every $t \in R_+$.

Let $x(t) = x(t;t_0,x_0)$ be a solution of system (4.2.1) starting at point (t_0,x_0).

We designate by $P_t = (t,x(t))$ the representative point at the phase space $R_+ \times R^n$. The set of all representative points for $t \in [0,+\infty)$ is referred to as a trajectory of the dynamical system.

The evolution of the process takes place as follows: the point P_t starts moving from the initial point $P_{t_0} = (t_0,x_0)$ and goes on in the curve $\{(t,x)\colon t \geq t_0, x = x(t)\}$ up to the time $t_1 > t_0$, when the point P_t meets the set $M(t)$. At the time $t = t_1$ the operator $A(t)$ transfers the point $P_{t_1} = (t_1,x(t_1))$ at the point $P_{t_1}^+ = (t_1,x_1^+) \in N(t_1)$, where $x_1^+ = A(t_1)x(t_1)$. Then the point P_t goes on along the trajectory with $x(t) = x(t;t_1,x_1^+)$

as a solution of system (4.2.1), started at the point $P_{t_1} = (t_1, x_1^+)$, till it meets the set $M(t)$ at time $t_2 > t_1$. As earlier at the point $t = t_2$ the point $P_{t_2} = (t_2, x(t_2))$ is transferred to the point $P_{t_2}^+ = (t_2, x_2^+) \in N(t_2)$, where $x_2^+ = A(t_2)x(t_2)$. After that, the point P_t keeps its motion in the trajectory with $x(t) = x(t; t_2, x_2^+)$ as the solution of system (4.2.1), started at the point (t_2, x_2^+). This lasts till the solution of system (4.2.1) exists, or till it is reasonable in a concrete engineering or scientific problem.

We define the system with impulsive effect as a procession of sets and spaces, the correlations between which are established by system (4.2.1), the sets $M(t)$ and $N(t)$ and the operator $A(t)$.

The trajectory made by the point P_t is refered to as an integral curve, and the function $x = x(t)$ that assigns this curve is called a solution with impulsive effect.

The solution to a system with impulsive effect can be

(a) a continuous function, if the integral curve does not intersect the set $M(t)$ or hits it at the fixed point of the operator $A(t)$;

(b) a piecewise continuous function having a finite number of discontinuities of the first kind if the integral curve meets $M(t)$ at a finite number of points which are not the fixed points of the operator $A(t)$;

(c) a piecewise continuous function having a countable number of discontinuities of the first kind if the integral curve encounters the set $M(t)$ at a countable number of points that are not the fixed points of the operator $A(t)$.

The times t_k when the representative point $P-t$ gets into the set $M(t)$ are called *impulsive effect moments*. We assume that the solutions $x(t)$ of a system with impulsive effect are continuous on the left at point t_k, $k \in \mathcal{Z}$, which means that

$$x(t_k^-) = \lim_{h \to 0^+} x(t_k - h) = x(t_k).$$

Under a certain choice of the correlations (a), (b) and (c) some types of systems with impulsive effect are obtained.

4.2.1 Systems with impulses at fixed times If in a real process described by the system (4.2.1) the impulses occur at fixed times, the mathematical model of this process will be given by the following impulsive system

(4.2.2)
$$\frac{dx}{dt} = f(t, x), \quad t \ne t_k, \quad k \in \mathcal{Z},$$
$$\Delta x = I_k(x), \quad t = t_k,$$

where $\Delta x(t_k) = x(t_k^+) - x(t_k)$ for $t = t_k$ and

$$x(t_k^+) = \lim_{h \to 0^+} x(t_k + h).$$

4.2.2 Systems with impulses at variable times Let a sequence of hypersurfaces $\{S_k\}$ be defined as $S_k : t = (\tau_k(x))$, $k \in \mathcal{Z}$, and moreover, $\tau_k(x) < \tau_{k+1}(x)$ and $\lim_{k \to \infty} \tau_k(x) = \infty$. Here it is reasonable to consider a system with impulsive effect in the form

(4.2.3)
$$\frac{dx}{dt} = f(t, x), \quad t \neq \tau_k(x),$$
$$\Delta x = I_k(x), \quad t = \tau_k(x), \quad k \in \mathcal{Z}.$$

Systems of the form (4.2.3) are more difficult for investigation then those of the type (4.2.2). The obvious difficulty is that the moments of impulse effects depend on the solution that remains unknown, i.e. $t_k = \tau_k(x(t_k))$, $k \in \mathcal{Z}$. The other difficulty is connected with the phenomena of beating of the solutions on the surfaces $t = \tau_k(x)$. It is the beating that frequently prevents the solution from being continuable on the time interval of interest for the application.

In the present Section we deal with the systems of equations where beating is absent. These are the systems the solutions of which intersect each one of the hypersurfaces $t = (\tau_k(x))$, $k \in \mathcal{Z}$ only once.

4.2.3 Autonomous systems with impulses Let the sets $M(t) \equiv M$, $N(t) \equiv N$ and the operator $A(t) \equiv A$ be independent of t and let $A \colon M \to N$ be defined by $Ax = x + I(x)$, where $I \colon \Omega \to \Omega$. Consider the autonomous impulsive differential system

(4.2.4)
$$x' = f(x), \quad x \notin M,$$
$$\Delta x = I(x), \quad x \in M.$$

When any solution $x(t) = x(t, 0, x_0)$ hits the set M at some time t, the operator A instantly transfers the point $x(t) \in M$ into point $y(t) = x(t) + I(x(t)) \in N$.

Since (4.2.4) is autonomous, the motion of the point $x(t)$ will be considered in Ω, along the trajectories of the system (4.2.4).

Currently the presented systems of differential equations describing various processes in real systems with impulsive perturbations are studied by means of both analytical and quantitative approaches. As for the system of ordinary differential equations, the problem of solutions existence (motion realizability, or conditions of some process going in time) is of importance.

4.3 Existence of Solutions and the Problem of "Beating"

The system of equations (4.2.3) is general enough to model a wide class of physical systems functioning in the presence of impulse perturbations. We shall present some existence conditions for solutions to this class of systems of equations.

Let $\Omega \in R^n$ be an open set and $D = R_+ \times \Omega$, where $R_+ = [0, \infty)$. Assume for each $k = 1, 2, \ldots$, $\tau_k \in C(\Omega, (0, +\infty))$, $\tau_k(x) < \tau_{k+1}(x)$, and $\lim_{k \to \infty} \tau_k(x) = \infty$ for all $x \in \Omega$. Assume further that $\tau_0(x) \equiv 0$ and k takes the values from 1 to $+\infty$.

Definition 4.3.1 *Function* $x: (t_0, t_0 + \alpha) \to R^n$, $t_0 \geq 0$, $\alpha > 0$ *is a solution of system (4.2.4) if:*

(a) $x(t_0^+) = x_0$ and $(t, x(t)) \in D$ for all $t \in [t_0, t_0 + \alpha)$,
(b) *the function* $x(t)$ *is continuously differentiable and satisfies the equation*

(4.3.1a) $$\frac{dx}{dt} = f(t, x(t))$$

for $t \in [t_0, t_0 + \alpha)$ and $t \neq \tau_k(x(t))$;
(c) if $t \in [t_0, t_0 + \alpha)$ and $t = \tau_k(x(t))$, then

(4.3.1b) $$x(t^+) = x(t) + I_k(x(t)).$$

For these values of t the function $x(t)$ is assumed to be continuous on the left and for some $\delta > 0$, $s \neq \tau_j(x(s))$ for any j, $t < s < \delta$.

Note that the use of the initial conditions $x(t_0^+) = x_0$ instead of the usual ones $x(t_0) = x_0$ for system (4.3.1) is reasonable because the values (t_0, x_0) may be such that for some k: $t_0 = \tau_k(x_0)$. Whenever $t_0 \neq \tau_k(x_0)$ for any k, the initial conditions are understood in the usual sense, i.e. $x(t_0) = x_0$.

Theorem 4.3.1 *Assume that*

(1) *in system (4.3.1) the vector-function* $f: R_+ \times \Omega \to R^n$ *is continuous for* $t \neq \tau_k(x)$, $k = 1, 2, \ldots$, *for all* $(t, x) \in R_+ \times \Omega$;

(2) there exists a locally integrable function $\kappa \in L^1$ such that in the arbitrary small neighborhood of the point $(t,x) \in R_+ \times \Omega$ the estimate $\|f(s,x)\| \leq \kappa(s)$ is satisfied;

(3) for each $k \in Z$ and $t_1 = \tau_k(x_1)$ there exists $\delta > 0$ such that $t \neq \tau_k(x)$ for any $0 < t - t_1 < \delta$ and $\|x - x_1\| < \delta$.

Then for any initial values $(t_0, x_0) \in R_+ \times \Omega$ a solution $x \colon [t_0, t_0 + \alpha) \to R^n$ exists for the initial problem (1.4.1) for some $\alpha > 0$.

For the *proof* of this theorem see the monograph by Lakshmikantham, Bainov and Simeonov [1].

Assume that system (4.3.1) is

(4.3.2)
$$\frac{dx}{dt} = A(t)x, \quad \text{for} \quad t \neq t_k,$$
$$\Delta x = B_k x, \quad \text{for} \quad t = t_k, \quad k \in Z,$$
$$x(t_0^+) = x_0.$$

Here $A(t)$ is $n \times n$ matrix, continuous on the interval $(a, b) \subset R_+$, B_k are $n \times n$ constant matrices, $t_k \in R_+$ are fixed instants of time such that $t_k < t_{k+1}$, $\lim_{k \to \infty} t_k = \infty$.

Corollary 4.3.1 Assume that the interval $[t_0, t_0 + \alpha] \subset R_+$ contains a finite number of points t_k. Then, for any $x_0 \in R^n$ the solution $x \colon [t_0, t_0 + \alpha] \to R^n$ of system (4.3.2) exists for all $t \in [t_0, t_0 + \alpha]$.

Moreover, if for all k, the instants $t_k \in [t_0, t_0 + \alpha]$ and the matrices $E + B_k$ are nonsingular, then $x(t; t_0, x_0) \neq x(t; t_0, y_0)$ for all $t \in [t_0, t_0 + \alpha]$ if $x_0 \neq y_0$.

We designate by $U(t, \tau)$ the solution to the Cauchy problem for the matrix equation

(4.3.3)
$$\frac{dU}{dt} = A(t)U, \quad U(t, \tau) = E.$$

Besides, any solution of the matrix system

(4.3.4)
$$\frac{dx}{dt} = A(t)x \quad \text{for} \quad t \neq t_k,$$
$$\Delta X = B_k X \quad \text{for} \quad t = t_k$$

can be presented as

$$X(t) = U(t, \tau_{j+k})(E + B_{j+k})U(t_{j+k}, t_{j+k-1})$$
$$\times (E + B_{j+k-1})\ldots(E + B_j)U(t_j, t_0)X(t_0),$$
$$t_{j-1} < t_0 \leq t_j < t_{j+k} < t \leq t_{j+k+1}.$$

For the matriciant of system (4.3.4) the following expression is known

$$X(t,t_0) = U(t,t_{j+k})(E + B_{j+k}) \prod_{v=k}^{1} U(t_{j+v},t_{j+v-1})(E + B_{j+v-1})U(t_j,t_0).$$

For the investigation of the inhomogeneous impulsive system

(4.3.5)
$$\frac{dx}{dt} = A(t)x + f(t) \quad \text{for} \quad t \neq t_k,$$
$$\Delta x = B_k x + a_k \quad \text{for} \quad t = t_k$$

the following result is known.

Theorem 4.3.2 *Let* $x = \varphi(t)$ *be a solution of system (4.3.2) and* $x = \psi(t)$ *be a solution of (4.3.5). Then the function* $x = \varphi(t) + \psi(t)$ *is a solution of system (4.3.5) and vice versa, if* $x = \varphi_1(t)$ *and* $x = \varphi_2(t)$ *are solutions of inhomogeneous system (4.3.5), then the function* $x = \varphi_1(t) - \varphi_2(t)$ *is a solution of system (4.3.2).*

Corollary 4.3.2 *Let* $X(t,t_0)$ *be a matriciant of system (4.3.2) and the matrices* $E + B_k$, $k \in Z$, *non-degenerated Then, any solution* $x(t,t_0) = x(t_0,x_0) = x_0$ *of system (4.3.5) can be presented as*

$$x(t,t_0) = X(t,t_0)x_0 + \int_{t_0}^{t} X(t,\tau)f(\tau)\,d\tau + \sum_{t_0 < t_k < t} X(t,t_k+0)a_k.$$

Assume that in system (4.3.2) matrices $A(t)$ and B_k are constant, i.e.

(4.3.6)
$$\frac{dx}{dt} = Ax \quad \text{for} \quad t \neq t_k,$$
$$\Delta x = Bx \quad \text{for} \quad t \neq t_k, \quad k \in Z.$$

Assume that $t_k \to +\infty$ as $k \to \infty$ and $t_1 > t_0$. Then, any solution of system (4.3.6) presented the function

(4.3.7)
$$x(t,t_0) = X(t,t_0)x_0,$$

where

(4.3.8)
$$X(t,t_0) = e^{A(t-t_k)} \prod_{t_0 < t_k < t} (E + B)e^{A(t_v - t_{v-1})}.$$

In the case when the matrices A and B in system (4.3.6) is commute with each other, solution (4.3.7) becomes

$$x(t, x_0) = e^{\lambda t}(E + B)^{i(0,t)} x_0, \qquad (4.3.9)$$

where λ is an eigenvalue of the matrix A and x_0 is an eigenvector corresponding to it.

In systems with impulsive effect the phenomenon of "beating" is of essential importance. In consists in one or multiple intersections by the solution $x(t, t_0, x_0)$ of the surfaces $S_k: t = \tau_k(x)$, $k = 1, 2, \ldots$.

Qualitative analysis of the impulsive system motions is simplified when the beating is absent. We shall present some conditions under which the solution $x(t, t_0, x_0)$ intersects the surfaces $S_k: t = \tau_k(x)$ only once.

Theorem 4.3.3 *Assume that system (4.3.1) satisfies the following conditions:*

(1) *the vector-function* $f \in C(R_+ \times \Omega, R^n)$, $\tau_k \in C^1(\Omega, (0, \infty))$, $\tau_k(x) < \tau_{k+1}(x)$ *for any* k, *and* $\lim_{k \to \infty} \tau_k(x) = \infty$ *uniformly in* $x \in \Omega$, $I_k \in C(\Omega, R^n)$;

(2) $\dfrac{\partial \tau_k(x)}{\partial x} f(t, x) \leq 0$ *for all* $(t, x) \in R_+ \times \Omega$;

(3) $x + I_k(x) \in \Omega$ *for* $x \in \Omega$ *and for any* k

$$\left(\dfrac{\partial \tau_k(x)}{\partial x}(x + \sigma I_k(x))\right)^T I_k(x) \leq 0, \quad 0 \leq \sigma \leq 1.$$

Then any solution of system (4.3.1) remaining in the domain Ω *for all* $t \in [t_0, t_0 +]$ *intersects the surface* $S_j: t = \tau_j(x)$ *only once.*

Finally, we shall present conditions for which the solution $x(t; t_0, x_0)$ intersects repeatedly the hypersurfaces $S_k: t = \tau_k(x)$, $k = 1, 2, \ldots$.

First, we consider the case of one hypersurface $S: t = \tau(x)$.

Theorem 4.3.4 *Assume that*

(1) *condition (1) of Theorem 4.3.3 is satisfied;*
(2) *function* $\tau(x)$ *is bounded;*
(3) *for all* $x \in \Omega$ *the inclusion* $x + I(x) \in \Omega$ *holds and for* $0 \leq \alpha \leq 1$

$$\left(\dfrac{\partial \tau}{\partial x}(x + \alpha I(x))\right)^T I(x) > 0.$$

is true.

Then, any solution $x(t, t_0, x_0)$ of system (4.3.1) for $0 \leq t_0 < \tau(x_0)$ intersects the hypersurface $S: t = \tau(x)$ several times.

When there are several hypersurfaces $t = \tau_k(x)$, $k = 1, 2, \ldots$, Theorem 4.3.4 is generalized as follows.

Theorem 4.3.5 *Assume that*

(1) *condition (1) of Theorem 4.3.3 is satisfied;*
(2) *for every fixed $k \in Z$, $\tau_k(x)$ are bounded and*

(a) $\left(\dfrac{\partial \tau_{k-1}(x)}{\partial x} \right)^{\mathrm{T}} f(t, x) \leq 1$ *for* $(t, x) \in R_+ \times \Omega$;

(b) *for all $x \in \Omega$ $x + I_k(x) \in \Omega$ and*

$$\left(\frac{\partial \tau_k}{\partial x} (x + \sigma I_k(x)) \right)^{\mathrm{T}} I_k(x) > 0;$$

(c) $\left(\dfrac{\partial \tau_{k-1}}{\partial x} (x + \sigma I_k(x)) \right)^{\mathrm{T}} I_k(x) \leq 0$ *for* $0 \leq \sigma \leq 1$.

Then any solution $x(t) = x(t; t_0, x_0)$ of system (4.3.1) for $\tau_{k-1}(x_0) < t_0 < \tau_k(x_0)$, $k \in Z$ intersects the surface S_k several times.

Proof For the proof of Theorem 4.3.3–4.3.5 see Lakshmikantham, Bainov and Simeonov [1].

4.4 Stability in the Sense of Liapunov

We go on with our discussion of the dynamical properties of the impulse system (4.3.1). Assume that the hypersurfaces S_k are determined by the equations $t = \tau_k(x)$, $k = 1, 2, \ldots$, and $\tau_k(x) \to \infty$ as $k \to \infty$. The notion of stability of solutions to the impulse system (4.3.1) should allow for two peculiarities characteristic of the systems of the type.

First, in the systems of (4.3.1) type the beating of solutions on the surfaces $S_k: t = \tau_k(x)$ is possible.

Second, the solutions of system (4.3.1) in general do not depend continuously on the initial conditions, and the continuity is uniform on some finite interval.

As a result, the classical definition of stability of arbitrary solution $x(t) = x(t; t_0, x_0)$ and/or zero solution $x(t) = 0$ (under the corresponding assumptions on system (4.3.1)) is somewhat modified.

Definition 4.4.1 Let $x_0(t) = x(t; t_0, y_0)$ be a prescribed solution of system (4.3.1) existing for all $t \geq t_0$. Assume that the solution $x_0(t)$ intersects the surfaces $S_k: t = \tau_k(x)$ at times τ_k such that $\tau_k(x) < \tau_{k+1}(x)$, $\tau_k(x) \to \infty$ as $k \to \infty$. We claim that the solution $x_0(t)$ of system (4.3.1) is:

(a) *stable*, if for arbitrary $\varepsilon > 0$, $\eta > 0$ and $t_0 \in R_+$ a $\delta = \delta(t_0, \varepsilon, \eta) > 0$ exists such that for any other solution $x(t)$ of system (4.3.1) the estimate $\|x(t) - x_0(t)\| < \varepsilon$ is satisfied for all $t \geq t_0$ whenever $\|x(t_0) - x_0(t_0)\| < \delta$ and $|t - t_k^0| > \eta$, where t_k^0 is the time when the solution $x(t)$ intersects the hypersurfaces $t = \tau_k(x)$;

(b) *uniformly stable*, if the value δ in Definition 4.4.1(a) does not depend on t_0;

(c) *attractive*, if for arbitrary $\varepsilon > 0$, $\eta > 0$ and $t_0 \in R_+$ there exists $\delta_0 = \delta_0(t_0)$ and some $T = T(t_0, \varepsilon, \eta) > 0$ such that $\|x(t) - x_0(t)\| < \varepsilon$ for $t \geq t_0 + T$ and $|t - t_k^0| > \eta$ whenever $\|x(t_0) - x_0(t_0)\| < \delta_0$;

(d) *uniformly attractive*, if the values δ_0 and T in Definition 4.4.1 (b) do not depend on t_0;

(e) *asymptotically stable*, if conditions of Definitions 4.4.2(a) and (c) are satisfied;

(f) *uniformly asymptotically stable*, if conditions of Definitions 4.4.1 (b) and (d) are satisfied.

Note that if $f(t, 0) \equiv 0$ and $I_k(0) \equiv 0$ for all $k \in Z$, then system (4.3.1) admits trivial solution. Besides, Definitions 4.4.1 (a)–(f) are simplified.

4.4.1 Stability analysis in the first approximation
We consider some partial cases of system (4.3.1) and present the corresponding stability conditions of the equilibrium state $x = 0$.

Assume that system (4.3.1) can be written in quasilinear form

(4.4.1)
$$\frac{dx}{dt} = A(t)x + g(t, x) \quad \text{for} \quad t \neq \tau_k(x),$$
$$\Delta x = B_k x + I_k^*(x) \quad \text{for} \quad t = \tau_k(x),$$

where $x \in R^n$, $A(t) \in C(R_+, R^{n \times n})$, the matrix $A(t)$ is bounded for all $t \geq t_0$, B_k are constant matrices for $k \in Z$, vector-functions $g(t, x)$ and $I_k^*(x)$ are definite for all $t \geq t_0$ and $x \in \Omega(H)$ and

(4.4.2)
$$\|g(t, x)\| \leq \alpha(t)\|x\|, \quad \|I_k^*(x)\| \leq \beta_k \|x\|,$$

for all $t \geq t_0$, $x \in \Omega(H)$, $k \in Z$, $\alpha(t) > 0$, for all $t \geq t_0$, $\beta_k > 0$, $k \in Z$. We assume that the functions $\tau_k(x)$ satisfy the conditions

(a) $|\tau_k(x') - \tau_k(x'')| \leq \|x' - x''\|$, $k = 1, 2, \ldots,$

(b) $\tau_k(x) \geq \tau_k(x + I_k(x))$, $x, x', x'' \in \Omega(H)$,

and, moreover, a constant $\theta > 0$, exists such that

(4.4.3) (c) $\sup_k \left(\min_{x \in \Omega(H)} \tau_{k+1}(x) - \max_{x \in \Omega(H)} \tau_k(x) \right) \geq \theta.$

Together with system (4.4.1) we consider the linear approximation

(4.4.4)
$$\frac{dx}{dt} = A(t)x \quad \text{for} \quad t \neq \tau_k(x), \quad k \in Z,$$
$$\Delta x = B_k x \quad \text{for} \quad t = \tau_k(x),$$

and assume that the functions $\tau_k(x)$ satisfy the condition

(4.4.5) $\qquad |\tau_k(x) - \tau_k(0)| \leq \delta^*(r),$

where $\delta^*(r) > 0$, $\delta^*(r) \to 0$ for $r \to 0$.

Theorem 4.4.1. *Assume that*

(1) *for system (4.4.1) conditions (4.4.2), (4.4.3) and*

$$\tau_k(x) \geq \tau_k[(E + B_k)x + I_k^*(x)]$$

are satisfied for all $x \in \Omega(H)$ and $k \in Z$;

(2) *conditions*

$$\int_{t_0}^{\infty} \alpha(t)dt < +\infty, \qquad \prod_{\tau_k > t_0} (1 + \beta_k) < +\infty;$$

are satisfied;

(3) *under condition (4.4.5) the solutions of system (4.4.4) are stable (exponentially stable).*

Then the equilibrium state $x = 0$ of system (4.4.1) is stable (exponentially stable).

Proof For the proof of this Theorem see Samoilenko and Perestyuk [1].

In the case, when $n \times n$ matrices $A(t)$ and B_k in system (4.4.1) are constant, the stability conditions for the state $(x = 0) \in R^n$ of system

(4.4.6)
$$\frac{dx}{dt} = Ax + g(t, x), \quad t \neq \tau_k(x),$$
$$\Delta x = B_k x + I_k^*(x), \quad t = \tau_k(x),$$

can be formulated as follows.

Designate $\gamma = \max_j \operatorname{Re} \lambda_j(A)$;

$$\alpha^2 = \max_j \lambda_j[(E + B)^{\mathrm{T}}(E + B)], \quad j = 1, 2, \ldots, n,$$

$$p = \lim_{\sigma \to \infty} \frac{k(t, t + \sigma)}{\sigma},$$

where $k(t, t + \sigma)$ is a number of points $\tau_k(x)$ on the interval $[t, t + \sigma]$.

Theorem 4.4.2 *Assume that*

(1) *there exists a sufficiently small constant value $a > 0$ such that*

(4.4.7)
$$\|g(t, x)\| \leq a\|x\|, \quad \|I_k^*(x)\| \leq a\|x\|$$

for all $t \geq t_0$ and $x \in \Omega(H)$;

(2) *functions $\tau_k(x)$ satisfy Lipschitz condition and $\tau_k(x) \geq \tau_k((E + B)x + I_k^*(x))$ for all $k = 1, 2, \ldots$ and $x \in \Omega(H)$;*

(3) *for any finite p the inequality $\gamma + p \ln \alpha < 0$ is satisfied.*

Then the equilibrium state $x = 0$ of system (4.4.6) is asymptotically stable.

Remark 4.4.1 Condition (4.4.3) employed in Theorems 4.4.1 and 4.4.2 can be replaced by the following one

$$0 < \theta_1 \leq \min_{x \in \Omega(h)} \tau_{k+1}(x) - \max_{x \in \Omega(h)} \tau_k(x) \leq \theta_2 \quad \text{for all} \quad k = 1, 2, \ldots.$$

Further we shall present some results of stability analysis of the state $x = 0$ of system

(4.4.8)
$$\frac{dx}{dt} = A(t)x + g(t, x), \quad t \neq t_k,$$
$$\Delta x = B_k x + I_k^*(x), \quad t = t_k,$$
$$x(t_0^+) = x_0$$

under the following assumptions

H_1: $0 < t_1 < t_2 < \cdots < t_k < \ldots \lim_{k \to \infty} t_k = \infty$;

H_2: $A(t)$ is an $n \times n$ matrix piece-wise continuous on R_+ with discontinuity points of the first kind at points $t = t_k$ and continuous on the left at these points.

H_3: vector function $g(t, x) \colon R_+ \times R^n \to R^n$ is continuous for $(t_{k-1}, t_k] \times R^n$ for any $x \in R^n$, $k = 1, 2, \ldots$, and $\lim_{(t,y) \to (t_k, x)} g(t, y)$ exists for $t > t_k$;

H_4: for any $k \in Z$ the matrices B_k are constant and the vector-functions $I_k^*(x) \colon R^n \to R$ are continuous for all $x \in R^n$, $k \in Z$.

Let $W_k(t, s)$ be a fundamental matrix of solutions of the linear impulsive system

(4.4.9)
$$\frac{dx}{dt} = A(t)x \quad \text{for} \quad t \neq t_k,$$
$$\Delta x = B_k x \quad \text{for} \quad t = t_k, \quad k \in Z,$$

where $x \in R^n$, and $t_{k+1} - t_k \geq \theta$, for some $\theta > 0$.

Assumption 4.4.1 There exist

(1) piece-wise continuous functions $\varphi(t)$ and $\psi(t)$ with the values in $(0, +\infty)$, $\varphi(t_k) > 0$, $\psi(t_k) > 0$ such that
$$\|W_k(t, s)\| \leq \varphi(t)\psi(s), \quad t_{k-1} < s \leq t \leq t_k;$$

(2) piece-wise continuous function $a(t) \geq 0$ and $m > 1$ such that
$$\|g(t, x)\| \leq a(t)\|x\|^m \quad \text{for all} \quad (t, x) \in R_+ \times \Omega(H);$$

(3) constants $\gamma_k, \rho_k > 0$ such that
$$\|E + B_k x\| \leq \gamma_k, \quad \|I_k^*(x)\| \leq \rho_k \|x\| \quad \text{for all} \quad x \in \Omega(H).$$

We introduce the designations
$$r_k = (\gamma_k + \rho_k)\varphi(t_k)\psi(t_k^+),$$
$$D(t_0, t) = \int_{t_0}^t \left(\prod_{t_0 < t_k < s} r_k \right)^{m-1} \varphi^m(s)\psi(s)a(s)\,ds$$

and formulate additional conditions.

Assumption 4.4.2 For system (4.4.8)

(1) there exists a function $N(t_0)$ such that

$$\varphi(t) \prod_{t_0 < t_k < t} r_k < N(t_0) \quad \text{for all} \quad t \geq t_0;$$

(2) inequality $D(0, \infty) < +\infty$ is satisfied;

(3) there exists a constant $N > 0$ such that

$$\varphi(t)\psi(t_0^+) \prod_{t_0 < t_k < t} r_k < N \quad \text{for all} \quad t \geq t_0, \quad \text{and}$$

$$\psi^{m-1}(t_0^+)D(t_0, \infty) < N \quad \text{for all} \quad t_0 > 0;$$

(4) for any $0 < \varepsilon < H$ and $t_0 \in R_+$ there exists $T = T(t_0, \varepsilon) > 0$ such that

$$\varphi(t) \prod_{t_0 < t_k < t} r_k < \varepsilon \quad \text{for all} \quad t \geq t_0 + T;$$

(5) for any $0 < \varepsilon < H$ there exists $T^* = T^*(\varepsilon) > 0$ such that

$$\varphi(t)\psi(t_0^+) \prod_{t_0 < t_k < t} r_k < \varepsilon \quad \text{for all} \quad t \geq t_0 + T^*, \quad t_0 \in R_+.$$

Note that for the known matrix of fundamental solutions of system (4.4.9) for the function $n(t) = \frac{\|x(t)\|}{\varphi(t)}$, where $x(t) = x(t; t_0, x_0)$ is a solution of system (4.4.8) the inequality

$$n(t) \leq \|x_0\|\varphi(t_0^+) \prod_{t_0 < t_k < t} \gamma_k \varphi(t_k)\psi(t_k^+)$$

(4.4.10)
$$+ \int_{t_0}^{t} \prod_{s < t_k < t} \gamma_k \varphi(t_k)\psi(t_k^+)\varphi^m(s)\psi(s)a(s)n^m(s)\, ds$$

$$+ \sum_{t_0 < t_k < t} \rho_k \varphi(t_k)\psi(t_k^+) \prod_{t_k < t_i < t} \gamma_i \varphi(t_i)\psi(t_i^+)n(t_k),$$

is valid for all $t > t_0$.

In terms of estimate (4.4.10) the validity of the following assertion is established.

Theorem 4.4.3 *Assume that*
(1) *system of equations (4.4.8) satisfies assumptions $H_1 - H_4$;*
(2) *conditions (1) - (3) of Assumption 4.4.1 are fulfilled;*
(3) *in Assumption 4.4.2*
  (a) *conditions (1), (2);*
  (b) *condition (3);*
  (c) *conditions (2), (4);*
  (d) *conditions (3), (5) are satisfied.*

Then the equilibrium state $x = 0$ of system (4.4.8) is
  (a) *stable;*
  (b) *uniformly stable;*
  (c) *asymptotically stable;*
  (d) *uniformly asymptotically stable,*

respectively.

Proof For the proof of this theorem see Lakshmikantham, Bainov and Simeonov [1].

4.5 Matrix Liapunov Functions Method

The direct Liapunov method (see Liapunov [1]) is an effective tool of qualitative analysis of solutions to impulsive systems. The application of multicomponent auxiliary functions makes this technique more versatile and fruitful. This is true for both the vector Liapunov function and the matrix-valued one. In this section we present the method of matrix Liapunov functions for systems with impulsive perturbations.

4.5.1 A scalar approach Now system (4.3.1) is considered under the following assumptions:

A_1: Vector-function $f\colon R_+ \times \Omega \to R^n$ is continuous on $R_+ \times \Omega$, $f(t,0) = 0$ for all $t \in R_+$ and there exists a constant $L > 0$ such that

$$\|f(t, x') - f(t, x'')\| \leq L\|x' - x''\|$$

for all $t \in R_+$ and $x', x'' \in \Omega$;

A_2: Vector-functions $I_k(x)\colon \Omega \to R^n$, $k = 1, 2, \ldots$ are continuous on Ω, $I_k(0) = 0$, $k = 1, 2, \ldots$;

A_3: There exists a constant $\mu \in (0, H)$ such that if $x \in \Omega(\mu)$, then $x + I_k(x) \in \Omega(H)$, $k = 1, 2, \ldots$;

A_4: Functions $\tau_k(x)\colon \Omega \to R_+$, $k = 1, 2, \ldots$, are continuous on Ω and for each $x \in \Omega$:

$$0 < \tau_1(x) < \cdots < \tau_k(x) < \ldots, \quad \lim_{k \to \infty} \tau_k(x) = \infty;$$

A_5: Integral curves of system (4.3.1) intersect each hypersurface $S_k\colon t = \tau_k(x)$, $k = 1, 2, \ldots$, only once.

A_6: For system (4.3.1) there exist a matrix-valued function $U(t, x)$, $U \in C(R_+ \times \Omega, R^{m \times m})$ and a vector $\alpha \in R_+^m$ such that the function

(4.5.1) $$v(t, x, \alpha) = \alpha^T U(t, x) \alpha$$

satisfies the conditions below.

Let $\tau_0(x) \equiv 0$ for all $x \in \Omega$, and

$$G_k = \{(t, k) \in R_+ \times \Omega\colon \tau_{k-1}(x) < t < \tau_k(x)\},$$

$$k = 1, 2, \ldots, \quad G = \bigcup_{k=1}^{\infty} G_k.$$

Definition 4.5.1 *Function $v(t, x, \alpha)$, $v \in C(R_+ \times \Omega \times R_+^m, R)$, belongs to the class SL_0 if:*

(1) *$v(t, x, \alpha)$ is continuous on each set G_k and $v(t, 0, \alpha) = 0$ for all $t \in R_+$;*

(2) *for each $k = 1, 2, \ldots$ and $(t_0, x_0) \in S_k$ there exist finite limits*

$$\lim [v(t, x, \alpha)\colon (t, x) \to (t_0, x_0);\ (t, x) \in G_k] = v(t_0 - 0,\ x_0, \alpha)$$
$$\lim [v(t, x, \alpha)\colon (t, x) \to (t_0, x_0);\ (t, x) \in G_{k+1}] = v(t_0 + 0,\ x_0, \alpha)$$

and the correlation $v(t_0 - 0, x_0, \alpha) = v(t_0, x_0, \alpha)$ is true.

If $(t_0, x_0) \notin S_k$ then $v(t_0 + 0, x_0, \alpha)$ is designated by $v(t_0, x_0, \alpha)$.

Function $v(t, x, \alpha)$ belongs to the class SL_1 if v is of class SL_0 and continuously differentiable on G.

For $(t, x) \in G$ the function

$$Dv(t, x, \alpha)|_{(4.3.1)} = \frac{\partial v}{\partial t} + (\operatorname{grad} v)^T f(t, x), \quad t \neq \tau_k(x), \quad k = 1, 2, \ldots$$

is a total derivative of function $v \in SL_1$ along solutions of system (4.3.1).

4. IMPULSIVE SYSTEMS

Theorem 4.5.1 *For system of equations (4.3.1) let conditions $A_1 - A_6$ be satisfied and for the function v of class SL_1*

(1) $a(\|x\|) \leq v(t, x, \alpha)$ *for all* $(t, x, \alpha) \in R_+ \times \Omega \times R_+^m$;

(2) $Dv(t, x, \alpha)|_{(4.3.1)} \leq \psi^{\mathrm{T}}(\|x\|) Q_1 \psi(\|x\|)$ *for all* $(t, x, \alpha) \in R_+ \times \Omega \times R_+^m$, $\psi^{\mathrm{T}}(\|x\|) = (\psi_1(\|x\|), \ldots, \psi_m(\|x\|))$ *for some constant $m \times m$ - matrix Q_1 and comparison function ψ_i is of class K, $i = 1, 2, \ldots, m$;*

(3) $v(t+0, x+I_k(x), \alpha) \leq v(t, x, \alpha)$ *for all* $(t, x, \alpha) \in (R_+ \times \Omega \times R_+^m) \cap S_k$, $k = 1, 2, \ldots$;

(4) *constant $m \times m$-matrix Q_1^* is negative semidefinite.*

Then the equilibrium state $x = 0$ of system (4.3.1) is stable.
If in addition to conditions (1)–(4) one more condition

(5) $v(t, x, \alpha) \leq b(\|x\|)$ *for all* $(t, x) \in R_+ \times \Omega$

is satisfied, where b is of class K, then the equilibrium state $x = 0$ of system (4.3.1) is uniformly stable.

Here the matrix Q_1^* is of the form $Q_1^* = \frac{1}{2}(Q_1 + Q_1^{\mathrm{T}})$. Since this matrix is negative semidefinite by condition (4) of Theorem 4.5.1, all its eigenvalues are real and satisfy the condition $\lambda_i(Q_1^*) \leq 0$, $i = 1, 2, \ldots, m$.

Theorem 4.5.2 *Assume that for system (4.3.1) Assumptions $A_1 - A_6$ and conditions (1), (2), (4) and (5) of Theorem 4.5.1 are satisfied and condition (3) is replaced by*

(3′) $v(\tau_k(x), x + I_k(x), \alpha) - v(\tau_k(x), x, \alpha) \leq -\psi(v(\tau_k, x, \alpha))$ *for all* $k = 1, 2, \ldots$,

where $\psi(s)$ is a continuous function, $s \geq 0$, $\psi(0) = 0$ and $\psi(s) > 0$ for $s > 0$.

Then the equilibrium state $x = 0$ of system (4.3.1) is uniformly asymptotically stable.

Let the set

$$A(t_0) = \{x_0 \in \Omega \colon \|x(t, t_0, x_0)\| \to 0 \text{ for } t \to +\infty\}$$

be an estimation of the attraction domain of the equilibrium state $x = 0$ of system (4.3.1). Together with the set $A(t_0)$ we shall consider the set

$$A(t, v, r) = \{x \in \Omega \colon v(t+0, x, \alpha) \leq a(r)\},$$

where $t \in R_+$, $r \in R_+$, function $v \in SL_1$ and a is of class K.

Theorem 4.5.3 *For system (4.3.1) let the conditions $A_1 - A_6$ be satisfied, the function v be of class SL_1, the comparison functions a, b and c be of class K, $i = 1, 2, \ldots, m$, so that*

(1) *for all $(t, x, \alpha) \in R_+ \times \Omega \times R_+^m$ the function $v(t, x, \alpha)$ satisfies the inequalities*

$$a(\|x\|) \leq v(t, x, \alpha) \leq b(\|x\|);$$

(2) *the inequality*

$$Dv(t, x, \alpha)|_{(4.3.1)} \leq c^{\mathrm{T}}(\|x\|) Q_2 r(\|x\|)$$

holds true for all $(t, x, \alpha) \in G \times R_+^m$;

(3) *for all $(t, x, \alpha) \in S_k \cap (R_+ \times \Omega(\mu) \times R_+^m)$, $k = 1, 2, \ldots$ the estimates*

$$v(t+0,\ x + I_k(x),\ \alpha) \leq v(t, x, \alpha)$$

are true;

(4) *the constant $m \times m$-matrix Q_2^* is negative definite.*

Then, if $0 < r < \mu$,, then for $(t_0, x_0) \in R_+ \times A(t_0, v, r)$ the limiting correlation $\lim\limits_{t \to \infty} \|x(t, t_0, x_0)\| = 0$ is valid and the equilibrium state $x = 0$ of system (4.3.1) is uniformly asymptotically stable.

Here the matrix Q_2^* is of the form $Q_2^* = \frac{1}{2}(Q_2 + Q_2^{\mathrm{T}})$. All its eigenvalues are real and satisfy the condition $\lambda_i(Q_2^*) < 0$, $i = 1, 2, \ldots, m$.

Theorem 4.5.4 *Assume that for system (4.3.1) the conditions $A_1 - A_6$ are satisfied, the function v is of class SL_1 and the comparison functions a, b, φ_i and ψ are of class K so that*

(1) *for all $(t, x, \alpha) \in R_+ \times \Omega \times R_+^m$ the estimates*

$$a(\|x\|) \leq v(t, x, \alpha) \leq b(\|x\|)$$

are valid;

(2) *for all $(t, x, \alpha) \in G \times R_+^m$ the inequality*

$$Dv(t, x, \alpha)|_{(4.3.1)} \leq \varphi^{\mathrm{T}}(v(t, x)) Q_3 \varphi(v(t, x, \alpha))$$

holds true for some constant $m \times m$-matrix Q_3;

(3) *for all $(t, x) \in S_k \cap (R_+ \times \Omega(\mu) \times R_+^m)$, $k = 1, 2, \ldots$, the estimates*

$$v(\tau_k(x),\ x + I_k(x),\ \alpha) \leq \psi(v(\tau_k(x), x, \alpha)), \quad k = 1, 2, \ldots,$$

are valid;
(4) there exists a constant $\theta > 0$ such that

(a) $\sup_{k} \left(\min_{\|x\| \leq H} \tau_{k+1}(x) - \max_{\|x\| \leq H} \tau_k(x) \right) = \theta > 0;$

(b) $\int_{a}^{\psi(a)} \dfrac{ds}{\varkappa(s)} \leq \theta;$

(c) for some $a_0 > 0$ when all $a \in (0, a_0]$,

$$\int_{a}^{\psi(a)} \dfrac{ds}{\varkappa(s)} \leq \theta - \gamma$$

for some $\gamma > 0;$

(5) constant $m \times m$-matrix Q_3^* is negative definite.

Then the equilibrium state $x = 0$ of system (4.3.1) is

(a) stable under conditions (1) – (4)(a), (b) and (5);
(b) asymptotically stable under conditions (1) – (4)(a), (c) and (5).

Here $\varkappa(s) = \lambda_M(Q_3^*) \Phi(s)$, where $\Phi(s)$ is of class K so that $\Phi(s) \geq \varphi^T(s)\varphi(s)$, $s = v(t, x, \alpha)$, $\lambda_M(Q_3^*)$ is a maximal eigenvalue of the matrix $Q_3^* = \frac{1}{2}(Q_3 + Q_3^T)$.

In instability theorems the auxiliary functions $v(t, x, \alpha)$ are applied with the following two properties

(A) the cross-section of the domain $\Pi(t, x) = \{(t, x) \in R_+ \times \Omega \times R_+^m : v(t, x, \alpha) > 0\}$ by the plane $t = \mathrm{const} > 0$ is a non-empty open set adjacent to the equilibrium state $x = 0$ for all $t \geq t_0$ and

(B) in the domain $\Pi(t, x)$ the function v of class SL_1 is bounded.

Theorem 4.5.5 Assume that for system (4.3.1) the conditions $A_1 - A_6$ are satisfied, the function v is of class SL_1 and possesses properties (A) and (B) and the comparison functions φ_i and ψ are of class K so that

(1) for all $(t, x, \alpha) \in G \times R_+^m$ the inequality

$$Dv(t, x, \alpha)|_{(4.3.1)} \geq \varphi^T(\|x\|) B_1 \varphi(\|x\|)$$

holds true for some constant $m \times m$-matrix B and vector $\varphi^T(\|x\|) = (\varphi_1(\|x\|), \ldots, \varphi_m(\|x\|));$

(2) for all $(t, x, \alpha) \in S_k \cap (\Pi(t, x) \times R_+^m)$, $k = 1, 2, \ldots$ the inequalities

$$v(\tau_k(x), x + I_k(x), \alpha) - v(\tau_k(x), x, \alpha) \geq \psi(v(\tau_k(x), x, \alpha)), \quad k = 1, 2, \ldots,$$

are satisfied;

(3) the constant $m \times m$-matrix B_1^* is positive semidefinite.

Then the equilibrium state $x = 0$ of system (4.3.1) is unstable.

Here $B_1^* = \frac{1}{2}(B_1 + B_1^T)$ and by condition (3) of Theorem 1.4.5 the condition $\lambda_i(B_1^*)$ is satisfied for all $i = 1, 2, \ldots, m$.

The following result is based on the conditions opposite to the ones of Theorem 4.5.4.

Theorem 4.5.6 *Assume that for system (4.3.1) the conditions $A_1 - A_6$ are satisfied, the function v is of class SL_1 with properties (A) and (B) and the comparison functions ψ and φ_i are of class K so that*

(1) *for all $(t, x, \alpha) \in G \cap (\Pi(t, x) \times R_+^m)$ the inequality*

$$Dv(t, x, \alpha)|_{(4.3.1)} \geq \varphi^T(v(t, x, \alpha))B_2\varphi(v(t, x, \alpha))$$

is true for some constant matrix B_2;

(2) *for all $(t, x, \alpha) \in S_k \cap (\Pi(t, x, \alpha) \times R_+^m)$ the inequalities*

$$v(\tau_k(x), x + I_k(x), \alpha) \geq \psi(v(\tau_k(x), x, \alpha)), \quad k = 1, 2, \ldots,$$

are satisfied;

(3) *there exists a constant $\theta > 0$ for which the condition (4)(a) of Theorem 4.5.4 is satisfied;*

(4) *for some $\gamma > 0$ when all $a \in (0, a_0]$*

$$\int_{\psi(a)}^{a} \frac{ds}{\chi(s)} \leq \theta - \gamma;$$

(5) *the constant $m \times m$-matrix B_2^* is positive definite.*

Then the equilibrium state $x = 0$ of system (4.3.1) is unstable.

Here $\chi(s) = \lambda_m(B_2^*)w(s)$, where $w(s) \leq \varphi^T(s)\varphi(s)$, $s = v(t, x, \alpha)$, w is of class K and $\lambda_m(B_2^*)$ is minimal eigenvalue of the matrix $B_2^* = \frac{1}{2}(B_2 + B_2^T)$.

Further we present the following result.

Theorem 4.5.7 *Assume that for system (4.3.1) the conditions $A_1 - A_6$ are satisfied, the function v is of class SL_1 with properties (A) and (B), φ_i and ψ_k are of class K and the continuous function $p(t): R_+ \to R_+$ so that*

(1) *for all $(t, x, \alpha) \in G \cap (\Pi(t, x, \alpha) \times R_+^m)$*

$$Dv(t, x, \alpha)|_{(4.3.1)} \geq -p(t)\varphi^T(v(t, x, \alpha))B_3\varphi(v(t, x, \alpha));$$

(2) for all $(t, x, \alpha) \in S_k \cap (\Pi(t, x, \alpha) \times R_+^m)$

$$v(\tau_k(x), x + I_k(x), \alpha) \geq \psi_k(v(\tau_k(x), x, \alpha)), \quad k = 1, 2, \ldots;$$

(3) the series $\sum_{k=1}^{\infty} \beta_k$ is divergent and for all $z \in (0, c]$, $c = \text{const} > 0$

$$-\int_{\tau_{k-1}(x)}^{\tau_k(x)} p(s)\,ds + \int_z^{\psi_k(z)} \frac{ds}{\sigma(s)} \geq \beta_k, \quad k = 1, 2, \ldots;$$

(4) the constant $m \times m$-matrix $B_3^* = \frac{1}{2}(B_3 + B_3^{\mathrm{T}})$ is positive definite.

Then the equilibrium state $x = 0$ of system (4.3.1) is unstable.

Here $\sigma(s) = \lambda_m(B_3^*)\gamma(s)$, $\gamma(s) \geq \varphi^{\mathrm{T}}(s)\varphi(s)$, $s = v(t, x, \alpha)$, $\gamma \in K$, $\lambda_m(B_3^*)$ is the minimal eigenvalue of the matrix B_3^*.

System (4.3.1) can be considered as a nonlinear approximation of the system

(4.5.2)
$$\frac{dx}{dt} = f(t, x) + g(t, x) \quad \text{for} \quad t \neq \tau_k(x),$$
$$\Delta x = I_k(x) + P_k(x) \quad \text{for} \quad t = \tau_k(x).$$

Here the functions f and g have the same properties as in system (4.3.1) and the functions g and P_k are such that

A_7: $g\colon R_+ \times \Omega \to R^n$; $P_k\colon \Omega \to R^n$, $k = 1, 2, \ldots$;
A_8: $g(t, 0) = P_k(0) = 0$ for all $k = 1, 2, \ldots$;
A_9: the zero solution of system (4.5.2) is unique on the right.

Theorem 4.5.8 *Assume that for system (4.5.2) the conditions $A_1 - A_9$ are satisfied and the function v is of class SL_1. If there exist a and b of K-class, locally integrable function $\beta(t)\colon R_+ \to R_+$ and non-negative constants b_k, $k = 1, 2, \ldots,$ such that*

(1) for all $(t, x, \alpha) \in R_+ \times \Omega \times R_+^m$

$$a(\|x\|) \leq v(t, x, \alpha) \leq b(\|x\|);$$

(2) for all $(t, x', \alpha), (t, x'', \alpha) \in G_k \cap (R_+ \times \Omega(p_1) \times R_+^m)$, $k = 1, 2, \ldots$

$$|v(t, x', \alpha) - v(t, x'', \alpha)| \leq L(t)\|x' - x''\|,$$

where the function $L(t)$ is piece-wise continuous;
(3) for all $(t, x, \alpha) \in G \cap (R_+ \times \Omega(p_1) \times R_+^m)$, $p_1 < H$

$$Dv(t, x, \alpha)|_{(4.3.1)} \leq -cv(t, x, \alpha), \quad 0 < c = \text{const};$$

(4) for all $(t, x, \alpha) \in S_k \cap (R_+ \times \Omega(p_1) \times R_+^m)$, $k = 1, 2, \ldots$

$$v(t+0, x + I_k(x), \alpha) \leq v(t, x, \alpha);$$

(5) for all $(t, x) \in R_+ \times \Omega(p_2)$, $p_2 < \varkappa$

$$\|g(t, x, \alpha)\| \leq \beta(t); \quad \|P_k(x)\| \leq b_k, \quad k = 1, 2, \ldots;$$

(6) the limiting correlations

$$\lim_{t \to \infty} \int_t^{t+1} L(s)\beta(s)\, ds = 0,$$

$$\lim_{t \to \infty} \sum_{t < \tau_k(x) < t+1} L(\tau_k(x) + 0)\, b_k = 0 \quad \text{for all} \quad x \in \Omega(p_2)$$

are satisfied;
(7) there exists p_0, $0 < p_0 < p^*$, such that the condition $x \in \Omega(p_0)$ implies

$$x + I_k(x) \in \Omega(p^*), \quad k = 1, 2, \ldots,$$

where $p^* = \min(p_1, p_2)$.

Then the equilibrium state $x = 0$ of system (4.5.2) is uniformly asymptotically stable.

Proof The proof of the Theorem 4.5.7 is similar to the proof of Theorem 3 by Kulev [1]. Analogous results on nonimpulsive system of differential equations are contained the papers by Chow and Yorke [1], and Furumochi [1].

4.5.2 Vector approach Further we shall present some results obtained for the impulsive system in terms of the vector approach in the method of matrix-valued functions.

We consider the impulsive system

(4.5.3)
$$\frac{dx}{dt} = f(t, x) \quad \text{for} \quad t \neq t_k, \quad k \in Z,$$
$$\Delta x = I_k(x) \quad \text{for} \quad t = t_k,$$
$$x(t_0^+) = x_0, \quad t_0 \geq 0, \quad k = 1, 2, \ldots$$

under the following assumptions

E_1: $0 < t_1 < t_2 < \ldots < t_k < \ldots$, $t_k \to \infty$ as $t \to \infty$;

E_2: the vector-function $f\colon R_+ \times R^n \to R^n$ and it is continuous on $(\tau_{k-1}, \tau_k] \times R^n$ for each $x \in R^n$, $k = 1, 2, \ldots$, and there exists the limit
$$\lim [f(t,y)\colon (t,y) \to (t_k^+, x)] = f(t_k^+, x);$$

E_3: vector-functions $I_k\colon R^n \to R^n$, $k = 1, 2, \ldots$;

E_4: for system (4.3.1) there exist a matrix-valued function $U(t,x)$, $U\colon R_+ \times \Omega \to R^{m \times m}$, a constant $m \times m$-matrix K and a vector $b \in R^m$ such that the vector-function

(4.5.4) $$L(t, x, b) = KU(t, x)b$$

satisfies the condition below.

Definition 4.5.2 The function $L(t, x, b)$, $L\colon R_+ \times \Omega \times R^m \to R^m$, is of class VL_0, if

(1) L is continuous in $(t_{k-1}, t_k] \times \Omega$ and for each $x \in \Omega$, $k = 1, 2, \ldots$,

$$\lim [L(t, x, b)\colon (t, x) \to (t_0, x_0)], \quad (t, x) \in G_k] = L(t - 0, x_0, b),$$
$$\lim [L(t, x, b)\colon (t, x) \to (t_0, x_0)], \quad (t, x) \in G_{k+1}] = L(t + 0, x_0, b),$$

and the correlation

$$L(t_0 - 0, x_0, b) = L(t_0, x_0, b)$$

is satisfied;

(2) L is locally Lipschitzian in x;

(3) $L(t, 0, b) = 0$ for all $t \in R_+$, $b \neq 0$.

The function $L(t, x, b)$ belongs to the class VL_1, if L belongs to class VL_0 and is continuously differentiable on G, i.e. determined is the total derivative of the function $L(t, x, b)$ along solutions of system (4.3.1)

$$DL_i(t, x, b)|_{(4.3.1)} = \frac{\partial L_i}{\partial t} + (\operatorname{grad} L_i)^T f(t, x)$$

for all $t \neq t_k$, $k = 1, 2, \ldots$, $i = 1, 2, \ldots, m$.

Assume that the solution $x(t, t_0, x_0)$ of system (4.3.1) exists on $[t_0, \infty)$ and intersects each hyperplane $S_k\colon t = t_k$, $k = 1, 2, \ldots$, only once.

Proposition 4.5.1 Assume that for system (4.3.1) the conditions E_1 – E_4 are satisfied and there exist functions L_i of class VL_1, $g\colon R_+ \times R_+ \to R^m$, $\psi_k\colon R_+ \to R_+^m$, $k = 1, 2, \ldots$, $g(t,u)$ and $\psi_k(u)$ are nondecreasing in u and so that

(1) $DL(t, x, b)|_{(4.3.1)} \leq g(t, L(t, x, b))$, $t \neq t_k$;

(2) $L(t, x + I_k(x), b) \leq \psi_k(L(t, x, b))$, $t = t_k$;

(3) for all $t \in [t_0, \infty)$ there exists a maximal solution of the comparison system

(4.5.5)
$$\frac{du}{dt} = g(t, u), \quad t \neq t_k,$$
$$u(t_k^+) = \psi_k(u(t_k)),$$
$$u(t_0^+) = u_0 \geq 0.$$

Then, if $L(t_0^+, x_0, b) \leq u_0$, then along solutions $x(t) = x(t; t_0, x_0)$ of system (4.3.1)

$$L(t, x(t), b) \leq r(t), \quad t \geq t_0.$$

The proof of this assertion is similar to that by Lakshmikantham, Leela and Martynyuk [1] conducted for auxiliary scalar function.

Theorem 4.5.9 *Assume that*

(1) *all conditions of Proposition 4.5.1 are satisfied;*

(2) *there exist functions a and b of class K such that*

$$a(\|x\|) \leq \sum_{i=1}^{m} L_i(t, x, b) \leq b(\|x\|) \quad \text{for all} \quad (t, x) \in R_+ \times \Omega(H);$$

(3) *for all* $(t, x) \in R_+ \times \Omega(r)$ *the estimates*

$$DL_i(t, x, b)|_{(4.3.1)} \leq g_i(t, L_1(t, x, b), \ldots, L_m(t, x, b)), \quad i = 1, 2, \ldots, m,$$

are fulfilled, where $g_i\colon R_+ \times R_+^m \to R$, $g_i(t, 0) = 0$ *for all* $t \in R_+$ *and* $g_i(t, u)$ *is quasimonotone nondecreasing in* u;

(4) *there exists* $r_0 > 0$, *such that the condition* $x_0 \in \Omega(r_0)$ *implies* $x_0 + I_k(x) \in \Omega(r)$ *and*

$$L_i(t, x + I_k(x), b) \leq \psi_k(L_i(t, x, b)), \quad t = t_k, \quad i = 1, 2, \ldots, m,$$

where $x \in \Omega(r_0)$, $\psi_{ik}\colon R_+ \to R_+$, $k = 1, 2, \ldots$;

(5) *the zero solution of system (4.5.5) possesses a certain type of stability.*

Then the equilibrium state $x = 0$ of system (4.3.1) possesses the same type of stability as the zero solution of system (4.5.5).

Assume that the matrix-valued function $U(t,x)$ contains only one component, which satisfies all conditions of Theorem 4.5.9. Then the following assertion takes place.

Corollary 4.5.1 *If in the conditions (3), (4) of Theorem 4.5.9, we choose:*

(a) $g(t,u) \equiv 0$, $\psi_k(u) = u$ for all $k = 1, 2, \ldots$, then $v(t, x(t))$ is nonincreasing in t and $v(t, x(t)) \leq v(t_0^+, x_0)$ for all $t \geq t_0$;

(b) $g(t,u) \equiv 0$, $\psi_k(u) = d_k u$, $d_k \geq 0$ for all $k = 1, 2, \ldots$, then

$$v(t, x(t)) \leq v(t_0^+, x_0) \prod_{t_0 < t_k < t} d_k, \quad t \geq t_0;$$

(c) $g(t,u) = -\alpha u$, $\alpha > 0$, $\psi_k(u) = d_k u$, $d_k \geq 0$ for all $k = 1, 2, \ldots$, then

$$v(t, x(t_1)) \leq \left(v(t_0^+, x_0) \prod_{t_0 < t_k < t} d_k \right) \exp(-\alpha(t - t_0))$$

for all $t \geq t_0$;

(d) $g(t,u) = \lambda'(t) u$, $\psi_k(u) = d_k u$, $d_k \geq 0$ for all $k = 1, 2, \ldots$, $\lambda \in C^1(R_+, R_+)$, then

$$v(t, x(t_1)) \leq \left(v(t_0^+, x_0) \prod_{t_0 < t_k < t} \right) \exp[\lambda(t) - \lambda(t_0)]$$

for all $t \geq t_0$;

(e) $g_0(t, u) = g(t, u) = -\dfrac{r'(t)}{(r(t))} u$, where $r(t) > 0$ is continuously differentiable on R_+, and $r(t) \to \infty$, and $\psi_i(u) = d_i$, $u, d_i \geq 0$ for all i, then

$$v(t, x(t)) \leq \left(v(t_0^+, x_0) \prod_{t_0 \leq t_0 < t} d_i \right) \frac{r(t_0)}{r(t)}, \quad t \geq t_0.$$

In particular, $r(t) = e^{\alpha t}$, $\alpha > 0$, is admissible;

(f) $g_0(t, u) = g(t, u) = -\gamma(u)$ where $\gamma \in K$, $\psi_i(u) = u$ for all i, then

$$v(t, x(t)) \leq J^{-1}\big(J(v(t_0^+, x_0)) - (t - t_0))\big), \quad t \geq t_0,$$

where J^{-1} is the inverse function of J and $J' = \dfrac{1}{\gamma(u)}$.

The method of matrix-valued functions allows the reduction of the complicate problem of stability of state $x = 0$ of system (4.3.1) to the system of algebraic conditions of the property of having a fixed sign of special matrices. This displays both the versatility of the direct Liapunov method and the existence of new ways for generalizing adequate classes of systems of equations modelling the real world phenomena.

4.6 Stability Analysis of Large Scale Impulsive Systems

To show stability for a large scale impulsive system, it is necessary to find a Liapunov function whose derivative along the trajectories is negative definite. Here we define a new class of auxiliary functions for impulsive large scale systems. The system of differential equations of general type (4.3.1) has the meaning of large-scale impulsive systems (LSIS), if it can be decomposed into m interconnected impulsive subsystems

(4.6.1)
$$\frac{dx_j}{dt} = f_j(t, x_j) + f_j^*(t, x), \quad t \neq \tau_k(x_j), \quad j = 1, 2, \ldots, m,$$
$$\Delta x_j = I_{kj}(x_j) + I_{kj}^*(x), \quad t = \tau_k(x_j), \quad k = 1, 2, \ldots.$$

We assume on system (4.6.1) that

(1) $x_j \in R^{n_j}$, $f_j(t, x_j) = 0$ iff $x_j = 0$;
(2) $0 < \tau_k(x_j) < \tau_{k+1}(x_j)$, $\tau_k(x_j) \to +\infty$ as $k \to \infty$;
(3) $I_{kj}\colon R^{n_j} \to R^{n_j}$ and $I_{kj} = 0$ iff $x_j = 0$;
(4) functions $f(t, x) = (f_1^T(t, x_1), \ldots, f_m^T(t, x_m))^T$ and $I_{kj}(x) = (I_{r1}^T(x_1), \ldots, I_{km}^T(x_m))^T$ are definite and continuous in the domain

(4.6.2) $R_+ \times S(\rho) = [t_0, \infty) \times \{x\colon \|x\| \leq \rho \leq \rho_0\}, \quad t_0 \geq 0;$

(5) functions $\tau_k(x_j) = \tau_k(x)$, $k = 1, 2, \ldots$, and number ρ satisfy conditions excluding beating of solutions of system (4.3.1) against the hypersurfaces $S_i\colon t = \tau_k(x)$, $k = 1, 2, \ldots$, $t \geq 0$.

We assume on system (4.6.1) that

(1) $x_j = (0^T, \ldots, 0^T, x_j^T, 0^T, \ldots, 0^T)^T \in R^n$, $x_j \in R^{n_j}$, $f_j^*(t, x) = f_j(t, x) - f_j(t, x_j)$;
(2) $I_{kj} = (I_{k1}^T, I_{k2}^T, \ldots, I_{km}^T)^T$, $I_{kj}^*(x) = I_{kj}(x) - I_{kj}(x_j)$, $n = n_1 + \cdots + n_m$.

4. IMPULSIVE SYSTEMS

The state of j-th noninteracting impulsive subsystem is described by the equations

(4.6.3)
$$\frac{dx_j}{dt} = f_j(t, x_j), \quad t \neq \tau_k(x_j);$$
$$\Delta x_j = I_{kj}(x_j), \quad t = \tau_k(x_j).$$

The problem on stability for LSIS (4.6.1) is formulated as follows:

To establish conditions under which stability of equilibrium state $x = 0$ of system (4.6.1) is derived from the properties of stability of impulsive subsystems (4.6.3) and properties of connection functions $f_j^*(t, x)$, $I_j^*(x)$.

4.6.1 Auxiliary estimations
Further we shall need some systematized conditions on functions similar to Liapunov functions for system (4.6.1).

Assumption 4.6.1 There exist

(1) open connected time-invariant neighborhoods $\mathcal{N}_{jx} \subseteq R^{n_j}$ of states $x_j = 0$, $j = 1, 2, \ldots, m$;
(2) functions $\varphi_{j1}, \psi_{j1}: \mathcal{N}_{jx} \to R_+$ (φ_{j1}, ψ_{j1} of class K);
(3) constants a_{jl}, b_{jl}, $j, l = 1, 2, \ldots, m$;
(4) matrix function $U(t, x) = [u_{jl}(t, \cdot)]$ with elements

$$v_{jj} = v_{jj}(t, x_j); \quad v_{jl} = v_{lj} = v_{jl}(t, x_j, x_l), \quad j \neq l,$$
$$v_{jj}(t, 0) = 0, \quad v_{jl}(t, 0, 0) = 0, \quad j, l = 1, 2, \ldots, m$$

of class SL_1 in the domain $\times S(\rho_0)$, where $\rho_0 = \text{const} > 0$, and satisfying estimates

(a) $a_{jj}\varphi_{j1}^2(\|x_j\|) \leq v_{jj}(t, x_j) \leq b_{jj}\psi_{j1}^2(\|x_j\|)$ for all $(t, x_j) \in R_+ \times \mathcal{N}_{jx}$, $j = 1, 2, \ldots, m$;

(b) $a_{jl}\varphi_{j1}(\|x_j\|)\varphi_{l1}(\|x_l\|) \leq v_{jl}(t, x_j, x_l) \leq b_{jl}\psi_{j1}(\|x_j\|)\psi_{l1}(\|x_l\|)$ for all $(t, x_j, x_l) \in R_+ \times \mathcal{N}_{jx} \times \mathcal{N}_{kx}$, $j, l = 1, 2, \ldots, m$.

Here $v_{jj}(t, x_j)$ correspond to subsystems (4.2.3) and $v_{jl}(t, x_j, x_l)$ take into account connections $f_j^*(t, x)$ and $I_{lj}^*(x)$ between them.

We introduce the scalar function

(4.6.4) $$v(t, x, \eta) = \eta^T U(t, x) \eta, \quad \eta \in R_+^m, \quad \eta > 0,$$

and its total derivative

(4.6.5) $$Dv(t, x, \eta) = \eta^T DU(t, x)\eta$$

due to system of equations (4.6.1).

Proposition 4.6.1 If all conditions of Assumption 4.6.1 are satisfied, then for function (4.6.4) the estimate

(4.6.6) $\quad u_1^T H^T A H u_1 \leq v(t, x, \eta) \leq u_2^T H^T B H u_2 \quad \text{for all} \quad (t, x) \in R_+ \times \mathcal{N}_x,$

is valid, where

$$u_1 = (\varphi_{11}(\|x_1\|), \varphi_{21}(\|x_2\|), \ldots, \varphi_{m1}(\|x_m\|))^T,$$
$$u_2 = (\psi_{11}(\|x_1\|), \psi_{21}(\|x_2\|), \ldots, \psi_{m1}(\|x_m\|))^T,$$
$$H = \text{diag}\{\eta_1, \eta_2, \ldots, \eta_m\}, \quad A = [a_{jl}], \quad B = [b_{jl}],$$
$$a_{jl} = a_{lj}, \quad b_{jl} b_{lj}, \quad j, l = 1, 2, \ldots, m,$$

$\mathcal{N}_x \subseteq \mathcal{N}_{1x} \times \mathcal{N}_{2x} \times \ldots \times \mathcal{N}_{mx}$ is an open connected neighborhood of state $x = 0$, such that $\mathcal{N}_x = \{x\colon \|x\| < \rho_0\}$, $\rho_0 = \text{const} > 0$.

The proof of estimate (4.6.6) is similar to that in Chapter 1 for a system without impulse effects.

Assumption 4.6.2 There exist

(1) open connected time invariant neighborhoods $\mathcal{N}_{jx} \subseteq R^{n_j}$ of states $x_j = 0$, $j = 1, 2, \ldots, m$, and open connected neighborhood $\mathcal{N}_x \subseteq \mathcal{N}_{1x} \times \ldots \times \mathcal{N}_{mx}$ of state $x = 0$;

(2) functions v_{jl}, $j, l = 1, 2, \ldots, m$, mentioned in Assumption 4.6.1 and functions φ_j, $j = 1, 2, \ldots, m$, φ_m, φ_M such that in domain $R_+ \times S(\rho_0)$ the conditions $\varphi_j(0) = \varphi_m(0) = \varphi_M(0) = 0$ hold,

(4.6.7) $\quad 0 < \varphi_m(v(t,x)) \leq \sum_{j=1}^{m} \varphi_j^2(v_{jj}(t, x_j)) \leq \varphi_M(v(t,x));$

(3) constants $\rho_j^{(1)}$, $\rho_j^{(2)}$, ρ_{jl}, $j \neq l$, $j, l = 1, 2, \ldots, m$, and the following conditions for all $t \neq \tau_k(x_j)$ are satisfied

(a) $\eta_j^2 \{D_t v_{jj} + (D_{x_j} v_{jj})^T f_j(t, x_j)\} \leq \rho_j^{(1)} \varphi_j^2(v_{jj}(t, x_j))$ for all $(t, x_j) \in R_+ \times \mathcal{N}_{jx0}$, $j = 1, 2, \ldots, m;$

(b) $\sum_{j=1}^{m} \eta_j^2 (D_{x_j} v_{jj})^T f_j^*(t, x) + 2 \sum_{j=1}^{m} \sum_{\substack{l=2 \\ l>j}}^{m} \eta_j \eta_l \{D_t v_{jl}$

$+ (D_{x_j} v_{jl})^T f(t, x) + (D_{x_l} v_{il})^T f(t, x)\} \leq \sum_{j=1}^{m} \rho_j^{(2)} \varphi_j^2(v_{jj}(t, x_j))$

$+ 2 \sum_{j=1}^{m} \sum_{\substack{l=2 \\ l>j}}^{m} \rho_{jl} \varphi_j(v_{jj}(t, x_j)) \varphi_l(v_{ll}(t, x_l)) \quad \text{for all} \quad (t, x_j, x_l) \in R_+ \times \mathcal{N}_{jx0} \times \mathcal{N}_{lx0}$, $j \neq l$.

Here $\mathcal{N}_{jx0} = \{x_j \colon x_j \in \mathcal{N}_{jx}, \ x_j \neq 0\}.$

Proposition 4.6.2 If all conditions of Assumption 4.6.2 are satisfied, then for expression (4.6.5) the inequality

(4.6.8) $$Dv(t,x) \leq u^T G u \quad \text{for all} \quad t \neq \tau_k(x_j), \quad x \in \mathcal{N}_{x0}$$

is valid, where

$$u^T = (\varphi_1(v_{11}(t,x_1)), \varphi_2(v_{22}(t,x_2)), \ldots, \varphi_m(v_{mm}(t,x_m))),$$
$$G = [\sigma_{jl}], \quad j,l = 1,2,\ldots,m, \quad \sigma_{jl} = \sigma_{lj},$$
$$\sigma_{jj} = \rho_j^{(1)} + \rho_j^{(2)}, \quad \sigma_{jl} = \rho_{jl}, \quad j \neq l, \quad j,l = 1,2,\ldots,m.$$

Let $\lambda_M(G)$ be the maximal eigenvalue of matrix G.

Corollary 4.6.1 If all conditions of Assumption 4.6.2 are satisfied and
(1) $\lambda_M(G) < 0$;
(2) $\lambda_M(G) > 0$,

then the following estimates hold true
(1) $Dv(t,x) \leq \lambda_M(G)\varphi_m(v(t,x))$ for all $t \neq \tau_k(x_j)$, $x \in \mathcal{N}_{x0}$;
(2) $Dv(t,x) \leq \lambda_M(G)\varphi_M(v(t,x))$ for all $t \neq \tau_k(x_j)$, $x \in \mathcal{N}_{x0}$

correspondingly.

Assumption 4.6.3 There exist

(1) functions v_{jl}, $j,l = 1,2,\ldots,m$, mentioned in Assumption 4.6.1, and functions ψ_j, $j = 1,2,\ldots,m$, ψ_m, ψ_M: $\psi_j(0) = \psi_m(0) = \psi_M(0) = 0$, such that in domain $R_+ \times S(\rho_0)$ the conditions

(4.6.9) $$0 < \psi_m(v(\tau_k(x),x)) \leq \sum_{j=1}^m \psi_j^2(v_{jj}(\tau_k(x_j),x_j)) \leq \psi_M(v(\tau_k(x),x)),$$
$$k = 1,2,\ldots$$

are satisfied;
(2) constants $\alpha_j^{(1)}$, $\alpha_j^{(2)}$, α_{jl} $(j \neq l)$, $j,l = 1,2,\ldots,m$, and the following inequalities are satisfied
(a) $\eta_j^2\{v_{jj}(\tau_k(x_j), x_j + I_{kj}(x_j)) - v_{jj}(\tau_k(x_j),x_j)\}$
$\leq \alpha_j^{(1)}\psi_j^{(2)}(v_{jj}(\tau_k(x_j),x_j))$ for all $x_j \in \mathcal{N}_{jx}$, $j = 1,2,\ldots,m$;
(b) $\sum_{j=1}^m \eta_j^2\{v_{jj}(\tau_k(x), x_j + I_{kj}(x))v_{jj}(\tau_k(x_j), x_j + I_{kj}(x_j))$
$+ v_{jj}(\tau_k(x_j),x_j) - v_{jj}(\tau_k(x),x_j)\} + 2\sum_{j=1}^m \sum_{\substack{l=2 \\ l>j}}^m \eta_j\eta_l\{v_{jl}(\tau_k(x),$

$$x_j + I_{kj}(x), x_l + J_{kl}(x)) - v_{jl}(\tau_k(x), x_j, x_l)\} \le$$
$$\sum_{j=1}^{m} \alpha_j^{(2)} \psi_j^2(v_{jj}(\tau_k(x_j), x_j)) + 2 \sum_{j=1}^{m} \sum_{\substack{l=2 \\ l>j}}^{m} \alpha_{jl} \psi_j(v_{jj}(\tau_j(x_j), x_j))$$
$$\times \psi_k(v_{ll}(\tau_k(x_l), x_l)) \quad \text{for all} \quad (x_j, x_l) \in \mathcal{N}_{jx} \times \mathcal{N}_{lx}, \ k = 1, 2, \ldots$$

Proposition 4.6.3 *If all conditions of Assumption 4.6.3 are satisfied, the estimate*

(4.6.10) $$v(\tau_k(x), x + I_k(x)) - v(\tau_k(x), x) \le u_k^{\mathrm{T}} C u_k$$

is valid, where

$$u_k = (\psi_1(v_{11}(\tau_k(x_1), x_1)), \psi_2(v_{22}(\tau_k(x_2), x_2)), \ldots, \psi_m(v_{mm}(\tau_k(x_m), x_m)))^{\mathrm{T}},$$
$$C = [c_{jl}], \quad j, l = 1, 2, \ldots, m, \quad c_{jl} = c_{lj}, \quad k = 1, 2, \ldots,$$
$$c_{jj} = \alpha_j^{(1)} + \alpha_j^{(2)}, \quad c_{jl} = \alpha_{jl}, \quad j \ne l, \quad j, l = 1, 2, \ldots, m.$$

Proof Under all conditions of Assumption 4.6.3 we have

$$v(\tau_k(x), x + I_k(x)) - v(\tau_k(x), x) = \eta^{\mathrm{T}}[U(\tau_k(x), x + I_k(x)) - U(\tau_k(x), x)]\eta$$
$$= \sum_{j=1}^{m} \eta_j^2 \{v_{jj}(\tau_k(x), x_j + I_{kj}(x)) - v_{jj}(\tau_k(x), x_j)\}$$
$$+ 2 \sum_{j=1}^{m} \sum_{\substack{l=2 \\ l>j}}^{m} \eta_j \eta_l \{v_{jl}(\tau_k(x), x_j + I_{kj}(x), x_l + I_{kl}(x))$$
$$- v_{jl}(\tau_k(x), x_j, x_k)\} = \sum_{j=1}^{m} \eta_j^2 \{v_{jj}(\tau_k(x_j), x_j + I_{kj}(x_j))$$
$$- v_{jj}(\tau_k(x_j), x_j)\} + \sum_{j=1}^{m} \eta_j^2 \{v_{jj}(\tau_k(x), x_j + I_{kj}(x_j)))$$
$$- v_{jj}(\tau_k(x_j), x_j + I_{kj}(x_j)) + v_{jj}(\tau_k(x_j), x_j - v_{jj}(\tau_k(x), x_j)\}$$
$$+ 2 \sum_{j=1}^{m} \sum_{\substack{l=2 \\ l>j}}^{m} \eta_j \eta_l \{v_{jl}(\tau_k(x), x_j + I_{kj}(x), x_l + I_{kl}(x))$$
$$- v_{jl}(\tau_k(x), x_j, x_l)\} \le \sum_{j=1}^{m} \alpha_j^{(1)} \psi_j^2(v_{jj}(\tau_k(x_j), x_j))$$

$$+ \sum_{j=1}^{m} \alpha_j^{(2)} \psi_j^2(v_{jj}(\tau_k(x_j)), x_j))$$

$$+ 2 \sum_{j=1}^{m} \sum_{\substack{l=2 \\ l>j}}^{m} \alpha_{jl} \psi_j(v_{jj}(\tau_k(x_j)), x_j)) \, \psi_l(v_{ll}(\tau_k(x_l)), x_l))$$

$$= \sum_{j=1}^{m} c_{kj} \psi_j(v_{jj}(\tau_k(x_j)), x_j)) \, \psi_l(v_{ll}(\tau_k(x_k)), x_l)) = u_k^T C u_k,$$

$$k = 1, 2, \ldots.$$

Corollary 4.6.2 If all conditions of Assumption 4.6.3 are satisfied and

(1) $\lambda_M(C) < 0$,
(2) $\lambda_M(C) > 0$,

then the following estimates hold true

(1) $v(\tau_k(x), x + I_k(x)) - v(\tau_k(x), x) \leq \lambda_M(C)\psi_m(v(\tau_k(x), x))$ for all $x \in \mathcal{N}_{x0}$;
(2) $v(\tau_k(x), x + I_k(x)) - v(\tau_k(x), x) \leq \lambda_M(C)\psi_m(v(\tau_k(x), x))$ for all $x \in \mathcal{N}_{x0}$

correspondingly.

Here $\lambda_M(C)$ is the maximal eigenvalue of matrix C.

Assumption 4.6.4 There exist

(1) functions v_{jl}, $j, l = 1, 2, \ldots, m$, mentioned in Assumption 4.6.1 and functions ψ_j, $j = 1, 2, \ldots, m$, ψ_m, ψ_M, mentioned in Assumption 4.6.3;
(2) constants $\beta_j^{(1)}$, $\beta_j^{(2)}$, β_{jl}, $j \neq l$, $j, l = 1, 2, \ldots, m$, and for all $k = 1, 2, \ldots$ the following conditions are satisfied

  (a) $\eta_j^2 v_{jj}(\tau_k(x_j), x_j + I_{kj}(x_j)) \leq \beta_j^{(1)} \psi_j^2(v_{jj}(\tau_k(x_j)), x_j))$ for all $x_j \in \mathcal{N}_{jx}$, $j = 1, 2, \ldots, m$;

  (b) $\sum_{j=1}^{m} \eta_j^2 \{v_{jj}(\tau_k(x), x_j + I_{kj}(x)) - v_{jj}(\tau_k(x_j), x_j + I_{kj}(x_j))\} +$

  $2 \sum_{j=1}^{m} \sum_{\substack{l=2 \\ l>j}}^{m} \eta_j \eta_l v_{jl}(\tau_k(x), x_j + I_{kj}(x), x_l + I_{kl}(x)) \leq$

  $\sum_{j=1}^{m} \beta_j^{(2)} \psi_j^2(v_{jj}(\tau_k(x_j)), x_j)) + 2 \sum_{j=1}^{m} \sum_{\substack{l=2 \\ l>j}}^{m} \beta_{jl} \psi_j(v_{jj}(\tau_k(x_j)), x_j)) \times$

  $\psi_l(v_{ll}(v_{ll}(\tau_k(x_l)), x_l)))$ for all $(x_j, x_l) \in \mathcal{N}_{jx} \times \mathcal{N}_{lx}$.

Proposition 4.6.4 If all conditions of Assumption 4.6.4 are satisfied, then the estimate

(4.6.11) $v(\tau_k(x), x + I_k(x)) \leq u_k^{\mathrm{T}} C^* u_k$ for all $x \in \mathcal{N}_x$, $k = 1, 2, \ldots$,

takes place where

$$C^* = [c_{jl}^*], \quad j, l = 1, 2, \ldots, m, \quad c_{jl} = c_{lj},$$
$$c_{jj}\beta_j^{(1)} + \beta_j^{(2)}, \quad c_{jl} = \beta_{jl}, \quad j \neq l, \quad j, l = 1, 2, \ldots, m.$$

The proof of Proposition 4.6.4 is similar to that of Proposition 4.6.3.

Corollary 4.6.3 If all conditions of Assumption 4.6.4 are satisfied and
(1) $\lambda_M(C^*) < 0$;
(2) $\lambda_M(C^*) > 0$,

then for all $k = 1, 2, \ldots$

(1) $v(\tau_k(x), x + I_k(x)) \leq \lambda_M(C^*)\psi_m(v(\tau_k(x), x))$ for all $x \in \mathcal{N}_{x0}$;
(2) $v(\tau_k(x), x + I_k(x)) \leq \lambda_M(C^*)\psi_M(v(\tau_k(x), x))$ for all $x \in \mathcal{N}_{x0}$;

correspondingly.

Here $\lambda_M(C^*)$ is the maximal eigenvalue of matrix C^*.

Assumption 4.6.5 Conditions (1) and (2) of Assumption 4.6.2 are satisfied and in the inequalities of condition (3) of Assumption 4.6.2 the inequality sign "$\leq$" is reversed.

Proposition 4.6.5 If conditions of Assumption 4.6.5 are satisfied, then for the total derivative (4.6.5) the estimate

(4.6.12) $Dv(t, x) \geq u^{\mathrm{T}} G u$ for all $t \neq \tau_k(x)$, $x \in \mathcal{N}_{x0}$,

is valid, where u and G are defined as in Proposition 4.6.2.

Corollary 4.6.4 If conditions of Assumption 4.6.5 are satisfied and
(1) $\lambda_m(G) < 0$;
(2) $\lambda_m(G) > 0$;

then

(1) $Dv(t, x) \geq \lambda_m(G)\varphi_M(v(t, x))$ for all $t \neq \tau_k(x)$, $x \in \mathcal{N}_{x0}$;
(2) $Dv(t, x) \geq \lambda_m(G)\varphi_m(v(t, x))$ for all $t \neq \tau_k(x)$, $x \in \mathcal{N}_{x0}$

correspondingly.

Here $\lambda_m(G)$ is a minimal eigenvalue of matrix G.

Assumption 4.6.6 Condition (1) of Assumption 4.6.3 is satisfied and in inequalities of condition (2) of Assumption 4.6.3 the inequality sign "$\leq$" is reversed.

Proposition 4.6.6 Under conditions of Assumption 4.6.6, for all $k = 1, 2, \ldots$ the estimate

$$(4.6.13) \qquad v(\tau_k(x), x + I_k(x)) - v(\tau_k(x), x) \geq u_k^T C u_k$$

takes place, where u_k, C are defined as in Proposition 4.6.3.

Corollary 4.6.5 If in inequality (4.6.13) $\lambda_m(C) > 0$, then for all $k = 1, 2, \ldots$ estimate

$$v(\tau_k(x), x + I_k(x)) - v(\tau_k(x), x) \geq \lambda_m(C) \psi_m(v(\tau_k(x), x)) \quad \text{for all} \quad x \in \mathcal{N}_{x0}$$

takes place.

Assumption 4.6.7 Condition (1) of Assumption 4.6.4 is satisfied and in inequalities of condition (2) of the same assumption the inequality sign "$\leq$" is reversed.

Proposition 4.6.7 Under conditions of Assumption 4.6.7 for all $k = 1, 2, \ldots$ the estimate

$$(4.6.14) \qquad v(\tau_k(x), x + I_k(x)) \geq u_k^T C^* u_k \quad \text{for all} \quad x \in \mathcal{N}_x$$

takes place, where u_k and C^* are defined as in Proposition 4.6.4.

Corollary 4.6.6 If in equality (4.6.14) $\lambda_m(C^*) > 0$, then for all $k = 1, 2, \ldots$ the inequality

$$v(\tau_k(x), x + I_k(x)) \geq \lambda_m(C^*) \psi_m(v(\tau_k(x), x)) \quad \text{for all} \quad x \in \mathcal{N}_{x0}$$

is valid.

Assumption 4.6.8 Let

$$\Pi_j = \{(t, x_j) \in R_+ \times \mathcal{N}_{jx} : \, v_{jj}(t, x_j) > 0\}$$

be domains of positiveness of functions $v_{jj}(t, x_j)$, $j = 1, 2, \ldots, m$, for any $t \geq t_0$ having non zero open crossection by plane $t = $ const adherent to the origin, and in this domain functions v_{jl}, $j, l = 1, 2, \ldots, m$, are bounded.

Proposition 4.6.8 *Under conditions of Assumptions 4.6.1 and 4.6.8 and if matrix A in estimate (4.6.6) is positive definite, i.e. $\lambda_m(H^{\mathrm{T}}AH) > 0$, then*

(1) *domain $\Pi = \{(t,x) \in R_+ \times \mathcal{S}(\rho) \colon v(t,x,\eta) > 0\}$ of function $v(t,x,\eta)$ positiveness for any $t \in R_+$ has non zero open crossection by plane $t = \mathrm{const}$ adherent to the origine;*

(2) *in domain Π function $v(t,x,\eta)$ is bounded.*

Proof Under conditions of Assumption 4.6.1 and $\lambda_m(H^{\mathrm{T}}AH) > 0$ the domain of function $v(t,x)$ positiveness is

$$\Pi = \{(t,x) \in R_+ \times \mathcal{S}(\rho) \colon x \neq 0\},$$

that has an open crossection by the plane $t = \mathrm{const}$ for any $t \in R_+$. Moreover, positiveness of function $v_{jj}(t,x)$ is a necessary condition for positiveness of function $v(t,x)$ and therefore, $\Pi \subseteq \bigcap_{j=1}^{m} \Pi_j$. Domains Π_j for any $j = 1, 2, \ldots, m$ have, by condition os Assumption 4.6.8, nonzero open crossection by plane $t = \mathrm{const}$ adherent to the origine. This proves assertion (1) of Proposition 4.6.8.

Boundedness of functions v_{jl}, $j, l = 1, 2, \ldots, m$, implies that matrix $U(t,x)$ is bounded, but then function $v(t,x,\eta)$ constructed according to (4.6.4) is also bounded.

4.6.2 Tests for stability Assumptions 4.6.1–4.6.8, Propositions 4.6.1–4.6.8 and Corollaries 4.6.1–4.6.6 allow us to formulate various stability and asymptotic stability conditions for zero solution of system (4.6.1).

Theorem 4.6.1 *If differential perturbed motion equations (4.6.1) are such that in domain $R_+ \times \mathcal{S}(\rho)$ all conditions of Assumptions 4.6.1, 4.6.2, and 4.6.3 are satisfied, and*

(1) *matrix A is positive definite ($\lambda_m(H^{\mathrm{T}}AH) > 0$);*
(2) *matrix G is negative semidefinite ($\lambda_M(G) \leq 0$);*
(3) *matrix C is negative semidefinite or equals to zero ($\lambda_M(C) \leq 0$),*

then the zero solution of (4.6.1) is stable.

If instead of condition (3) the following condition is satisfied

(4) *matrix C is negative definite ($\lambda_M(C) < 0$),*

then zero solution of (4.6.1) is asymptotically stable.

Proof Under Assumption 4.6.1, Proposition 4.6.1 and condition (1) of Theorem 4.6.1 function $v(t, x, \eta)$, constructed by (4.6.4), is positive definite. Conditions of Assumption 4.6.2, Proposition 4.6.2 and condition (2) of Theorem 4.6.1 imply

$$Dv(t, x) \leq 0, \quad t \neq \tau_k(x), \quad (t, x) \in R_+ \times S(\rho). \tag{4.6.15}$$

From conditions of Assumption 4.6.3, Proposition 4.2.3 and condition (3) of Theorem 4.6.1 it follows that on hypersurfaces $S_k: t = \tau_k(x)$, $x \in S(\rho)$, the inequalities

$$v(\tau_k(x), x + I_k(x)) \leq v(\tau_k(x), x), \quad k = 1, 2, \ldots, \tag{4.6.16}$$

are satisfied.

Conditions (4.6.15) and (4.6.16) are sufficient for zero solution of (4.6.1) to be stable.

If instead of conditions (3) of Theorem 4.6.1 condition (4) of Theorem 4.6.1 is satisfied, then by Proposition 4.6.3 and Corollary 4.6.2 we find the estimate
(4.6.17)
$$v(\tau_k(x), x + I_k(x)) - v(\tau_k(x), x) \leq \lambda_M(C)\psi_m(v(\tau_k(x), x), \quad k = 1, 2, \ldots$$

By virtue of conditions (4.6.15) and (4.6.17) the zero solution of LSIS (4.6.1) is asymptotically stable.

Theorem 4.6.1 is proved.

Example 4.6.1 Consider the system of equations

$$\tag{4.6.18} \begin{aligned} \frac{dx}{dt} &= (4.5 + 0.9 \sin^2 x) y^3, \\ \frac{dy}{dt} &= -(5 + \sin^2 x) y^3, \quad t \neq \tau_k(x, y); \\ \Delta x &= -x + \sigma y, \quad t = \tau_k(x, y); \\ \Delta y &= \sigma x - y, \quad t = \tau_k(x, y). \end{aligned}$$

For system (4.6.18) we construct a matrix function $U(x, y)$ with the elements

$$v_{11}(x) = x^2, \quad v_{22}(y) = y^2,$$
$$v_{12}(x, y) = v_{21}(x, y) = 0.9\, xy,$$

for which estimates

$$v_{11}(x) \geq |x|^2, \quad v_{22}(y) \geq |y|^2,$$
$$v_{12}(x,y) \geq -0.9|x|\,|y|$$

are valid.

The matrix

$$A = \begin{pmatrix} 1 & -0.9 \\ -0.9 & 1 \end{pmatrix}$$

is positive definite.

If $\eta^T = (1, 1)$, then matrices G and C from (4.6.8) and (4.6.12) are

$$G = \begin{pmatrix} 0 & 0 \\ 0 & -0.28 \end{pmatrix}, \quad C = \begin{pmatrix} \sigma^2 - 1 & \frac{9}{10}|\sigma^2 - 1| \\ \frac{9}{10}|\sigma^2 - 1| & \sigma^2 - 1 \end{pmatrix}.$$

We note that $\lambda_M(G) = 0$, i.e. matrix G is negative semidefinite, for $\sigma = \pm 1$ matrix C is equal to zero, and for $|\sigma| < 1$ we have $\lambda_m(C) < 0$, i.e. matrix C is negative definite.

Thus all conditions of Theorem 4.6.1 are satisfied, and zero solution of system of equations (4.6.18) is stable for $\sigma = \pm 1$, and asymptotically stable for $|\sigma| < 1$.

Theorem 4.6.2 *Let differential perturbed motion equations (4.6.1) be such that in domain $R_+ \times S(\rho)$ the conditions of Assumptions 4.6.1, 4.6.2 and 4.6.4 are satisfied, and*

(1) *matrix A is positive definite $(\lambda_m(H^T A H) > 0)$;*
(2) *matrix G is negative definite $(\lambda_M(G) < 0)$;*
(3) $\lambda_M(C^*) > 0$;
(4) *functions $\tau_k(x)$ and a constant ϑ satisfy the inequality*

$$\sup_i \left(\min_{x \in S(\rho)} \tau_{k+1}(x) - \max_{x \in S(\rho)} \tau_k(x) \right) = \vartheta > 0, \quad \text{where} \quad \rho < \rho_0;$$

(5) *functions $\varphi_m(y)$ and $\psi_M(y)$ and a constant $a_0 > 0$ are such that for all $a \in (0, a_0]$ the estimate*

(4.6.19) $$-\frac{1}{\lambda_M(G)} \int_a^{\lambda_M(C^*)\psi_M(a)} \frac{dy}{\varphi_m(y)} \leq \vartheta$$

is valid.

Then the zero solution of (4.6.1) is stable.

If instead of inequality (4.6.19) for some $\gamma > 0$ the inequality

$$(4.6.20) \qquad -\frac{1}{\lambda_M(G)} \int_a^{\lambda_M(C^*)\psi_M(a)} \frac{dy}{\varphi_m(y)} \leq \vartheta - \gamma$$

is satisfied, the zero solution of (4.6.1) is asymptotically stable.

Proof of Theorem 4.6.2 is similar to that of Theorem 4.6.1.

Theorem 4.6.3 *Let differential perturbed motion equations (4.6.1) be such that in domain $R_+ \times S(\rho)$ the conditions of Assumptions 4.6.1, 4.6.2, and 4.6.4 are satisfied, and*

(1) *matrix A is positive definite, i.e. $\lambda_m(H^T A H) > 0$;*
(2) $\lambda_M(G) > 0$;
(3) $\lambda_M(C^*) > 0$;
(4) *functions $\tau_k(x)$ and a constant $\vartheta_1 > 0$ are such that for all $k = 1, 2, \ldots$ the inequality*

$$\max_{x \in S(\rho)} \tau_k(x) - \min_{x \in S(\rho)} \tau_{k-1}(x) \leq \vartheta_1, \quad \text{where} \quad \rho < \rho_0$$

is satisfied;

(5) *functions $\varphi_M(y)$ and $\psi_M(y)$ and a constant a_0 are such that for all $a \in (0, a_0]$ the estimate*

$$(4.6.21) \qquad \frac{1}{\lambda_M(G)} \int_{\lambda_M(C^*)\psi_M(a)}^a \frac{dy}{\varphi_M(y)} \geq \vartheta_1$$

is satisfied.

Then zero solution of (4.6.1) is stable.

If instead of (4.6.21) the inequality

$$(4.6.22) \qquad \frac{1}{\lambda_M(G)} \int_{\lambda_M(C^*)\psi_M(a)}^a \frac{dy}{\varphi_M(y)} \geq \vartheta_1 + \gamma$$

is satisfied, for which $\gamma > 0$, then the zero solution of (4.6.1) is asymptotically stable.

Proof of this theorem is similar to that of Theorem 4.6.1.

4.6.3 The conditions of instability We establish sufficient instability conditions for the zero solution of (4.6.1) in terms of Assumptions 4.6.5–4.6.8 and Propositions 4.6.6–4.6.8.

Theorem 4.6.4 *If differential perturbed motion equations (4.6.1) are such that conditions of Assumptions 4.6.1, 4.6.5, 4.6.6, and 4.6.8 are satisfied and in the domain* Π

(1) *matrix A is positive definite (i.e.* $\lambda_m(H^{\mathrm{T}}AH) > 0$);
(2) *matrix G is positive semidefinite or equals to zero (i.e.* $\lambda_m(G) \geq 0$);
(3) *matrix C is positive definite (i.e.* $\lambda_m(C) > 0$),

then zero solution of (4.6.1) is unstable.

Proof Under conditions of Assumptions 4.6.1 and 4.6.8 and Propositions 4.6.1, 4.6.8 and condition (1) of Theorem 4.6.4, the function $v(t, x, \eta)$ is positive definite and possesses properties (A) and (B). By conditions of Assumption 4.6.5, Proposition 4.6.5 and condition (2) of Theorem 4.6.4

(4.6.23) $\qquad Dv(t, x, \eta) \geq 0, \quad t \neq \tau_k(x) \quad \text{for all} \quad x \in \Pi.$

From Assumption 4.6.6, Proposition 4.6.6, Corollary 4.6.5 and condition (3) of Theorem 4.6.4 it follows

(4.6.24)
$$v(\tau_k(x), \, x + I_k(x)) - v(\tau_k(x), x) \geq \lambda_m(C)\psi_m(v(\tau_k(x), x)),$$
$$t = \tau_k(x), \quad x \in \Pi, \quad k = 1, 2, \ldots.$$

Conditions (4.6.23) and (4.6.24) are sufficient for the zero solution of (4.6.1) be unstable.

Theorem 4.6.5 *Let differential perturbed motion equations (4.6.1) be such that conditions of Assumptions 4.6.1, 4.6.5, 4.6.7, and 4.6.8 are satisfied and in the domain* Π

(1) *matrix A is positive definite (i.e.* $\lambda_m(H^{\mathrm{T}}AH) > 0$);
(2) $\lambda_m(G) < 0$;
(3) *matrix C^* is positive definite (i.e.* $\lambda_m(C^*) > 0$);
(4) *functions $\tau_k(x)$ and a constant $\vartheta_1 > 0$ are such that for all $k = 1, 2, \ldots$ the inequality*

$$\max_{x \in S(\rho)} \tau_k(x) - \min_{x \in S(\rho)} \tau_{k-1}(x) \leq \vartheta_1$$

holds, where $\rho < \rho_0$;
(5) *functions $\varphi_m(y)$ and $\psi_m(y)$ and a constant $\gamma > 0$ for all $a \in (0, a_0]$ satisfies inequality*

$$-\frac{1}{\lambda_m(G)} \int_a^{\lambda_m(C^*)\psi_m(a)} \frac{dy}{\varphi_M(y)} \geq \vartheta_1 + \gamma.$$

Then the zero solution of (4.2.2) is unstable.

Proof of Theorem 4.6.5 is similar to that of Theorem 4.6.4.

Theorem 4.6.6 *Let differential perturbed motion equations (4.6.1) be such that conditions of Assumptions 4.6.1, 4.6.5, 4.6.7, and 4.6.8 are satisfied and in the domain* Π

(1) *matrix A is positive definite (i.e. $\lambda_m(H^{\mathrm{T}}AH) > 0$);*
(2) *matrix G is positive definite (i.e. $\lambda_m(G) > 0$);*
(3) *matrix C^* is positive definite (i.e. $\lambda_m(C^*) > 0$);*
(4) *functions $\tau_k(x)$ and a constant ϑ are such that*

$$\sup_i \Big(\min_{x \in S(\rho)} \tau_{k+1}(x) - \max_{x \in S(\rho)} \tau_k(x) \Big) = \vartheta > 0, \quad \rho < \rho_0;$$

(5) *functions $\varphi_m(y)$ and $\psi_m(y)$ and a constant $\gamma > 0$ are such that for all $a \in (0, a_0]$ the inequality*

$$\frac{1}{\lambda_m(G)} \int_{\lambda_m(C^*)\psi_m(a)}^{a} \frac{dy}{\varphi_m(y)} \leq \vartheta - \gamma$$

is satisfied.
Then the zero solution of (4.6.1) is unstable.

Proof of Theorem 4.6.6 is similar to that of Theorem 4.6.4.

4.7 Novel Approach to Stability Analysis of Uncertain Impulsive Systems

In this section we set out a new approach to the problem of stability investigation of impulsive systems whose parameters are known imprecisely. We propose to realize the approach via the application of the block-diagonal matrix Liapunov function and comparison principle.

4.7.1 Statement of the problem

Let $0 < t_1 < t_2 < \ldots < t_k < \ldots$ and $g_k \to \infty$ as $k \to \infty$. The vector-function $w \in PC(R_+ \times R^n, R^m)$, if $w \colon (t_{k-1}, t_k] \times R^n \to R^m$ is continuous on $(t_{k-1}, t_k] \times R^n$ and for any $x \in R^n$ there exists the limit

$$\lim_{(t,y) \to (t_k^+, x)} w(t, y) = w(t_k^+, x), \quad k = 1, 2, \ldots$$

An uncertain impulsive system of general form is the system of equations

(4.7.1)
$$\frac{dx}{dt} = f(t, x, \alpha), \quad t \neq t_k,$$
$$\Delta x(t) = I_k(x(t), \alpha), \quad t = t_k,$$
$$x(t_0) = x_0.$$

Here $x(t) \in R^n$ is a state vector of the system at the time $t \in R_+$, $f \in PC(R_+ \times R^n \times R^d, R^n)$, $I_k \colon R^n \times R^d \to R^n$, $k = 1, 2, \ldots$, denotes an instant value of the state vector of system (4.7.1) at the instants of the impulse effect t_k, $\alpha \in \mathcal{S} \subseteq R^d$, where $d \geq 1$, $\mathcal{S}$ is a compact set, is an "uncertainty" parameter of system (4.7.1).

Further impulse system (4.7.1) in the absence of the uncertainty parameter is called a nominal impulse system.

The totality of all instants of the impulse effect on the constant component of system (4.7.1) forms the set $E = \{t_1, t_2, \ldots : t_1 < t_2 < \ldots\} \subset R_+$, which is unbounded and closed on R_+.

We assume on system (4.7.1) that the motion subjected to equations (4.7.1) is described by the function $x(t, t_0, x_0, \alpha) \triangleq x(t, \alpha)$ with the following properties in any open neighborhood D of the state $x = 0$, $D \subseteq R^n$:

(a) the motion $x(t, \alpha)$ of system (4.7.1) is continuous on the left on $[t_0, \infty)$ for some $t_0 \geq 0$ and such that $x(t_0, \alpha) = x_0$ for any $(t_0, x_0, \alpha) \in \mathcal{T}_i \times D \times \mathcal{S}$;

(b) the motion $x(t, \alpha)$ of system (4.7.1) is differentiable with respect to t and
$$\frac{dx(t, \alpha)}{dt} = f(t, x(t, \alpha), \alpha)$$
almost everywhere on (t_0, ∞) excluding the set of values $E \subset R_+$ for any $\alpha \in \mathcal{S}$;

(c) for any $t = t_k \in E$ the condition
$$x(t^+, \alpha) = \lim_{t^* \to t, t^* > t} x(t^*, \alpha) = x(t, \alpha) + I_k(x(t, \alpha), \alpha)$$
for any $\alpha \in \mathcal{S}$ and $k = 1, 2, \ldots$.

It is assumed that for any $\alpha \in \mathcal{S}$ the order of system (4.7.1) remains unchanged all the time of system functioning, and for all $t \in \mathcal{T}$ the equilibrium state $x = 0$ is unique for all values of $\alpha \in \mathcal{S}$, i.e. $f(t, 0, \alpha) = I_k(0, \alpha) = 0$.

4. IMPULSIVE SYSTEMS

The strict stability of system (4.7.1) is investigated further under different assumptions on the dynamical properties of continuous and discrete components of nominal impulse system.

System (4.7.1) is presented as

(4.7.2)
$$\frac{dx}{dt} = f(t, x, 0) + \Delta f(t, x, \alpha), \qquad t \neq t_k,$$
$$\Delta x(t) = I_k(x(t), 0) + \Delta I_k(x(t), \alpha), \qquad t = t_k,$$
$$x(t_0) = x_0.$$

where the designations Δf and ΔI_k are obvious. Further the nominal impulsive system

(4.7.3)
$$\frac{dx}{dt} = f(t, x, 0), \qquad t \neq t_k,$$
$$\Delta x(t) = I_k(x(t), 0), \qquad t = t_k,$$
$$x(t_0) = x_0.$$

is decomposed into m interconnected subsystems

(4.7.4)
$$\frac{dx_i}{dt} = f_i(t, x_i, 0) + r_i(t, \hat{x}, 0), \qquad t \neq t_k,$$
$$\Delta x_i(t) = I_{ik}(x_i(t), 0) + a_{ik}(\hat{x}(t), 0), \qquad t = t_k,$$
$$x_i(t_0) = x_{i0}, \quad i = 1, 2, \ldots, m.$$

It is obvious that

$$x(t) = (x_1^T(t), \ldots, x_m^T(t))^T \in R^n, \quad n = \sum_{i=1}^{m} n_i,$$
$$r_i(t, \hat{x}, 0) = f_i(t, x_1, \ldots, x_m, 0) - f_i(t, x_i, 0),$$
$$a_{ik}(\hat{x}(t), 0) = I_{ik}(x_1(t), \ldots, x_m(t), 0) - I_{ik}(x_i(t), 0),$$
$$f_i \colon R_+ \times R^{n_i} \to R^{n_i}, \quad r_i \colon R_+ \times R^{n-n_i} \to R^{n_i},$$
$$I_{ik} \colon R^{n_i} \to R^{n_i}, \quad a_{ik} \colon R^{n-n_i} \to R^{n_i}.$$

We shall make some assumptions on the dynamical properties of the independent nominal subsystems

(4.7.5)
$$\frac{dx_i}{dt} = f_i(t, x_i, 0), \qquad t \neq t_k,$$
$$\Delta x_i(t) = I_{ik}(x_i(t), 0), \qquad t = t_k,$$
$$x_i(t_0) = x_{i0}, \quad i = 1, 2, \ldots, m,$$

of system (4.7.4).

Assumption 4.7.1 Among m independent continuous components (4.7.5) of the impulsive system there are r asymptotically stable and $m - r \geq 1$ unstable.

Assumption 4.7.2 Among m independent discrete components (4.7.5) of the impulsive system there exist r unstable and $m-r \geq 1$ asymptotically stable.

Remark 4.7.1 Stability of impulsive systems (4.7.1) with Assumptions 4.7.1, 4.7.2 described dynamical properties of continuous and discrete components was not analysed before in the qualitative theory of impulse systems (see Halanay and Weksler [1], Pandit and Deo [1], Lakshmikantham, Bainov and Simeonov [1], Samoilenko and Perestyuk [1], etc.).

4.7.2 Comparison principle with the block-diagonal matrix function To study system (4.7.1) we construct the matrix function $U(t,x)$ in the form

(4.7.6) $$U(t,x) = \text{diag}\{U_1(t,x), U_2(t_k,x)\},$$

where $U_1 \in PC(R_+ \times R^n, R^{m' \times m'})$, $m' = \sum_{i=1}^{r} n_i$, $U_2 \colon E \times R^{n-m'} \to R^{(n-m') \times (n-m')}$.

The auxiliary functions of (4.7.6) type are called the block-diagonal matrix functions. Two-component vector functions are partial cases of these functions. One component of the functions characterizes the continuous component and the other is characteristic of the discrete component.

The block $U_1(t,x)$ of the matrix-valued function is constructed in terms of r stable subsystems of the continuous part of system (4.7.4) and the block $U_2(t_k,x)$ is constructed in terms of $m-r$ asymptotically stable components of the discrete part of system (4.7.4).

Assumption 4.7.3 The elements $v_{ij}(t,\cdot)$ of the matrix-valued function $U_1(t,x)$ satisfy the following conditions

(a) v_{ij} are continuous on $(t_{k-1}, t_k] \times R^{m'}$ for every $x \in R^{m'}$ and there exist

$$\lim_{(t,y) \to (t_k^+, x)} v_{ij}(t,y) = v_{ij}(t_k^+, x), \quad i,j = 1,2,\ldots,r;$$

(b) v_{ij} are locally Lipschitzian with respect to x;

(c) for any $(t,x) \in (t_{k-1}, t_k] \times R^n$ the derivatives
$$D_- v_{ij}(t,x) = \liminf\{[v_{ij}(t+\theta, x+\theta f(t,x,\alpha)) - v_{ij}(t,x)]\theta^{-1}: \theta \to 0^-\}, \quad i,j = 1,2,\ldots,m',$$
along solutions of the corresponding subsystems (of system (4.7.1)) are determined;

(e) for $(t, x_p) \in E \times R^{n-m'}$ the first forward differences
$$\Delta v_{pq}(t_k, x_p) = v_{pq}(t_{k+1}, x(t_{k+1})) - v_{pq}(t_k, x(t_k)),$$
$$p,q = 1,2,\ldots,m', \quad k = 1,2,\ldots$$
are determined.

In terms of the matrix-valued function $U(t,x)$ we construct the auxiliary scalar function

(4.7.7) $$v(t,x,\eta) = \eta^T U(t,x)\eta, \quad \eta \in R_+^m.$$

Consider the comparison equations

(4.7.8) $$\frac{du}{dt} = g_1(t,u,\mu(\alpha)), \quad t \neq t_k,$$
$$u(t_k^+) = \psi_k(u(t_k), \mu(\alpha)),$$
$$u(t_0) = u_0 \geq 0,$$

and

(4.7.9) $$\frac{dv}{dt} = g_2(t,v,\mu(\alpha)), \quad t \neq t_k,$$
$$v(t_k^+) = \phi_k(v(t_k), \mu(\alpha)),$$
$$v(t_0) = v_0 \geq 0,$$

Here $g_1, g_2 \in PC(R_+^3, R)$, $g_2(t,v,\mu(\alpha) \leq g_1(t,u,\mu(\alpha)))$ for all $\alpha \in S$, $\mu(\alpha) \geq 0$, $\psi_k, \phi_k \colon R_+^2 \to R$ are nondecreasing in the first argument and moreover $\phi_k(u,\mu(\alpha)) \leq \psi_k(u,\mu(\alpha))$ for every k, $\alpha \in S$ and $\mu(\alpha) \geq 0$.

Let $u^+(t,t_0,u_0)$ and $v^-(t,t_0,v_0)$ be upper and lower solutions of equations (4.7.8) and (4.7.9) respectively, defined for all $t \geq t_0$. Then by the theory of differential inequalities it follows that

(4.7.10) $$v^-(t,t_0,v_0) \leq u^+(t,t_0,u_0)$$

for all $t \geq t_0$, $\mu(\alpha) \geq 0$, whenever

(4.7.11) $$v_0 \leq u_0.$$

Theorem 4.7.1 *Assume that*

(1) *all conditions of Assumptions 4.7.1–4.7.3 are satisfied;*
(2) *there exist functions $g_1, g_2 \in PC(R_+^3, R)$, $\psi_k, \phi_k \colon R_+^2 \to R$ such that*

   (a) $g_2(t, v(t,x,\eta), \mu(\alpha)) \leq D_- v(t,x,\eta) \leq g_1(t, v(t,x,\eta), \mu(\alpha))$ *for $t \neq t_k$ and all $(t,x) \in R_+ \times D(\rho)$;*

   (b) $\phi_k(v(t,x,\eta), \mu(\alpha)) \leq v(t, x+I_k(x,\alpha), \eta) \leq \psi_k(v(t,x,\eta), \mu(\alpha))$ *for $t = t_k$ for some $\rho_0 = \rho_0(\rho) > 0$ such that the condition $x \in D(\rho_0)$ implies the inclusion $x + I_k(x,\alpha) \in D(\rho)$ for all k and $\mu(\alpha) \geq 0$;*

(3) *the vector $\eta \in R_+^m$ and initial values (x_0, u_0, v_0) are taken so that $v_0 \leq v(t, x_0, \eta) \leq u_0$ for $t = t_0$;*
(4) *under the initial conditions mentioned in condition (3) the motions $x(t, \alpha)$ of system (4.7.1) are defined for all $t \geq t_0$.*

Then the bilateral estimate

$$(4.7.12) \qquad v^-(t, t_0, v_0) \leq v(t, x(t,\alpha), \eta) \leq u^+(t, t_0, u_0)$$

holds true for all $t \geq t_0$ when any $\alpha \in S$ and $\mu(\alpha) \geq 0$.

Proof The assertion of Theorem 4.7.1 follows from Theorem 4.9.1 by Lakshmikantham, Leela and Martynyuk [1] (see also Lakshmikantham and Devi [1])

4.7.3 Test for strict stability We shall cite some definitions necessary for subsequent presentation.

Definition 4.7.1 The equilibrium state $x = 0$ of impulsive system (4.7.1) is

(a) strictly equistable, if given $t_0 \in R_+$, $\varepsilon_1 > 0$ and $\Delta > 0$ one can indicate positive functions $\delta_1 = \delta_1(t_0, \varepsilon_1, \Delta)$, $\delta_2 = \delta_2(t_0, \varepsilon_1, \Delta)$ and $\varepsilon_2 = \varepsilon_2(t_0, \varepsilon_1, \Delta)$ which are continuous in t_0 for fixed (ε_1, Δ) and ordered by the inequalities $(\varepsilon_2 < \delta_2 < \delta_1 < \varepsilon_1)$, so that $\varepsilon_2 < \|x(t,\alpha)\| < \varepsilon_1$ for all $t \geq t_0$ and any $\alpha \in S$, whenever $\delta_2 < \|x_0\| < \delta_1$ and $|t - t_k| > \Delta$;
(b) strictly uniformly stable, if in Definition 4.7.1(a) functions δ_1, δ_2 and ε_2 do not depend on t_0.

Definition 4.7.2 The zero solution of comparison equations (4.7.8), (4.7.9) is strictly equistable, if, given $t_0 \in R_+$, $\varepsilon_1 > 0$, $\widetilde{\Delta} > 0$, it is possible to indicate positive functions $\tilde{\delta}_1 = \tilde{\delta}_1(t_0, \tilde{\varepsilon}_1, \widetilde{\Delta})$, $\tilde{\delta}_2 = \tilde{\delta}_2(t_0, \tilde{\varepsilon}_1, \widetilde{\Delta})$ and $\tilde{\varepsilon}_2 = \tilde{\varepsilon}_2(t_0, \tilde{\varepsilon}_1, \widetilde{\Delta})$ which are continuous in t_0 and ordered by the inequalities $\tilde{\varepsilon}_2 < \tilde{\delta}_2 < \tilde{\delta}_1 < \tilde{\varepsilon}_1$ such that

(4.7.13) $$\tilde{\varepsilon}_2 < v^-(t, t_0, v_0) \le u^+(t, t_0, u_0) < \tilde{\varepsilon}_1$$

for all $t \ge t_0$ and $\mu(\alpha) \ge 0$ whenever $\tilde{\delta}_2 < v_0 \le u_0 < \tilde{\delta}_1$ and $|t - t_k| > \widetilde{\Delta}$.

Sufficient conditions of strict stability for the equilibrium state $x = 0$ of system (4.7.1) are established by the following result.

Theorem 4.7.2 *Assume that*

(1) *all conditions of Theorem 4.7.1 are satisfied;*
(2) *for all $t \ge t_0$ the vector-functions $f(t, x, \alpha) = 0$ and $I_k(x, \alpha) = 0$, iff $x = 0$ for every k and any $\alpha \in \mathcal{S}$;*
(3) *there exist functions a and b of class K such that for the function $v(t, x, \eta) = \eta^{\mathrm{T}} U(t, x) \eta$*

$$b(\|x\|) \le v(t, x, \eta) \le a(\|x\|)$$

for all $(t, x) \in R_+ \times D(\rho)$.

Then the equilibrium state $x = 0$ of uncertain impulse system (4.7.1) possesses the same type of strict stability as the zero solution of comparison equations (4.7.8) and (4.7.9).

Proof Consider the property of strict equistability of motion $x = 0$ of system (4.7.1). By condition (2) of Theorem 4.7.2, system (4.7.1) has a unique equilibrium state $x = 0$ for any change of $\alpha \in \mathcal{S}$ and for all $k = 1, 2, \ldots$. Assume that $t_0 \in R_+$ and $0 < \varepsilon_1 < \rho^* = \min(\rho_o, \rho)$ are given.

Let the zero solution of comparison equations (4.7.8), (4.7.9) be strictly equistable. In this case, given $\varepsilon_1 > 0$, we take $\tilde{\varepsilon}_1$ from the correlation $\tilde{\varepsilon}_1 = b(\varepsilon_1)$. For $\varepsilon_1 > 0$ we take $\tilde{\delta}_1 = \tilde{\delta}_1(t_0, \varepsilon_1, \Delta)$, $\tilde{\delta}_2 = \tilde{\delta}_2(t_0, \varepsilon_1, \Delta)$ and $\tilde{\varepsilon}_2 = \tilde{\varepsilon}_2(t_0, \varepsilon_1, \Delta)$ continuous in t_0 and satisfying the inequalities

(4.7.14) $$\tilde{\varepsilon}_2 < \tilde{\delta}_2 < \tilde{\delta}_1 < b(\varepsilon_1),$$
$$\tilde{\varepsilon}_2 < v^-(t, t_0, v_0) < u^+(t, t_0, u_0) < b(\varepsilon_1), \quad t \ge t_0, \quad \mu(\alpha) \ge 0,$$

whenever $\tilde{\delta}_2 < v_0 \leq u_0 < \tilde{\delta}_1$, $|t - t_k| > \Delta$. Under condition (3) of Theorem 4.7.1 (see condition (1) of Theorem 4.7.2) we choose the vector $\eta \in R_+^m$ and values (x_0, u_0, v_0) so that

(4.7.15) $$v_0 = v(t, x_0, \eta) = u_0 \quad \text{for} \quad t = t_0.$$

the fact that the function $v(t, x, \eta)$ is continuous in the neighborhood of point (t_0, x_0) and $v(t, 0, \eta) = 0$ for all $t \in R_+$, implies that, given $\tilde{\delta}_2 < \tilde{\delta}_1$, there exist $\delta_2 < \delta_1 < \varepsilon_1$ such that

(4.7.16) $$\tilde{\delta}_2 < v(t_0, x_0, \eta) < \tilde{\delta}_1,$$

whenever $\delta_2 < \|x_0\| < \delta_1$, where $\delta_2 = \delta_2(t_0, \varepsilon_1, \Delta)$, $\delta_1 = \delta_1(t_0, \varepsilon_1, \Delta)$.

Then we choose the value $\varepsilon_2 = \varepsilon_2(t_0, \varepsilon_1, \Delta) > 0$ so that $a(\varepsilon_2) \leq \tilde{\varepsilon}_2$ and $\varepsilon_2 < \delta_2$. As a result we have a sequence of inequalities $\varepsilon_2 < \delta_2 < \delta_1 < \varepsilon_1$ (see Definition 4.7.1). Let us show that for this choice of the values ε_2, δ_2 and δ_1 the equilibrium state $x = 0$ of system (4.7.1) is strictly equistable whenever all conditions of Theorem 4.7.2 are satisfied. This means that the motion $x(t, \alpha)$ of system (4.7.1) for any $\alpha \in \mathcal{S}$ satisfies the estimate

$$\varepsilon_2 < \|x(t, \alpha)\| < \varepsilon_1 \quad \text{for all} \quad t \geq t_0$$

whenever $\delta_2 < \|x_0\| < \delta_1$.

If this is not true, there is an alternative between two cases of motion $x(t, \alpha)$ of system (4.7.1).

Case 1 For some k there exists $t_2 \in (t_k, t_{k+1}]$ such that $\varepsilon_2 \geq \|x(t_2, \alpha)\|$ for any $\alpha \in \mathcal{S}$. It is clear that for $t_0 \leq t \leq t_2$ the motion $x(t, \alpha)$ remains in the domain $D(\rho)$, and by conditions of Theorem 4.7.1 estimate (4.7.12) is true. Hence, in view of condition (3) of Theorem 4.7.2 and estimate (4.7.12), we get

$$a(\varepsilon_2) \geq a(\|x(t_2, \alpha)\|) \geq v(t, x(t, \alpha), \eta) \geq v^-(t_0, t_0, v_0) > \tilde{\varepsilon}_2 \geq a(\varepsilon_2).$$

The contradiction obtained proves that $t_2 \overline{\in} (t_k, t_{k+1}]$ and $\|x(t, \alpha)\| > \varepsilon_2$ for all $t \geq t_0$ and $\alpha \in \mathcal{S}$.

Case 2 For some p there exists $t_1 \in (t_p, t_{p+1}]$ such that $\varepsilon_1 < \|x(t_1, \alpha)\|$ and $\|x(t, \alpha)\| < \varepsilon_1$ for all $t_0 \leq t < t_1$. By condition 2(b) of Theorem 4.7.2 $\|x(t_p, \alpha)\| < \varepsilon_1 < \rho_0$. Hence we have

(4.7.17) $$\|x(t_p^+, \alpha)\| = \|x(t_p, \alpha) + I_k(x(t_p, \alpha), \alpha)\| < \rho.$$

Thus, a $\tilde{t} \in (t_p, t_1]$ must exist such that

(4.7.18) $$\varepsilon_1 \leq \|x(\tilde{t}, \alpha)\| < \rho.$$

This means that for $t_0 \leq t \leq \tilde{t}$ estimate $\|x(t, \alpha)\| < \rho$ is valid, and therefore, by Theorem 4.7.1 estimate (4.7.12) is true. By condition (3) of Theorem 4.7.2 and correlations (4.7.16) and (4.7.17) we have

(4.7.19) $$b(\varepsilon_1) \leq b(\|x(\tilde{t}, \alpha)\|) \leq v(\tilde{t}, x(\tilde{t}, \alpha), \eta) \leq u^+(\tilde{t}, t_0, u_0) < b(\varepsilon).$$

Contradiction (4.7.19) proves that $\tilde{t} \in (t_p, t_1]$, and consequently $\|x(t, \alpha)\| < \varepsilon_1$, for all $t \geq t_0$ and all $\alpha \in \mathcal{S}$. This proves the assertion of Theorem 4.7.2.

Further we shall apply an auxiliary vector function for the stability analysis of uncertain system.

4.7.4 Vector approach For system (4.7.1) let the matrix function $U(t, x)$ be constructed in the form

(4.7.20) $$U(t, x) = \operatorname{diag}\{U_1(t, x), U_2(t_k, x)\},$$

where $U_1 \in PC(R_+ \times R^n, R^{m' \times m'})$, $m' = \sum_{i=1}^{r} n_i$, $U_2 \colon E \times R^{n-m'} \to R^{(n-m') \times (n-m')}$.

By means of the matrix-valued function $U(t, x)$ we construct the auxiliary vector function

(4.7.21) $$L(t, x, \eta) = AU(t, x)\eta, \quad \eta \in R_+^m,$$

where $A = \operatorname{diag}(A_1, A_2)$, A_i are constant matrices whose dimensions correspond to those of the blocks of matrix function (4.7.20).

For system (4.7.1) and function (4.7.21) we shall consider the comparison function

(4.7.22) $$\frac{du}{dt} = g(t, u, \mu(\alpha)), \quad t \neq t_k,$$
$$u(t_k^+) = \psi_k(u(t_k), \mu(\alpha)),$$
$$u(t_0) = u_0 \geq 0.$$

Here $g \in PC(R_+^3, R^m)$, $g(t, u, \mu(\alpha)))$ is quasimonotone in u for all $\alpha \in \mathcal{S}$, $\mu(\alpha) \geq 0$, $\psi_k, \phi_k \colon R_+^2 \to R$ are nondecreasing with respect to the first argument.

Let $u^+(t, t_0, u_0)$ be the upper solution of system of equations (4.7.22) determined for all $t \geq t_0$. According to the theory of differential inequalities we have the following result.

Theorem 4.7.3 *Assume that*
(1) *all conditions of Assumptions 4.7.1, 4.7.2 and 4.7.3 are satisfied;*
(2) *there exist functions $g \in PC(R_+^3, R)$, $\psi_k, \phi_k \colon R_+^2 \to R$ such that*

  (a) $D^+ L(t, x, \eta) \leq g(t, L(t, x, \eta), \mu(\alpha))$ *for $t \neq t_k$ and for all $(t, x) \in R_+ \times D(\rho)$;*

  (b) $L(t, x + I_k(x, \alpha), \eta) \leq \psi_k(L(t, x, \eta), \mu(\alpha))$ *for $t = t_k$ for some $\rho_0 = \rho_0(\rho) > 0$ such that the condition $x \in D(\rho_0)$ implies the inclusion $x + I_k(x, \alpha) \in D(\rho)$ for all k and $\mu(\alpha) \geq 0$;*

(3) *the matrix A, vector $\eta \in R_+^m$ and initial values (x_0, u_0, v_0) are taken so that $L(t, x_0, \eta) \leq u_0$ for $t = t_0$;*
(4) *under the initial conditions, mentioned in condition (3), the motions $x(t, \alpha)$ of system (4.7.1) are determined for all $t \geq t_0$.*

Then the change of components of vector function (4.7.21) along solutions of system (4.7.1) is estimated by the inequality

(4.7.23) $$L(t, x(t, \alpha), \eta) \leq u^+(t, t_0, u_0)$$

for all $t \geq t_0$ for any $\alpha \in \mathcal{S}$ and $\mu(\alpha) \geq 0$.

Proof Assertion of this theorem is verified in the same way as that of comparison Theorem 3.4.1 by Lakshmikantham, Leela, and Martynyuk [3].

Stability conditions for the equilibrium state $x = 0$ of system (4.7.1) are established in the following result.

Theorem 4.7.4 *Let perturbed motion equations (4.7.1) be such that*
(1) *all conditions of Theorem 4.7.3 are fulfilled;*
(2) *for all $t \geq t_0$ the vector functions $f(t, x, \alpha) = 0$ and $I_k(x, \alpha) = 0$, iff $x = 0$ for every k and any $\alpha \in \mathcal{S}$;*
(3) *there exist functions a and b of class K and vector $c \in R^m$ such that for the function $L(t, x, \eta)$*

(4.7.24) $$b(\|x\|) \leq c^{\mathrm{T}} L(t, x, \eta) \leq a(\|x\|)$$

  for all $(t, x) \in R_+ \times D(\rho)$.

Then the equilibrium state $x = 0$ of system (4.7.1) possesses the same type of stability as the zero solution of comparison system (4.7.22).

Proof Let the values $t_0 \in R_+$ and $0 < \varepsilon < \rho^* = \min(\rho_0, \rho)$ be given. Assume that the zero solution of comparison system (4.7.22) is equistable. Then, given $t_0 \in R_+$ and $b(\cdot) > 0$, there exist $\delta_1(t_0, \varepsilon) > 0$ such that

(4.7.25) $$\|u(t; t_0, u_0)\| < b(\varepsilon) \quad \text{for all} \quad t \geq t_0$$

whenever $\|u_0\| < \delta_1$, where $u(t;t_0,u_0)$ is any solution of comparison system (4.7.22). By condition (3) of Theorem 4.7.3, given $L(t,x,b)$, we have for $t = t_0^+$ the estimate $e^{\mathrm{T}}L(t,x_0,b) \leq u_0$, where $u_0 = a(\|x_0\|)$ is admissible. We take $\delta_2 = \delta_2(\varepsilon)$ so that $a(\delta_2) < b(\varepsilon)$ and determine $\delta = \min(\delta_1,\delta_2) > 0$.

Given $\varepsilon > 0$ and δ mentioned above, we shall show that $\|x(t,\alpha)\| < \varepsilon$ for all $t \geq t_0$ and $\alpha \in \mathcal{S}$ whenever $\|x_0\| < \delta$, where $x(t,\alpha)$ is a solution of system (4.7.1), determined for all $t \geq t_0$. Let this be not true. Then a solution $x(t,\alpha)$ must exist at least for one $\alpha \in \mathcal{S}$ such that $\|x_0\| < \delta$ and for $t^* > t_0$ ($t_k < t^* < t_{k+1}$ for some k)

(4.7.26) $\quad\quad \|x(t^*,\alpha)\| \geq \varepsilon \quad \text{and} \quad \|x(t,\alpha)\| < \varepsilon \quad \text{for} \quad t_0 \leq t \leq t_k.$

Since $0 < \varepsilon < \rho^*$, it follows from condition (2)(b) of Theorem 4.7.3 that $|x_k^+| = |x_k + I_k(x_k,\alpha)| < \rho$, where $x_k = x(t_k,\alpha)$ and $|x_k| < \varepsilon$. Hence it follows that a t^0 can be found such that $t_k < t^0 < t^*$ for which $\varepsilon \leq \|x(t^0,\alpha)\| < \rho^*$ for all $\alpha \in \mathcal{S}$.

We designate $m(t) = c^{\mathrm{T}}L(t,x(t),b)$ for $t_0 \leq t \leq t^0$ and in view of conditions (2)–(4) of Theorem 4.7.3 we obtain

(4.7.27) $\quad\quad e^{\mathrm{T}}L(t,x(t),b) \leq e^{\mathrm{T}}u(t;t_0,u_0), \quad t_0 \leq t \leq t^0,$

where $u(t;t_0,u_0)$ is the maximal solution of comparison system (4.7.21). Inequality (4.7.27) yields

$$b(\varepsilon) \leq b(\|x(t^0,\alpha)\|) \leq e^{\mathrm{T}}L(t^0,x(t^0,\alpha),b) \leq e^{\mathrm{T}}u(t^0,t_0,a(\|x_0\|)) < b(\varepsilon).$$

Hence it follows that $\|x(t,\alpha)\| < \varepsilon$ for all $t \geq t_0$ and all $\alpha \in \mathcal{S}$. This proves stability of the state $x = 0$ of system (4.7.1).

4.7.5 Uncertain linear impulsive systems

Consider a linear impulsive system

(4.7.28) $\quad\quad \begin{aligned} \frac{dx}{dt} &= Ax, \quad t \neq t_i, \\ x(t^+) &= Bx(t), \quad t = t_i, \quad x(t_0) = x_0, \end{aligned}$

where $x \in R^n$ is a state vector of the system, $x(t^+)$ is the value of function $x(t)$ on the right, $A \in R^{n \times n}$ is a constant matrix, $B \in R^{n \times n}$ is a matrix from which a constant matrix B_0 can be taken, i.e. the representation $B = B_0 + \Delta B$ is valid, where ΔB is the uncertainty matrix whose elements belong to a compact set.

We make some general assumptions on system (4.7.28).

Assumption 4.7.4 Let for impulsive system (4.7.28) the following conditions be satisfied

(1) matrix A possesses the property of exponential dichotomy, i.e. there exists a projection P and positive constants K, L, α, β such that

$$|X(t)PX^{-1}(s)| \leq Ke^{-\alpha(t-s)} \quad \text{for} \quad t \geq s,$$
$$|X(t)(I-P)X^{-1}(s)| \leq Le^{-\beta(s-t)} \quad \text{for} \quad s \geq t,$$

where $X(t)$ is a fundamental matrix for linear differential equation (4.7.28);

(2) moments of the impulsive effect $\{t_i\}_{i=1}^{\infty}$ are unknown, but such that

$$0 < \theta_1 \leq t_{i+1} - t_i \leq \theta_2 < +\infty \quad \text{for all} \quad i = 1, 2, \ldots.$$

For the "nominal" system

(4.7.29)
$$\frac{dx}{dt} = Ax, \quad t \neq t_i,$$
$$x(t^+) = B_0 x(t), \quad t = t_i, \quad x(t_0) = x_0.$$

sufficient stability conditions for the state $x = 0$ are obtained via the vector Liapunov function and the second order comparison system

(4.7.30)
$$\dot{u}_1 = a_{11}u_1 + a_{12}u_2,$$
$$\dot{u}_2 = a_{21}u_1 + a_{22}u_2, \quad t \neq t_i,$$
$$u_1(t^+) = b_{11}u_1 + b_{12}u_2,$$
$$u_2(t^+) = b_{21}u_1 + b_{22}u_2, \quad t = t_i$$

By means of the linear nonsingular transformation

(4.7.31)
$$y_1 = w_{11}u_1 + w_{12}u_2,$$
$$y_2 = w_{21}u_1 + w_{22}u_2,$$

where

$$w_{11} = \frac{a_{11} - a_{22} + \sqrt{(a_{22} - a_{11})^2 + 4a_{12}a_{21}}}{2a_{12}}, \quad w_{12} = 1,$$

$$w_{21} = \frac{a_{11} - a_{22} - \sqrt{(a_{11} - a_{22})^2 + 4a_{12}a_{21}}}{2a_{12}}, \quad w_{22} = 1,$$

system (4.7.30) is reduced to the form

(4.7.32)
$$\begin{aligned}
\dot{y}_1 &= \lambda_1 y_1, \\
\dot{y}_2 &= \lambda_2 y_2, \quad t \neq t_i, \\
y_1(t^+) &= \bar{b}_{11} y_1(t) + \bar{b}_{12} y_2(t), \\
y_2(t^+) &= \bar{b}_{21} y_1(t) + \bar{b}_{22} y_2(t), \quad t = t_i.
\end{aligned}$$

Here

$$\lambda_1 = \frac{a_{11} + a_{22} + \sqrt{(a_{11} - a_{22})^2 + 4 a_{12} a_{21}}}{2},$$

$$\lambda_2 = \frac{a_{11} + a_{22} - \sqrt{(a_{11} - a_{22})^2 + 4 a_{12} a_{21}}}{2},$$

$$\bar{b}_{11} = \frac{b_{11} w_{11} w_{22} - b_{12} w_{11} w_{21} + b_{21} w_{12} w_{22} - b_{22} w_{12} w_{21}}{w_{11} w_{12} - w_{12} w_{21}},$$

$$\bar{b}_{12} = \frac{b_{12} w_{11}^2 - b_{11} w_{11} w_{12} + b_{22} w_{11} w_{12} - b_{21} w_{12}^2}{w_{11} w_{22} - w_{12} w_{21}},$$

$$\bar{b}_{21} = \frac{b_{11} w_{21} w_{22} - b_{12} w_{21}^2 + b_{21} w_{22}^2 - b_{22} w_{22} w_{21}}{w_{11} w_{22} - w_{12} w_{21}},$$

$$\bar{b}_{22} = \frac{b_{12} w_{21} w_{11} - b_{11} w_{21} w_{12} + b_{22} w_{22} w_{11} - b_{21} w_{22} w_{12}}{w_{11} w_{22} - w_{12} w_{21}}.$$

Sufficient stability conditions for the state $y = 0$ of system (4.7.32) are established by the following result.

Theorem 4.7.5 *Assume that the parameters of system (4.7.32) are such that*

(1) $\lambda_1 + \dfrac{\ln\left(|\bar{b}_{11}| + |\bar{b}_{21}|\right)}{\theta_2} < 0,$

(2) $\lambda_2 + \dfrac{\ln\left(|\bar{b}_{22}| + |\bar{b}_{12}|\right)}{\theta_k} < 0,$

where

$$k = \begin{cases} 1, & \text{for } |\bar{b}_{22}| + |\bar{b}_{12}| \geq 1 \\ 2, & \text{for } |\bar{b}_{22}| + |\bar{b}_{12}| < 1. \end{cases}$$

Then its equilibrium state $y_1 = y_2 = 0$ is asymptotically stable.

Proof of this assertion consists in the immediate calculation of the matriciant $X(t, t_0)$ of system (4.7.32) and establishing the estimate $\|X(t, t_0)\|_1 < c(t_0) q^{N(t,t_0)}$, where $N(t, t_0)$ is the largest integer satisfying the inequality

$N(t,t_0)\theta_1 \le t - t_0$, $q \in (0,1)$, $\|\cdot\|_1$ is the matrix norm conforming with the vector norm $\|y\|_1 = |y_1| + |y_2|$.

Further we employ linear nondegenerate transform $y = Wx$ to reduce system (4.7.28) to block-diagonal form

(4.7.33)
$$\dot{y}_1 = A_1 y_1,$$
$$\dot{y}_2 = A_2 y_2, \quad t \ne t_i,$$
$$y_1(t^+) = \overline{B}_{11} y_1(t) + \overline{B}_{12} y_2(t),$$
$$y_2(t^+) = \overline{B}_{21} y_1(t) + \overline{B}_{22} y_2(t), \quad t = t_i,$$

where A_1 is a totally unstable matrix, A_2 is the Hurwitz matrix, $y = (y_1^T, y_2^T)^T$, $y_1 \in R^r$, $y_2 \in R^{n-r}$, $A_1 \in R^{r \times r}$, $A_2 \in R^{(n-r) \times (n-r)}$, $\overline{B}_{11} \in R^{r \times r}$, $\overline{B}_{12} \in R^{r \times (n-r)}$, $\overline{B}_{21} \in R^{(n-r) \times r}$ and $\overline{B}_{22} \in R^{(n-r) \times (n-r)}$.

For system (4.7.33) we construct the vector Liapunov function $V: R^n \to R_+^2$ with the elements

(4.7.34)
$$v_1(y_1) = y_1^T P_1 y_1, \quad y_1 \in R^r,$$
$$v_2(y_2) = y_2^T P_2 y_2, \quad y_2 \in R^{n-r},$$

where P_1, P_2 are symmetric positive definite matrices determined by the matrix Liapunov equations

(4.7.35)
$$\overline{B}_{11}^T P_1 \overline{B}_{11} - P_1 = -Q_1,$$
$$A_2^T P_2 + P_2 A_2 = -Q_2.$$

Here Q_1 and Q_2 are some positive definite matrices.

It is easy to show that the estimates of total derivatives of components (4.7.34) along solutions of system (4.7.33) are as follows

$$\frac{dv_1}{dt} \le \frac{\lambda_M(G_1)}{\lambda_m(P_1)} v_1,$$
$$\frac{dv_2}{dt} \le -\frac{\lambda_m(Q_2)}{\lambda_M(P_2)} v_2, \quad \text{for} \quad t \ne t_i.$$

Besides, the following estimates are valid

$$v_1(\overline{B}y) \leq \frac{\lambda_M(P_1 - Q_1) + \|\overline{B}_{11}^T P_1 \overline{B}_{12}\|}{\lambda_m(P_1)} v_1$$

$$+ \frac{\lambda_M(\overline{B}_{12}^T P_1 \overline{B}_{12}) + \|\overline{B}_{11}^T P_1 \overline{B}_{12}\|}{\lambda_m(P_2)} v_2,$$

$$v_2(\overline{B}y) \leq \frac{\lambda_M(\overline{B}_{22}^T P_2 \overline{B}_{22}) + \|\overline{B}_{22}^T P_2 \overline{B}_{21}\|}{\lambda_m(P_2)} v_2$$

$$+ \frac{\lambda_M(\overline{B}_{21}^T P_2 \overline{B}_{21}) + \|\overline{B}_{22}^T P_2 \overline{B}_{21}\|}{\lambda_m(P_1)} v_1, \quad \text{for} \quad t = t_i.$$

Here $G_1 = A_1^T P_1 + P_1 A_1$ and $\lambda_M(\cdot)$ is the largest eigenvalue and $\lambda_m(\cdot)$ is the smallest eigenvalue of the corresponding matrices.

By comparison Theorem 1.11.3 from Lakshmikantham, Leela and Martynyuk [1] stability of solution $u_1 = u_2 = 0$ of the comparison system

(4.7.36)
$$\frac{du_1}{dt} = \frac{\lambda_M(G_1)}{\lambda_m(P_1)} u_1,$$

$$\frac{du_2}{dt} = -\frac{\lambda_m(Q_2)}{\lambda_M(P_2)} u_2, \quad t \neq t_i,$$

$$u_1(t^+) = \frac{\lambda_M(P_1 - Q_1) + \|\overline{B}_{11}^T P_1 \overline{B}_{12}\|}{\lambda_m(P_1)} u_1(t)$$

$$+ \frac{\lambda_M(\overline{B}_{12}^T P_1 \overline{B}_{12}) + \|\overline{B}_{11}^T P_1 \overline{B}_{12}\|}{\lambda_m(P_2)} u_2(t),$$

$$u_2(t^+) = \frac{\lambda_M(\overline{B}_{22}^T P_2 \overline{B}_{22}) + \|\overline{B}_{22}^T P_2 \overline{B}_{21}\|}{\lambda_m(P_2)} u_2(t)$$

$$+ \frac{\lambda_M(\overline{B}_{21}^T P_2 \overline{B}_{21}) + \|\overline{B}_{22}^T P_2 \overline{B}_{21}\|}{\lambda_m(P_1)} u_1(t), \quad t = t_i,$$

implies stability of the state $y_1 = y_2 = 0$ of system (4.7.33) and the state $x = 0$ of nominal system (4.7.29).

Further we shall use the designations

$$\gamma_1 = \frac{\lambda_M(P_1 - Q_1) + \|\overline{B}_{11}^T P_1 \overline{B}_{12}\| + \lambda_M(\overline{B}_{21}^T P_2 \overline{B}_{21}) + \|\overline{B}_{22}^T P_2 \overline{B}_{21}\|}{\lambda_m(P_1)},$$

$$\gamma_2 = \frac{\lambda_M(\overline{B}_{22}^{\mathrm{T}} P_2 \overline{B}_{22}) + \|\overline{B}_{22}^{\mathrm{T}} P_2 \overline{B}_{21}\| + \lambda_M(\overline{B}_{12}^{\mathrm{T}} P_1 \overline{B}_{12}) + \|\overline{B}_{11}^{\mathrm{T}} P_1 \overline{B}_{12}\|}{\lambda_m(P_2)},$$

$$\Gamma_1 = \frac{\lambda_M(G_1)}{\lambda_m(P_1)} + \frac{1}{\theta_2} \ln \gamma_1,$$

$$\Gamma_2 = -\frac{\lambda_m(Q_2)}{\lambda_M(P_2)} + \frac{1}{\theta_k} \ln \gamma_2,$$

where

$$k = \begin{cases} 1, & \text{for } \gamma_2 \geq 1, \\ 2, & \text{for } \gamma_2 < 1. \end{cases}$$

Stability conditions for the state $x = 0$ of nominal system (4.7.29) are established by the following result.

Theorem 4.7.6 *Let nominal impulsive system (4.7.29) be such that for the reduced system (4.7.33) corresponding to it*

(1) *there exist symmetric positive definite matrices P_1 and P_2, satisfying equations (4.7.35);*

(2) *inequalities $\Gamma_1 < 0$ and $\Gamma_2 < 0$ are satisfied.*

Then the solution $x = 0$ of nominal system (4.7.29) is asymptotically stable.

This theorem is proved by application of Theorem 4.7.5 to comparison system (4.7.36) and comparison principle (see Theorem 1.11.3 in Lakshmikantham, Leela and Martynyuk [1]).

Then we use the designations

$$k_1 = \frac{\theta_2 \lambda_M(G_1)}{\lambda_m(P_1)}, \quad k_2 = -\frac{\theta_1 \lambda_m(Q_2)}{\lambda_M(P_2)},$$

$$a_1 = \frac{2\|P_1\| + 2\|P_2\|}{\lambda_m(P_1)}, \quad a_2 = \frac{2\|P_1\| + 2\|P_2\|}{\lambda_m(P_2)},$$

$$b_1 = \frac{3\|\overline{B}_{11}^{\mathrm{T}} P_1\| + \|P_1 \overline{B}_{12}\| + \|P_2 \overline{B}_{21}\| + 3\|\overline{B}_{22}^{\mathrm{T}} P_2\|}{\lambda_m(P_1)},$$

$$b_2 = \frac{\|P_2 \overline{B}_{22}\| + 3\|P_2 \overline{B}_{21}\| + 3\|P_1 \overline{B}_{12}\| + \|P_1 \overline{B}_{11}\|}{\lambda_m(P_2)}.$$

Similar to Theorem 4.7.6 we proved the following result.

Theorem 4.7.7 *Let impulsive system (4.7.28) be such that*

(1) *there exist symmetric positive definite matrices P_1 and P_2 satisfying equations (4.7.35);*

(2) $\Gamma_1 < 0$, $\Gamma_2 < 0$;
(3) $0 \leq \mu < \|B_0^{-1}\|^{-1}$;
(4) $0 \leq \mu < \dfrac{\mu_0}{\|W\|\|W^{-1}\|}$,

where

$$\mu_0 = \min(\mu_1, \mu_2),$$

$$\mu_1 = \frac{-b_1 + \sqrt{b_1^2 - 4a_1(\gamma_1 - e^{-k_1})}}{2a_1}, \quad \mu_2 = \frac{-b_2 + \sqrt{b_2^2 - 4a_2(\gamma_2 - e^{-k_2})}}{2a_2}.$$

Then, under the condition $\|\Delta B\| < \mu$ the solution $x = 0$ of system (4.7.28) is asymptotically stable.

Example 4.7.1 Consider the fourth order impulsive system

(4.7.37)
$$\begin{aligned}
\dot{x}_1 &= 0.5x_1 - 0.1x_2, \\
\dot{x}_2 &= -0.2x_1 + 0.5x_2, \\
\dot{x}_3 &= -2x_3 + x_4, \\
\dot{x}_4 &= x_3 - 2x_4, \quad \text{for} \quad t \neq t_i, \\
x_1(t^+) &= 0.1x_1 + 0.05x_2 + 0.5x_3 + 0.5x_4, \\
x_2(t^+) &= 0.05x_1 + 0.05x_2 + 0.5x_3 + 0.2x_4, \\
x_3(t^+) &= 0.05x_1 + 0.1x_2 + 0.5x_3 + 0.1x_4, \\
x_4(t^+) &= 0.01x_1 + 0.1x_2 + 0.5x_3 + 0.2x_4, \quad \text{for} \quad t \neq t_i,
\end{aligned}$$

where $(x_1, x_2, x_3, x_4) \in R^4$, $t_1 < t_2 < \cdots < t_n < \ldots$, $0.25 \leq t_{i+1} - t_i \leq 0.3$.
Applying the vector Liapunov function

$$V(x) = (v_1(x_1, x_2), v_2(x_3, x_4))$$

with the components $v_1(x_1, x_2) = x_1^2 + x_2^2$ and $v_2(x_3, x_4) = x_3^2 + x_4^2$ and Theorem 4.7.6, we find that

(4.7.38) $\qquad \Gamma_1 = -3.1699 < 0, \quad \Gamma_2 = -0.30308 < 0.$

Together with the other conditions of this Theorem, inequalities (4.7.38) allow us to conclude on asymptotic stability of the state $x = 0$ of system (4.7.37).

Theorem 4.7.7 provides an estimation of the robustness boundary of system (4.7.37). Namely, for any matrices $\|\Delta B\|$, satisfying the condition

$\|\Delta B\| < 0.029869$, asymptotic stability of the state $x = 0$ of system (4.7.37) is preserved.

Remark 4.7.2 It is easy to verify that sufficient asymptotic stability conditions cited in Theorems 15.1–15.4 by Samoilenko and Perestyuk [1] are not applicable to system (4.7.37).

The proposed motion stability criteria for impulse systems with uncertain parameter values are a development of the known sufficient stability conditions for nominal impulse systems.

4.8 Problems for Investigation

4.8.1 System (4.7.1) includes two classes of uncertain impulsive systems being of interest for applications.

Class 1 of uncertain impulsive systems (UIS-1) comprises all systems of the type

(4.8.1)
$$\frac{dx}{dt} = f(t, x), \quad t \neq t_k,$$
$$\Delta x(t_k) = I_k(x(t_k), \alpha),$$

where $x \in R^n$, $f \in C(R_+ \times R^n, R^n)$, $I_k \colon R^n \times R^d \to R^n$, $d \geq 1$, $k = 1, 2, \ldots$. It is clear that system (4.8.1) describes an impulsive system evolution for which the state vector values at impulse effect instants are known imprecisely.

Class 2 of uncertain impulse systems (UIS-2) includes all systems of the type

(4.8.2)
$$\frac{dx}{dt} = f(t, x, \alpha), \quad t \neq t_k,$$
$$\Delta x(t_k) = I_k(x(t_k)),$$

where $x \in R^n$, $f \in C(R_+ \times R^n \times R^d, R^n)$ and $I_k \colon R^n \to R^n$, $k = 1, 2, \ldots$. System (4.8.2) describes the situation when the discrete component of the system does not depend on "uncertainties" whereas on the continuity intervals uncertainty of the system state vector is taken into account.

The problem of analysis of dynamical properties of systems (4.8.1) and (4.8.2) is an open area of investigations in the theory of impulsive dynamical systems.

To establish conditions under which:

(a) solutions of systems (4.8.1) and (4.8.2) are stable with respect to the moving surface

$$(4.8.3) \qquad \Delta(r) = \{x \in R^n : \|x\| = r(\alpha)\},$$

where $r(\alpha) > 0$, $r(\alpha)$ is a function characteristic of uncertainties in the impulsive system;

(b) solutions of systems (4.8.1) and (4.8.2) exponentially converge to moving surface (4.8.3);

(c) solutions of systems (4.8.1) and (4.8.2) are oscillating with respect to moving surface (4.8.3);

(d) set (4.8.3) is a minimal or minimal set of almost periodic motions of systems (4.8.1) or (4.8.2).

Note According to Nemytskii and Stepanov [1] for dynamical system definite in space R, the set Ω is called minimal if it is nonempty, closed and invariant and does not have a proper subset possessing these three properties.

4.9 Notes and References

Section 4.1 There are a variety of good and, by now, classic textbooks on the use of the accepted model of impulsive perturbation. Chief among these is the text by Samoilenko and Perestyuk [1]. Other exellent books are by Lakshmikantham, Bainov and Simeonov [1] and Halanay and Weksler [1]. Pandit and Deo [1] book is short but to the point.

Unfortunately there are no set rules, and no understood "right" way to model systems under impulsive perturbations. For a detailed discussion of this point see Barbashin [1], Blaquiere [1], Bromberg [1], Mil'man and Myshkis [1], Samoilenko and Perestyuk [1], etc. The impulsive processes that consists of between two impulses non-autonomous, non-liner continuous processes subjected to time-varing impulsive constraints see Shu [1]. The greatest current interest in such systems arises in the area of impact mechanics (see, e.g. Brogliato [1], and Brach [1]). The book by Fowler [1] discusses the methodological problems of modelling of real world phenomena in general.

Sections 4.2–4.4 In these sections we present a short survey of some results for impulsive systems. For the details see Mil'man and Myshkis [1],

Samoilenko and Perestyuk [1], Lakshmikantham, Bainov and Simeonov [1], etc.

Section 4.5 Theorems 4.5.1–4.5.7 are new. The direct Liapunov method via scalar auxiliary functions is developed in many papers (see Amatov [1], Kulev [1], Leela [1], Simeonov and Bainov [1], Hui Ye, Michel, and Ling Hou [1], Martynyuk and Slyn'ko [1], etc.)

Section 4.6 This section is based on the results by Martynyuk and Miladzhanov [1] and Martynyuk and Begmuratov [1]. For the application of vector Liapunov function to stability investigation of large scale impulsive system see Liu [1], Liu and Willms [1], etc.

Section 4.7 This section encorporates the results by Martynyuk [2], Martynyuk-Chernienko [1], and Dvirnyj [1]. Impulsive systems with structural perturbations were studied by Martynyuk and Miladzhanov [2], Martynyuk and Stavroulakis [1], *et al.* On the control of uncertain impulsive systems see Gao Yan, *et al.* [1], Guang Zhi Hong, *et al.* [1], Yang Tao [1].

Note that if in conditions of Theorem 4.7.4 Assumptions 4.7.1 and 4.7.2 are omitted, and an ordinary vector Liapunov function is considered, then Theorem 4.7.4 becomes the known Theorem 3.4.2 from Lakshmikantham, Matrosov, and Sivasundaram [1].

An extension of the method of vector Liapunov function for impulsive fuzzy dynamic systems see Devi and Vatsala [1]. For the impulsive differential inclusions, in general, see Watson [1].

References

Amatov, M.A.
[1] On stability of motion of impulsive systems. *Diff. Uravn.* **21**(9) (1977) 21–28. [Russian]

Bainov, D.D. and Simeonov, P.S.
[1] *Systems with Impulse Effect Stability: Theory and Applications.* New York: Halsted Press, 1989.

Barbashin, Ye.A.
[1] On stability with respect to impulsive perturbations. *Diff. Uravn.* **2** (1966) 863–871.

Blaquiere, A.
[1] Differential games with piecewise continuous trajectories. In: *Lecture Notes in Control and Information Sciences.* New York: Springer-Verlag, 1977, P.34–69.

Brach, R.M.
[1] *Mechanical Impact Dynamics.* New York: John Wiley, 1991.

Brogliato, B.
[1] *Nonsmooth Impact Mechanics.* Berlin: Springer-Verlag, 1996.

Bromberg, P.V.
[1] *Matrix Methods in the Theory of Impulsive Control.* Nauka, Moscow, 1967. [Russian]

Chow, S.-N. and Yorke, J.A.
[1] Lyapunov theory and perturbation of stable and asymptotically stable systems. *J. Diff. Eqns* **15** (1974) 308–321.

Devi, J.Vasundhara and Vatsala, A.S.
[1] Method of vector Lyapunov functions for impulsive fuzzy systems. *Dynamic Systems and Appl.* **13** (2004) 531–531.

Dvirnyj, A.I.
[1] On estimation of boundary of the robustness of linear impulsive system. *Reports of the National Academy of Sciences of Ukraine.* (9) (2003) 34–39. [Russian]

Fowler, A.C.
[1] *Mathematical Models in the Applied Science.* Cambridge: Cambridge University Press, 1997.

Furumochi, T.
[1] Uniform asymptotic stability in perturbed differential equations. *J. Math. Anal. Appl.* **137** (1989) 207–220.

Gao Yan, Lygeros, J., Quincampoix, M. and Seube, N.
[1] On the control of uncertain impulsive systems: Approximate stabilization and controlled invariance. *Int. J. Control* **77**(16) (2004) 1393–1407.

Guan Zhi Hong, Liao Jun Feng and Liao Rui Quan
[1] Robust H_∞ control of uncertain impulsive systems. *Control Theory Appl.* **19**(4) (2002) 623–626. [Chinese]

Halanay, A. and Wexler, D.
[1] *Qualitative Theory of Impulsive Systems.* Moscow: Mir, 1971. [Russian]

Hui, Ye., Michel, A.N. and Hou, Ling
[1] Stability analysis of systems with impulse effects. *IEEE Trans. on Automatic Control* **43**(12) (1998) 1719–1723.

Kulev, G.K.
[1] Uniform asymptotic stability in impulsive perturbed systems of differential equations. *J. Comput. and Appl. Mathematics* **41** (1992) 49–55.

Lakshmikantham, V. and Devi, J.Vasundhara
[1] Strict stability criteria for impulsive differential systems, *Nonlin. Anal.* **21**(10) (1993) 785–794.

Lakshmikantham, V., Bainov, D.D. and Simeonov, P.S.
[1] *Theory of Impulsive Differential Equations.* Singapore: World Scientific, 1989.

Lakshmikantham, V., Leela, S. and Martynyuk, A.A.
[1] *Stability Analysis of Nonlinear Systems.* New York: Marcel Dekker, 1988.

Lakshmikantham, V., Matrosov, V.M. and Sivasundaram, S.
[1] *Vector Liapunov Functions and Stability Analysis of Nonlinear Systems.* Amsterdam: Kluwer Academic Publishers, 1991.

Leela, S.
[1] Stability of differential systems with impulsive perturbations in terms of two measures. *Nonlin. Anal.* **1** (1977) 667–677.

Liapunov, A.M.
[1] *General Problem of Stability of Motion.* Kharkov: Kharkov Math. Soc., 1892. (Published in: *Collected Papers.* Ac. Sci. USSR, Moscow-Leningrad, **2**, 1956 5–263). [Russian]

Liu, Xinzhi
[1] Stability of large scale impulsive systems. In: *Proceedings of the Fifth International Colloquium on Differential Equations*, VSP Utrecht, 1995, P.193–203.

Liu, Xinzhi and Willms, A.
[1] Stability analysis and applications to large scale impulsive systems: A new approach. *Canad. Appl. Math. Quart.* **3** (1995) 419–444.

Martynyuk, A.A.
[1] *Qualitative Methods in Nonlinear Dynamics: Novel Approaches to Liapunov's Matrix Functions.* New York: Marcel Dekker, 2002.
[2] Block-diagonal matrix-valued Lyapunov function and stability of an uncertain impulsive system. *Intern. Appl. Mech.* **40**(3) (2004) 322–327.
[3] Matrix Liapunov function and stability with respect to two measures of impulsive systems *Dokl. Acad. Nauk* **338**(6) (1994) 728–730.

Martynyuk-Chernienko, Yu.A.
[1] Stability theory of solutions to an impulsive system with uncertain parameters. *Dokl. Acad. Nauk* **395**(2) (2004) 1–4. [Russian]

Martynyuk, A.A. and Begmuratov, K.
[1] Analytical construction of the hierarchical matrix Lyapunov function for impulsive systems. *Ukr. Mat. Zh.* **49** (1997) 548–557. [Russian]

Martynyuk, A.A. and Miladzhanov, V.G.
[1] Stability analysis of solutions of large-scale impulsive systems, (to appear).
[2] Stability Theory of Large Scale Dynamical Systems under Nonclassical Structural Perturbations, (to appear).

Martynyuk, A.A. and Slyn'ko, V.I.
[1] Stability of a nonlinear impulsive system. *Int. Appl. Mech.* **40**(2) (2004) 231–239.

Martynyuk, A.A. and Stavroulakis, I.P.
[1] Direct Liapunov's matrix function method for impulsive systems under structural perturbations. *Canad. Appl. Math. Quart.* **7**(2) (1999) 159–184.

Mil'man, V.D. and Myshkis, A.D.
[1] On the stability of motion presence of impulses. *Sib. Math. J.* **1**(20) (1960) 233–237.

Nemytskii, V.V. and Stepanov, V.V.
[1] *Qualitative Theory of Differential Equations.* Moscow-Leningrad: GITTL, 1949. [Russian]

Pandit, S.G. and Deo, S.G.
[1] *Differential Systems Involving Impulses.* Berlin: Springer-Verlag, 1980.

Pavlidis, T.
[1] Stability of systems described by differential equations containing impulses. *IEEE Trans. Automatic Control* **AC-12** (1967) 43–45.

Samoilenko, A.M. and Perestyuk, N.A.
[1] *Impulsive Differential Equations.* Singapore: World Scientific, 1995.

Simeonov, P.S. and Bainov, D.D.
[1] The second method of Liapunov for systems with an impulse effect. *Tamkang J. of Math.* **16**(4) (1985) 19–40.

Watson, P.J.
[1] Impulsive differential inclusions. *Nonlinear World* **4**(4) (1997) 395–402.

Yang Tao
[1] *Impulsive Control Theory.* Berlin: Springer-Verlag, 2001.

Zhongzhou Shu
[1] On impulsive processes. *Int. J. of Non-Linear Mechanics* **37** (2002) 213–224.

5

APPLICATIONS

5.1 Introduction

The strict stability theory developed by Liapunov [1] has numerous applications. The possibilities of the direct method (method of Liapunov functions) depend on a properly constructed auxiliary function which is a key point of the process of qualitative analysis of system motion. Knowledge of the Liapunov function allows us to solve many problems of importance for applications both in engineering and natural sciences. We note only some of them: estimation of the domain of system parameter variation ensuring prescribed motion; estimation of transient process time; estimation of the domain of initial motion perturbation; establishment of existence conditions for periodic motion; estimation of the chaotic motion existence domain.

In this chapter we present new applications of the method of multicomponent Liapunov function in the investigation of some urgent problems of nonlinear dynamics.

In Section 5.2 stabilization conditions are established for the motion of solid bearing moving material points in a prescribed direction.

Section 5.3 deals with stability analysis of solutions to discrete-time neural system with unprecise parameter values.

In Section 5.4 a method of stability analysis is proposed for quasilinear systems in the case when the application of vector Liapunov function is awkward.

Section 5.5 dwells on the solution of the problem of estimation of robust stability boundary of linear mechanical system.

In the final Section 5.6 we analyse boundedness and stability of growth of populations whose behavior is described by the Kolmogorov model.

In all the problems mentioned above, new approaches are proposed to qualitative analysis of solutions for systems of the type. Their efficiency is demonstrated by some numerical examples.

5.2 Stabilization of Motion of a Rigid Body

5.2.1 Statement of the problem Consider a mechanical system S, consisting of an absolutely rigid body T_0, with a fixed point O in the center of inertia of the connected with the body material points M_j, $j = 1, 2, \ldots, k$, with masses m_j. It is assumed that the points perform a prescribed motion with respect to the system of coordinates $Oxyz$, rigidly connected with the rigid body T_0.

We designate by $O\xi\eta\zeta$ an absolute system of coordinates which is assumed to be right Cartezian, similarly to system $Oxyz$. Let two unit vectors s and r be given. The unit vector s takes a constant place in the system $O\xi\eta\zeta$, while the unit vector r takes a constant place in the system $Oxyz$.

The problem of motion orientation of the mechanical system S in a given direction s consists in constructing a control moment M which, acting on system S, stabilizes the unit vector r in the direction s.

Note that in the case when the material points M_j are absent and system S consists only of a rigid body rotating around a fixed point, the problem is solved by Zubov [1]. In particular, Zubov showed that the control moment M_0, solving the problem on stabilization is determined by the formula

$$(5.2.1) \qquad M_0 = -\omega + [r, s],$$

where ω is a vector of angular velocity.

Further we shall investigate the action of control moment M_0 in more general form as compared with (5.2.1), namely

$$(5.2.2) \qquad M_0 = -\omega + K(s - r),$$

where K is a tensor whose components are constant with respect to a moving system of coordinates.

It is assumed that the rank of the matrix, whose elements are components of the tensor K with respect to the system of coordinates $Oxyz$, is not lower than 2. In the special case when $K = R$, where

$$R = \begin{pmatrix} 0 & r_1 & -r_3 \\ -r_1 & 0 & r_2 \\ r_3 & -r_2 & 0 \end{pmatrix},$$

r_1, r_2, r_3 are components of vector r with respect to the moving system of coordinates $Oxyz$, the moment specified by formula (5.2.2) coincides with moment (5.2.1). This follows from the fact that in three-dimensional space the vector multiplication operation is $[r, x] = Rx$. We can also make sure that the rank $R = 2$, i.e. control (5.2.2) is a generalization of control (5.2.1). Since the control moment of (5.2.2) type has a quite simple structure, further we shall assume that the control moment of (5.2.2) type acts on the system S under consideration and establish conditions under which it stabilizes the system motion in the prescribed direction. Note that the considered problem of stabilization of system S in the prescribed direction in the context of Liapunov stability notion makes sense only when the motion of the material points is bounded. In the cases when these motions are unbounded, it is reasonable to consider the problem on "technical stabilization", i.e. motion stabilization on a finite time interval.

5.2.2 Motion equations of mechanical system S We designate by r a radius-vector of the point which belongs to the rigid body T_0, and r_j is a radius-vector of point M_j. Applying the theorem on variation of moment of momentum of a mechanical system we compile the equation

$$\frac{dK_0}{dt} = M_0. \tag{5.2.3}$$

Here K_0 is the moment of momentum of system S with respect to the point O, M_0 is the moment of external forces applied to the system S, d/dt is the derivative in the fixed system of coordinates $O\xi\eta\zeta$.

It is known that

$$K_0 = \int_{(T_0)} [r, v]\, dm + \sum_{j=1}^{k} m_j [r_j, v_j],$$

where v is a motion velocity of a point of rigid body T_0 with a radius-vector r, and v_j is a motion velocity of point M_j with respect to the system of coordinates $O\xi\eta\zeta$.

In view of the formula of velocity distribution in a rigid body the expression for K_0 is reduced as follows

$$K_0 = \int_{(T_0)} [r, v]\, dm + \sum_{j=1}^{k} m_j \left[r_j, \frac{\widetilde{d r_j}}{dt} \right] + \sum_{j=1}^{k} m_j [r_j, [\omega, r_j]], \tag{5.2.4}$$

where $\widetilde{d}/dt$ is a local derivative in the system of coordinates $Oxyz$.

The first term in formula (5.2.4) can be represented as

$$\int_{(T_0)} [r, v]\, dm = \Theta_0 \omega,$$

where Θ_0 is an inertia tensor of the rigid body T_0 with respect to the fixed point O. We transform the third term in (5.2.4) using the known formula for the double vector product $[a, [b, c]] = b(ac) - c(ab)$. This term becomes

$$\sum_{j=1}^{k} m_j [r_j, [\omega, r_j]] = \sum_{j=1}^{k} m_j r_j^2 \omega - \sum_{j=1}^{k} m_j (r_j \omega) r_j = \sum_{j=1}^{k} (m_j r_j^2 I - m_j r_j r_j) \omega.$$

From here on we designate by (a, b) a scalar product of vectors a and b, the notation ab means a dyad product of vectors a and b, and I is a unit tensor.

Thus, for the expression of K_0 we get

$$K_0 = \left(\Theta_0 + \sum_{j=1}^{k} (m_j r_j^2 I - m_j r_j r_j) \right) \omega + \sum_{j=1}^{k} m_j \left[r_j, \frac{\widetilde{dr_j}}{dt} \right].$$

Taking into account the relationship between the vector derivatives in the moving and fixed systems of coordinates

$$\frac{da}{dt} = \frac{\widetilde{da}}{dt} + [\omega, a],$$

we get

(5.2.5) $$\frac{\widetilde{dK_0}}{dt} + [\omega, K_0] = M_0.$$

The first term in the left-hand part of (5.2.5) is

$$\frac{\widetilde{dK_0}}{dt} = 2 \sum_{j=1}^{k} \left(m_j \left(r_j \frac{\widetilde{dr_j}}{dt} \right) I - m_j r_j \frac{\widetilde{dr_j}}{dt} \right) \omega$$

$$+ \left\{ \Theta_0 + \sum_{j=1}^{k} (m_j r_j^2 I - m_j r_j r_j) \right\} \frac{\widetilde{d\omega}}{dt} + \sum_{j=1}^{k} m_j \left[r_j, \frac{\widetilde{d^2 r_j}}{dt^2} \right].$$

Transformation of the second term in the left-hand part of (5.2.5) yields

$$[\omega, K_0] = \left[\omega, \left(\Theta_0 - \sum_{j=1}^{k} m_j r_j r_j\right)\omega\right] + \sum_{j=1}^{k} m_j \left[\omega, \left[r_j, \frac{\widetilde{dr_j}}{dt}\right]\right]$$

$$= \left[\omega, \left(\Theta_0 - \sum_{j=1}^{k} m_j r_j r_j\right)\omega\right] + \sum_{j=1}^{k} m_j \left(\omega \frac{\widetilde{dr_j}}{dt}\right) r_j - \sum_{j=1}^{k} m_j (\omega r_j) \frac{\widetilde{dr_j}}{dt}$$

$$= \left[\omega, \left(\Theta_0 - \sum_{j=1}^{k} m_j r_j r_j\right)\omega\right] + \sum_{j=1}^{k} m_j r_j \frac{\widetilde{dr_j}}{dt} \omega - \sum_{j=1}^{k} m_j \frac{\widetilde{dr_j}}{dt} r_j \omega.$$

Thus, the motion equations of the mechanical system S are

$$M_0 = \left(\Theta_0 + \sum_{j=1}^{k} m_j (r_j^2 I - r_j r_j)\right) \frac{\widetilde{d\omega}}{dt} + \left[\omega, \left(\Theta_0 - \sum_{j=1}^{k} m_j r_j r_j\right)\omega\right]$$

$$+ \sum_{j=1}^{k} \left[2m_j \left(r_j \frac{\widetilde{dr_j}}{dt}\right) I - m_j r_j \frac{\widetilde{dr_j}}{dt} - m_j \frac{\widetilde{dr_j}}{dt} r_j\right] \omega + \sum_{j=1}^{k} m_j \left[r_j, \frac{\widetilde{d^2 r_j}}{dt^2}\right].$$

5.2.3 Orientation stabilization of mechanical system S in a prescribed direction

Assume that the motion of the material points M_j with respect to the moving system of coordinates $Oxyz$ is linear, i.e. for the radius-vector of point M_j the representation

(5.2.6) $$r_j(t) = a_j(t) e_j,$$

is admissible, where e_j is a unit vector which is constant in the system $Oxyz$ and $a_j \in C^2(J, R)$, $J \subseteq J_t^+ = \{t: t \geq 0\}$.

The problem of orientation stabilization of the mechanical system S is studied in the context of the problem of Liapunov stability under some restrictions on motion of the material points M_j. Namely, we formulate the following assumption.

Assumption 5.2.2 Functions $a_j(t)$ in expression (5.2.6) satisfy the conditions

(1) $a_j \in C^2(J, R)$, $J = J_t^+ = \{t : t \geq 0\}$, $j = 1, 2, \ldots, k$;

(2) $|a_j(t)| \leq c_1$, $|da_j(t)/dt| \leq c_2$, where c_1 and c_2 are some constants for all $t \in J_t^+$, $j = 1, 2, \ldots, k$.

Under conditions of Assumption 5.2.2, which specify the motion of points M_j, the following assertion is true.

Proposition 5.2.1 Assume that

$$\Theta(t) = \Theta_0 + \sum_{j=1}^{k} m_j a_j^2(t)(I - e_j e_j^{\mathrm{T}})$$

is the tensor matrix $\Theta(t) = \Theta_0 + \sum_{j=1}^{k} m_j a_j^2(t)(I - e_j e_j^{\mathrm{T}})$ in the system of coordinates $Oxyz$. Then, there exist constants $\gamma_1, \gamma_2 > 0$ such that for all $t \in J_t^+$

(1) $\|\Theta(t)\| \leq \gamma_1$;
(2) $\|\Theta^{-1}(t)\| \leq \gamma_2$.

Proof Note that the matrix $I - e_j e_j^{\mathrm{T}}$ is positive semidefinite. In fact, in view of this matrix symmetry we have $(I - e_j e_j^{\mathrm{T}})(I - e_j e_j^{\mathrm{T}}) = I - 2e_j e_j^{\mathrm{T}} + e_j e_j^{\mathrm{T}} e_j e_j^{\mathrm{T}} = I - e_j e_j^{\mathrm{T}}$. Hence it follows that the eigenvalues of this matrix are 0 or 1, and there must be 1 among them. Therefore $I - e_j e_j^{\mathrm{T}}$ is positive semidefinite and besides, $\|I - e_j e_j^{\mathrm{T}}\| = 1$. Hence the correctness of the estimate

$$\|\Theta(t)\| \leq \|\Theta_0\| + \sum_{j=1}^{k} m_j c_1^2 = \gamma_1.$$

The correlation

$$\|\Theta^{-1}(t)\| = \sup_{\|x\| \neq 0} \frac{x^{\mathrm{T}} \Theta^{-1}(t) x}{\|x\|^2}$$

takes place by definition.

As matrix $\Theta^{-1}(t)$ is symmetric and positive definite, we shall introduce variable $y = \Theta^{-1/2} x$. As a result of simple transformations we have

$$\|\Theta^{-1}(t)\|^2 = \sup_{\|x\| \neq 0} \frac{x^{\mathrm{T}} \Theta^{-1}(t) x}{\|x\|^2} = \sup_{\|y\| \neq 0} \frac{y^{\mathrm{T}} y}{y^{\mathrm{T}} \Theta(t) y}$$

$$\leq \sup_{\|y\| \neq 0} \frac{y^{\mathrm{T}} y}{y^{\mathrm{T}} \Theta_0 y} \leq \frac{1}{\lambda_m(\Theta_0)} = \gamma_2^2.$$

This completes the proof of Proposition 5.2.1.

Taking into account (5.2.6) we reduce motion equations of the mechanical system S to the form

(5.2.7)
$$\left[\Theta_0 + \sum_{j=1}^{k} m_j a_j^2(t)(I - e_j e_j)\right] \frac{d\widetilde{\omega}}{dt} + \left[\omega, \left(\Theta_0 - \sum_{j=1}^{k} m_j a_j^2(t) e_j e_j\right) \omega\right]$$

$$+ 2 \sum_{j=1}^{k} m_j a_j(t) \frac{a_j(t)}{dt}(I - e_j e_j)\omega = -\omega + K(s - r).$$

5. APPLICATIONS

We add to equations (5.2.7) a kinematic group of motion equations taking account of the fact that
$$\frac{ds}{dt} = 0.$$
Switching to the system of coordinates $Oxyz$, we get
$$\frac{\widetilde{ds}}{dt} + [w, s] = 0.$$

Further we designate
$$L(t) = \Theta_0 - \sum_{j=1}^{k} m_j a_j^2(t) e_j e_j,$$

$$N(t) = I + 2 \sum_{j=1}^{k} m_j a_j(t) \frac{da_j(t)}{dt}(I - e_j e_j).$$

As a result of these transformations a complete system of motion equations of the mechanical system S becomes

(5.2.8)
$$\Theta(t) \frac{\widetilde{dw}}{dt} + [w, L(t)w] + N(t)w + K(r - s) = 0,$$
$$\frac{\widetilde{ds}}{dt} + [w, s] = 0.$$

Let p, q and r be projections of the vector w on the coordinate axes of system $Oxyz$, s_1, s_2, s_3 be those of the vector s and r_1, r_2, r_3 be those of the vector r respectively.

Designate $w = (p, q, r)^T$, $s = (s_1, s_2, s_3)^T$, $r = (r_1, r_2, r_3)^T$, and, if $a = (a_1, a_2, a_3)^T$, $b = (b_1, b_2 b_3)^T$, then $[a, b] = (a_2 b_3 - b_2 a_3, a_1 b_3 - b_1 a_3, a_1 b_2 - a_3 b_1)^T$.

The following result is valid.

Proposition 5.2.2 Let $\Theta(t)$ be a matrix of tensor Θ in the moving system of coordinates $Oxyz$, and K be a matrix of tensor K in the system $Oxyz$. Assume that there exist values $T \in [t_0, \infty)$ such that the matrix $\left.\frac{d\Theta}{dt}\right|_{t=T}$ is of definite sign.

Then, if the matrix K is nondegenerate, system (5.2.8) possesses a unique equilibrium state $w = 0$, $s = r$. If the matrix K is degenerate, system (5.2.8) possesses two equilibrium states $w = 0$, $s = r$, and $w = 0$, $s = r - 2(kr)k$, where $k \neq 0$ and $k \in \text{Ker } K$.

Proof Let $\omega = \omega_0$, $s = s_0$ be the equilibrium state of system (5.2.8). Projecting the motion equations of system (5.2.8) on the moving axes $Oxyz$ one gets

$$[\omega_0, L(t)\omega_0] + \frac{d\Theta}{dt}\omega_0 = -\omega_0 + K(s_0 - r),$$
(5.2.9)
$$[\omega_0, s_0] = 0.$$

Assume that $\omega \neq 0$, then there exists a constant $\lambda \neq 0$ such that $s_0 = \lambda\omega_0$. We multiply equality (5.2.9) on the left by ω_0^T

(5.2.10)
$$\omega_0^T \frac{d\Theta}{dt}\omega_0 = -\omega_0^T\omega_0 + \omega_0^T K(\lambda\omega_0 - r).$$

Equality (5.2.10) is integrated from 0 to T, $T > 0$, and divided by T. We get

$$\omega_0^T \frac{\Theta(T) - \Theta(0)}{T}\omega_0 = -\omega_0^T\omega_0 + \omega_0^T K(\lambda\omega_0 - r).$$

For $T \to \infty$ one obtains the equation

$$-\omega_0^T\omega_0 + \omega_0^T K(\lambda\omega_0 - r) = 0.$$

Thus $\omega_0^T \dfrac{d\Theta}{dt}\omega_0 = 0$, and taking into account the conditions of Proposition 5.2.2 we conclude that $\omega_0 = 0$. Equality (5.2.9) yields

$$2K(s_0 - r) = 0.$$

If K is a nondegenerate matrix, then $s_0 = r$ is a unique equilibrium state. In the case when the matrix K is degenerate there exists one more equilibrium state $s_0 - r \in \mathrm{Ker}\, K$ in addition to the equilibrium state $s_0 = r$. Since Ker has the dimension one, we get the presentation $s_0 - r = \lambda k$, $k \in \mathrm{Ker}\, K$. As $\|s_0\| = \|r\| = 1$ we obtain $s_0 = r - 2(k^T r)k$. This completes the proof of Proposition 5.2.2.

Then we study stability of the state $\omega = 0$, $s = r$. To this end we introduce new variables $x = \omega$, $y = s - r$, and, by projecting the vectors ω, s, r on the axes of coordinates Ox, Oy, Oz, we present the system of perturbed motion equations corresponding to (5.2.8) as

$$\Theta(t)\frac{dx}{dt} + [x, L(t)x] + N(t)x - Ky = 0,$$
(5.2.11)
$$\frac{dy}{dt} + [x, r] + [x, y] = 0.$$

5. APPLICATIONS

In subsequent presentation we shall need the following designations

$$A(t) = -\Theta^{-1}(t)N(t), \quad B(t) = \Theta^{-1}(t)K,$$
$$A_{21} = R, \quad A_{22} = 0.$$

Now system (5.2.11) is rewritten as

(5.2.12)
$$\frac{dx}{dt} = A(t)x + B(t)y - \Theta^{-1}(t)[x, L(t)x],$$
$$\frac{dy}{dt} = Rx - [x, y].$$

For system (5.2.12) a Cauchy problem is considered with the initial conditions $x(t_0) = x_0$, $y(t_0) = y_0$, where $(x_0, y_0) \in R^3 \times R^3$. Together with system (5.2.12) we consider its linear approximation

(5.2.13)
$$\frac{dx}{dt} = A(t)x + B(t),$$
$$\frac{dy}{dt} = Rx.$$

According to the results of Section 1.3 we construct for system (5.2.13) the matrix-valued function $U(t, x, y)$. The diagonal elements of the function $U(t, x, y)$ are taken as the quadratic forms

(5.2.14) $$v_{11}(x) = x^{\mathrm{T}} P(t) x, \quad v_{22}(y) = y^{\mathrm{T}} y,$$

where $x \in R^3$, $y \in R^3$, $P \in C^1(J, R^{3 \times 3})$ is a symmetric positive definite matrix. The non-diagonal element $v_{12}(t, x, y)$ of the matrix-valued function $U(t, x, y)$ is constructed as a bilinear form

(5.2.15) $$v_{12}(t, x, y) = v_{21}(t, x, y) = x^{\mathrm{T}} P_{12}(t) y,$$

where the matrix $P_{12} \in C^1(J, R^{3 \times 3})$ is determined in terms of the matrix differential equation

(5.2.16) $$\frac{dP_{12}}{dt} + A^{\mathrm{T}}(t) P_{12} = -P(t)\Theta^{-1}(t)K + R.$$

Under the condition that the fundamental matrix of the homogeneous equation corresponding to (5.2.16) has the estimate

$$\|\Phi(t, t_0)\| \leq K e^{\gamma(t - t_0)},$$

where $K > 0$, $\gamma > 0$, the solution of equation (5.2.16) can be presented in the form

$$(5.2.17) \qquad P_{12}(t) = -\int_t^\infty \Phi(t,\tau)[R - P(\tau)\Theta^{-1}(\tau)K]\,d\tau.$$

It can be easily verified that under conditions of Assumption 5.2.2 matrix (5.2.17) is bounded for all $t \in [0, \infty)$.

Applying estimate (1.3.7) for the function $v(t, x, y, \eta) = \eta^T U(t, x, y)\eta$, where for the sake of simplicity we set $\eta = (1,1)^T$, we obtain the bilateral estimate

$$u^T \underline{C} u \leq v(t,x,y,\eta) \leq u^T \overline{C} u \quad \text{for all} \quad (t,x,y) \in J_t^+ \times R^3 \times R^3.$$

Here

$$\overline{C} = \begin{pmatrix} \overline{c}_{11} & \mu \\ \mu & 1 \end{pmatrix}, \quad \underline{C} = \begin{pmatrix} \underline{c}_{11} & -\mu \\ -\mu & 1 \end{pmatrix},$$

$$u = (\|x\|, \|y\|)^T, \quad \mu = \sup_{t \geq 0} \|P_{12}(t)\|,$$

$$\overline{c}_{11} = \sup_{t \geq 0} \lambda_M(P(t)), \quad \underline{c}_{11} = \inf_{t \geq 0} \lambda_m(P(t)).$$

The estimate of the total derivative of function $v(t, x, y, \eta)$ along solutions of system (5.2.12) in view of boundedness of the matrices $\Theta^{-1}(t)$, $N(t)$ and $L(t)$ is

$$(5.2.18) \qquad \left.\frac{dv}{dt}\right|_{(5.2.12)} \leq u^T S(t) u + R_1(\|x\|, \|y\|).$$

Here $S(t) = [\sigma_{ij}(t)]_{i,j=1}^2$, $R_1(\cdot)$ is a polynomial of order higher then 2. The elements of the matrix $S(t)$ in estimate (5.2.18) have the following structure

$$\sigma_{11}(t) = -\lambda_m(G(t)) + \lambda_M \left[\int_t^\infty \Phi(t,\tau)(R^2 - P(\tau)\Theta^{-1}(t)KR) \right.$$

$$\left. + (R^2 - RK^T P(\tau)\Theta^{-1}(\tau))\Phi(t,\tau)\,d\tau \right],$$

$$\sigma_{12}(t) = 0,$$

$$\sigma_{22}(t) = \lambda_M \left[\int_t^\infty (K^T \Theta^{-1}(t)\Phi(t,\tau)(P(\tau)\Theta^{-1}(\tau)K - R) \right.$$

$$\left. + (R + K^T P(\tau)\Theta^{-1}(\tau)\Phi^T(t,\tau))\Theta^{-1}(t)K)d\tau \right],$$

where $G(t) = \dfrac{dP}{dt} + A^{\mathrm{T}}(t)P(t) + P(t)A(t)$.

Now we can formulate the following result.

Theorem 5.2.1 *Assume that the mechanical system S is such that*

(1) *all conditions of Assumption 5.2.2 are satisfied;*
(2) *for all $t \in J_t^+$ and some $\delta > 0$*

$$\mu = \sup_{t \geq 0} \|P_{12}(t)\| \leq \sqrt{\lambda_m(P(t)) + \delta};$$

(3) *the matrix $S(t)$ satisfies generalized Silvester conditions.*

Then control (5.2.2) stabilizes the motion of system S in the prescribed direction, and the equilibrium state $\omega = 0$, $s = r$ is uniformly asymptotically stable in the sense of Liapunov.

Proof By condition (2) the function $v(t, x, y) = \eta^{\mathrm{T}} U(t, x, y)\eta$ is positive definite, and its total derivative along solutions of system (5.2.12) is negative definite in the arbitrary small neighborhood of the equilibrium state $x = y = 0$. By Theorem 1.2.4 the equilibrium state $x = y = 0$ of system (5.2.12) is uniformly asymptotically stable.

To go on with the stability investigation of the equilibrium state $x = y = 0$ of system (5.2.12) we need the following auxiliary results.

Proposition 5.2.3 *If the equilibrium state $x = y = 0$ of system (5.2.12) is uniformly x-polystable, then it is asymptotically stable.*

Proof Let us show that the equilibrium state $x = y = 0$ is attractive, i.e. the following conditions are satisfied: there exists a value $\sigma > 0$ such that for arbitrary initial condition $(x_0, y_0) \in R^3 \times R^3$ such that $\|x_0\| + \|y_0\| < \sigma$ and arbitrary $\varepsilon > 0$ there exists a time $T \in [0, \infty)$ such that for solutions of system (5.2.12) the estimate

$$\|x(t, t_0, x_0, y_0)\| + \|y(t, t_0, x_0, y_0)\| < \varepsilon \quad \text{for all} \quad t \geq t_0 + T, \quad t_0 \geq 0.$$

holds true.

Due to stability of the equilibrium state $x = y = 0$ and asymptotic x-stability there exists $\sigma = \sigma(K) > 0$ such that the inequality $\|x_0\| + \|y_0\| < \sigma(K)$ implies $\|x(t)\| + \|y(t)\| < l(K)$ for all $t \geq t_0$, where $l(K) = 1$ in the case when $|K| \neq 0$ or $(r, k) = 0$, and $l(K) = |(r, k)|$ in the other cases, and $\lim\limits_{t \to \infty} x(t) = 0$, where $x(t) = x(t, t_0, x_0, y_0)$, $y(t) = y(t, t_0, x_0, y_0)$.

Assume that $\|x_0\| + \|y_0\| < \sigma(K)$ and $\varepsilon > 0$. We show that in this case there exists $T \in [0, \infty)$ such that $\|x(t)\| + \|y(t)\| < \varepsilon$ for all $t \geq t_0 + T$. We multiply equation (5.2.11) by $(Ky)^{\mathrm{T}}$ and present it in the form

$$(5.2.19) \quad \frac{d}{dt}[y^{\mathrm{T}} K^{\mathrm{T}} \Theta(t) x] = y^{\mathrm{T}} K^{\mathrm{T}} \frac{d\Theta}{dt} x - y^{\mathrm{T}} K^{\mathrm{T}}[x, L(t)x] - y^{\mathrm{T}} K^{\mathrm{T}} N(t) x \\ + [x, r]^{\mathrm{T}} K^{\mathrm{T}} \Theta(t) x + [x, y]^{\mathrm{T}} K^{\mathrm{T}} \Theta(t) x + \|Ky\|^2.$$

Assume that there exists $\alpha > 0$ such that $\|Ky\|^2 > \alpha$ for all $t \geq t_0$. By Assumption 5.2.2 the matrices $\Theta(t)$, $\dfrac{d\Theta}{dt}$, $N(t)$ and $L(t)$ are bounded for all $t \in [t_0, \infty)$. Therefore, in view of the fact that $\|y(t)\| < 1$ and $\lim_{t \to \infty} \|x(t)\| = 0$, inequality (5.2.19) implies that there exists an instant of time $T^* \in [0, \infty)$ such that for $t \geq T^*$

$$(5.2.20) \quad \frac{d}{dt}\left[y^{\mathrm{T}} K^{\mathrm{T}} \Theta(t) x\right] \geq \frac{\alpha}{2}.$$

Integrating (5.2.20) from T^* to t and setting $t \to \infty$ one can easily find out that $y^{\mathrm{T}} K \Theta(t) x \to \infty$ as all $t \to \infty$, but this contradicts boundedness of $\|x\|$, $\|y\|$ and matrix $\Theta(t)$.

Consequently, there exists a sequence $\{t_n\}$, $t_n \to \infty$, such that $Ky(t_n) \to 0$ as $n \to \infty$. In the case when K is a nondegenerate matrix we get $y(t_n) \to 0$ as $n \to \infty$.

Let the matrix K be degenerated, then $y(t_n) = \lambda_n k + \varepsilon_n$, where $\|\varepsilon_n\| \to 0$ as $n \to \infty$, $\|k\| = 1$, $k \in \operatorname{Ker} K$. We shall show that $\lambda_n \to 0$ as $n \to \infty$. Actually, the equality $1 = \|s(t_n)\|^2 = 1 + \lambda_n^2 + \lambda_n k^{\mathrm{T}} \varepsilon_n + 2r^{\mathrm{T}} \varepsilon_n + 2\lambda_n r^{\mathrm{T}} k + \|\varepsilon_n\|^2$ implies that $\lambda_n(\lambda_n + 2r^{\mathrm{T}} k) \to 0$ as $n \to \infty$. If $r^{\mathrm{T}} k = 0$ then $\lambda_n \to 0$ as $n \to \infty$.

Let $r^{\mathrm{T}} k \neq 0$ and n be taken so large that $\|\varepsilon_n\| < \dfrac{|r^{\mathrm{T}} k|}{2}$. Then

$$|\lambda_n| = \|\lambda_n k\| = \|y(t_n) - \varepsilon_n\| \leq \|y(t_n)\| + \|\varepsilon_n\| < \frac{3}{2}|r^{\mathrm{T}} k|,$$
$$|\lambda_n + 2r^{\mathrm{T}} k| \geq 2|r^{\mathrm{T}} k| - |\lambda_n| > 2|r^{\mathrm{T}} k| - \frac{3}{2}|r^{\mathrm{T}} k| = \frac{1}{2}|r^{\mathrm{T}} k|$$

and, consequently, $\lambda_n \to 0$ for $n \to \infty$ and $\|y(t_n)\| \to \infty$ for $n \to \infty$.

Due to uniform stability of the equilibrium state $x = y = 0$ of system (5.2.12) there exists $\delta(\varepsilon) > 0$ such that the inequality $\|x_0\| + \|y_0\| < \delta(\varepsilon)$ implies $\|x(t)\| + \|y(t)\| < \varepsilon$. Therefore, the above consideration and properties of asymptotic x-stability of the equilibrium state $x = y = 0$ imply the existence of an instant of time t_n such that $\|x(t_n)\| + \|y(t_n)\| < \delta(\varepsilon)$ and then for $t \geq t_n$ the estimate $\|x(t) + y(t)\| < \varepsilon$ is valid for all $t \geq t_0 + t_n$.

Now we shall prove the following result.

Theorem 5.2.2 *Assume that the mechanical system S is such that*

(1) *all conditions of Assumption 5.2.2 are satisfied;*
(2) *for all $t \in J_t^+$ and some $\delta > 0$*

$$\mu = \sup_{t \geq 0} \|P_{12}(t)\| \geq \sqrt{\lambda_m(P(t))} + \delta;$$

(3) *there exists a constant $\varkappa > 0$ such that*

$$\sigma_{11}(t) + \varkappa < 0, \quad \sigma_{22}(t) \leq 0.$$

Then control (5.2.2) stabilizes the motion of system S in the prescribed direction and its equilibrium state $\omega = 0$, $s = r$ is asymptotically stable in the sense of Liapunov.

Proof Under condition (2) of the theorem the scalar function $v(t, x, y, \eta)$ is positive definite. For the total derivative of this function along solutions of system (5.2.12) we have estimate

$$\left.\frac{dv}{dt}\right|_{(5.2.12)} \leq \sigma_{11}(t)\|x\|^2 + \sigma_{22}(t)\|y\|^2 + 2x^{\mathrm{T}}P_{11}[x, L(t)]$$
$$- 2y^{\mathrm{T}}[x, y] - 2x^{\mathrm{T}}P_{12}(t)[x, y] + 2y^{\mathrm{T}}P_{12}^{\mathrm{T}}\Theta^{-1}(t)[x, L(t)x].$$

In view of boundedness of $L(t)$, $P_{12}(t)$ and $\Theta^{-1}(t)$ the estimate of the expression $\left.\dfrac{dv}{dt}\right|_{(5.2.12)}$ is presented as

$$\left.\frac{dv}{dt}\right|_{(5.2.12)} \leq \sigma_{11}(t)\|x\|^2 + \sigma_{22}(t)\|y\|^2 + 2c(\|x\| + \|y\|)\|x\|^2,$$

where $c > 0$ is a constant. By condition (3) of the theorem the inequality

$$\left.\frac{dv}{dt}\right|_{(5.2.12)} \leq -\varkappa\|x\|^2 + 2c\|x\|^2(\|x\| + \|y\|).$$

is satisfied.

Thus, in the open set $D = \{(x,y) \in R^3 \times R^3 \colon \|x\| + \|y\| < \dfrac{\varkappa}{4c}\}$, containing the equilibrium state $x = y = 0$, the total derivative of function $v(t,x,y,\eta)$ is estimated by the inequality

$$\left.\dfrac{dv}{dt}\right|_{(5.2.12)} \leq -\dfrac{\varkappa}{2}\|x\|^2.$$

Hence it follows that the function $v(t,x,y,\eta)$ satisfies all conditions of Theorem 1.2.1 on stability (see also Liapunov [1], and Rumjantsev and Oziraner [1]). Thus, applying Proposition 5.2.3 we establish asymptotic stability of the equilibrium state $x = y = 0$ of system (5.2.12).

Further we shall consider the case of control (5.2.2), when $K = R$. To study stability of the equilibrium state $x = y = 0$, we construct the matrix-valued function $U(t,x,y) = [v_{ij}(\cdot)]$, $i,j = 1,2$. Diagonal elements of this function are taken in the form

$$v_{11}(x) = x^{\mathrm{T}}\Theta(t)x, \quad v_{22}(y) = y^{\mathrm{T}}y.$$

To construct a non-diagonal element of the matrix-valued function we use equation (5.2.16) whose solution is

$$P_{12}(t) = -\int_t^\infty \Phi(t,\tau)(R - \Theta(\tau)\Theta^{-1}(\tau)R)\,d\tau = 0.$$

Consequently, $v_{12}(t,x,y) = 0$ and the corresponding scalar function becomes $v(t,x,y) = \eta^{\mathrm{T}}U(t,x,y)\eta = x^{\mathrm{T}}\Theta x + y^{\mathrm{T}}y$.

We note that in this case the function $v(t,x,y)$ has a simple physical sense, namely it is the double total energy of system S, since the control is performed by the potential forces. Therefore, the method of constructing the Liapunov function proposed for the problem generalizes the energetic approach for the case of nonconservative positional forces.

Later we shall need estimates of the function $v(t,x,y)$ and its total derivative along solutions of (5.2.1). It is easy to obtain

(5.2.21) $\qquad \gamma_2\|x\|^2 + \|y\|^2 \leq v(t,x,y) \leq \gamma_1\|x\|^2 + \|y\|^2$

and

(5.2.22) $\qquad \dfrac{dv}{dt} = -2x^{\mathrm{T}}\left(I + \displaystyle\sum_{j=1}^k m_j a_j(t)\dfrac{da_j(t)}{dt}(I - e_j e_j^{\mathrm{T}})\right)x.$

Estimates (5.2.21) and (5.2.22) enable us to formulate the following result.

Theorem 5.2.3 *Control (5.2.2) stabilizes motion of the mechanical system S in the prescribed direction, if the matrix $I + \sum_{j=1}^{k} m_j a_j(t) \frac{da_j(t)}{dt}(I - e_j e_j^T)$ is positive definite, i.e. for it the generalized Silvester conditions are satisfied.*

Proof Estimate (5.2.21) implies that the function $v(t, x, y)$ is positive definite and decreasing. According to estimate (5.2.22) and Theorem 1.2.3 we establish stability of the equilibrium state $x = y = 0$ of the system of perturbed motion equations and by Theorem 6.1 by Rumiantsev and Oziraner [1] obtain asymptotic x-stability of the equilibrium $x = y = 0$.

Therefore, basing on Proposition 5.2.3 we can confirm the asymptotic stability of the equilibrium state $x = y = 0$. The proof is completed.

We note that in the case when motions of the material points of system S are absent ($m_j = 0$) conditions of Theorem 5.2.3 are satisfied and controls (5.2.2) stabilize motion of the rigid body T_0, i.e. Theorem 5.2.3 generalizes Theorem 30 by Zubov [1].

5.2.4 Stabilization of dynamically symmetric rigid body in a prescribed direction

In the case under consideration $\Theta(t) = AI$, where I is a unit tensor, A is a value of the principle moment of inertia. The system of perturbed motion equations is autonomous and it can be presented as

(5.2.23)
$$A\frac{dx}{dt} = -x + Ky,$$
$$\frac{dy}{dt} = Rx - [x, y].$$

For system (5.2.23) we construct the matrix function $U(x, y)$, whose diagonal elements are taken as the quadratic forms

$$v_{11}(x) = Ax^T x, \quad v_{22}(y) = y^T y.$$

A off-diagonal element is constructed as a bilinear form $v_{12}(x, y)$, where the matrix P_{12} is determined by the equation (5.2.16) in which it is accepted $\eta = (1, 1)^T$

$$-\frac{1}{A} P_{12} = R - K.$$

Therefore we have $v_{12}(x, y) = Ax^T (K - R)y$. For the scalar function $v(x, y) = \eta^T U(x, y)\eta$ the estimate

(5.2.24)
$$v(x, y) \geq u^T Cu,$$

is valid, where

$$u = (\|x\|, \|y\|)^T, \quad C = \begin{pmatrix} A & A\|K-R\| \\ A\|K-R\| & 1 \end{pmatrix}.$$

We estimate the variation of total derivative of function $v(x,y)$ along solutions of system (5.2.23)

(5.2.25)
$$\left.\frac{dv}{dt}\right|_{(5.2.23)} \leq x^T(-2I + A(KR - RK^T - 2R^2))x$$
$$+ y^T(2K^TK + RK - K^TR)y + 2A\|K-R\|(\|x\| + \|y\|)\|x\|^2.$$

Estimates (5.2.24) and (5.2.25) enable us to establish the following result.

Theorem 5.2.4 *Controls (5.2.2) stabilize orientation of a rigid body in prescribed direction, if:*

(1) $\|K - R\| \leq \dfrac{1}{\sqrt{A}}$;

(2) *the matrix* $2K^TK + RK - K^TR$ *is negative semidefinite;*

(3) *the matrix* $-2I + A(KR - RK^T - 2R^2)$ *is negative definite.*

Proof Condition (1) of Theorem 5.2.4 implies that the function $v(x, y)$ is positive definite. Consider the neighborhood

$$H = \left\{(x,y) \in R^3 \times R^3 \colon \|x\| + \|y\| < \frac{\gamma}{2c}\right\},$$

where $\gamma = -\lambda_M(-2I + A(KR - RK^T - 2R^2))$, $\gamma > 0$, $c = 2A\|K-R\| + 1$, of the equilibrium state $x = y = 0$. In this neighborhood the estimate

$$\left.\frac{dv}{dt}\right|_{(5.2.23)} \leq -\gamma x^T x + c(\|x\| + \|y\|)\|x\|^2 \leq -\frac{\gamma}{2}\|x\|^2.$$

By Theorem 1.2.1 the equilibrium state $x = y = 0$ is stable. Moreover, Theorem 6.1 by Rumiantsev and Oziraner [1] ensures asymptotic x-stability of the equilibrium $x = y = 0$ of system (5.2.23). Asymptotic stability of the equilibrium state $x = y = 0$ of system (5.2.23) follows by conditions of Theorem 5.2.4 and Assumption 5.2.3.

5.2.5 Some comments on Theorem 5.2.4 It is of interest to investigate the structure of the matrices which satisfy conditions (1)–(3) of Theorem 5.2.4. With this in mind we shall use classical Schur theorem on unitary triangulation of matrices (see Gantmacher [1]).

Proposition 5.2.5 The matrix K, satisfying conditions (1)–(3) of Theorem 5.2.4, possesses one real eigenvalue equal to zero; its other two eigenvalues are complex-conjugated numbers.

Proof Let λ_1, λ_2, λ_3 be real eigenvalues of the matrix K. According to Gantmacher [1] there exists an orthogonal matrix U such that

$$K = USU^\mathrm{T},$$

where

$$S = \begin{pmatrix} \lambda_1 & s_1 & s_2 \\ 0 & \lambda_2 & s_3 \\ 0 & 0 & \lambda_3 \end{pmatrix}$$

is an upper triangular matrix, s_1, s_2, s_3 are real numbers. Taking into account condition (2) of Theorem 5.2.4, we conclude that the matrix

$$D = 2S^\mathrm{T}S + \overline{R}S - S^\mathrm{T}\overline{R}$$

is negative definite, where $\overline{R} = URU^\mathrm{T}$ is a skew-symmetric matrix, i.e.

$$\overline{R} = \begin{pmatrix} 0 & \overline{r}_1 & -\overline{r}_3 \\ -\overline{r}_1 & 0 & \overline{r}_2 \\ \overline{r}_3 & -\overline{r}_2 & 0. \end{pmatrix}$$

Calculating the element d_{11}, we find that $d_{11} = 2\lambda_1^2$, and therefore condition (2) of Theorem 5.2.4 can be satisfied only for $\lambda_1 = 0$. Similar speculations results in the conclusion that $\lambda_2 = \lambda_3 = 0$. Consequently, the matrix K is reduced by the orthogonal transformation to the triangular form

$$S = \begin{pmatrix} 0 & s_1 & s_2 \\ 0 & 0 & s_3 \\ 0 & 0 & 0 \end{pmatrix}.$$

Besides, the matrix D becomes

$$D = \begin{pmatrix} 0 & 0 & s_3\overline{r}_1 \\ 0 & 2s_1^2 - 2s_1\overline{r}_1 & 2s_1s_2 - s_2\overline{r}_1 + s_1\overline{r}_3 \\ s_3\overline{r}_1 & 2s_1s_2 - s_2\overline{r}_1 + s_1\overline{r}_3 & 2(s_2^2 + s_3^2) + 2(s_2\overline{r}_3 - s_3\overline{r}_2) \end{pmatrix}.$$

Since the matrix D is negative semidefinite, $s_3\overline{r}_1 = 0$. If $s_3 = 0$, then rank K = rank $S < 2$, which contradicts general assumptions on the structure of controls (5.2.2). If $\overline{r}_1 = 0$, then $s_1 = 0$, which is also a contradiction.

The proved Proposition 5.2.5 implies that the matrix K can be presented in the form $K + USU^T$, where U is an orthogonal matrix, while the matrix S has real elements and is of the form

$$S = \begin{pmatrix} \lambda_1 & \alpha & \beta \\ 0 & a & b \\ 0 & -b & a \end{pmatrix}.$$

Proceeding in the same way as above we can easily show that $\lambda_1 = 0$.

Theorem 5.2.5 *For the matrix $K \in R^{3\times 3}$ to satisfy conditions (1) – (3) of Theorem 5.2.4, it is necessary and sufficient that real numbers α, β, a, b exist and $a^2 + b^2 \neq 0$ so that*

(1) *the norm of the matrix*

$$\begin{pmatrix} \alpha^2 + a^2 + (-b \pm 1)^2 & \alpha\beta \\ \alpha\beta & \beta^2 + a^2 + (b \mp 1)^2 \end{pmatrix}$$

does not exceed $1/A$;

(2) *the matrix*

$$\begin{pmatrix} \alpha^2 + a^2 + b^2 \mp b & \alpha\beta \\ \alpha\beta & \beta^2 + a^2 + b^2 \mp b \end{pmatrix}$$

is negative semi-definite;

(3) *the following inequalities are fulfilled*

$$-1 + A(1 \mp b) < 0,$$
$$4(1 - A(1 \mp b)) > A^2(\alpha^2 + \beta^2)$$

and the orthogonal matrix U is such that

$$Ur = (0, \pm 1, 0)^T$$

and

$$K = USU^T,$$

where

$$S = \begin{pmatrix} 0 & \alpha & \beta \\ 0 & a & b \\ 0 & -b & a \end{pmatrix}.$$

5. APPLICATIONS

Proof. Necessity By Theorem Gantmacher [1] and Proposition 5.2.5 we find that there exists an orthogonal matrix U such that $K + USU^T$, where S is of the form

$$S = \begin{pmatrix} 0 & \alpha & \beta \\ 0 & a & b \\ 0 & -b & a \end{pmatrix}.$$

We note that conditions (1)–(3) of Theorem 5.2.4 are equivalent to the conditions

(i) $\|S - \overline{R}\| < 1/A$;
(ii) the matrix $2S^T S + \overline{R}S - S^T \overline{R}$ is negative definite;
(iii) the matrix $-2I + A(S\overline{R} - \overline{R}S - 2R^2)$ is negative semidefinite.

As the matrix

$$2S^T S + \overline{R}S - S^T \overline{R}$$

$$= \begin{pmatrix} 0 & a\bar{r}_1 + b\bar{r}_3 & b\bar{r}_1 - a\bar{r}_3 \\ a\bar{r}_1 + b\bar{r}_3 & 2(a^2 + \alpha^2 + \beta^2 - \alpha\bar{r}_1 - b\bar{r}_2 & 2\alpha\beta + a\bar{r}_3 - \beta\bar{r}_1 \\ b\bar{r}_1 - a\bar{r}_3 & 2\alpha\beta + a\bar{r}_3 - \beta\bar{r}_1 & 2(a^2 + \beta^2 + b^2 + \beta\bar{r}_3 - b\bar{r}_2) \end{pmatrix}$$

is negative semi-definite, we get

$$a\bar{r}_1 + b\bar{r}_3 = 0,$$
$$b\bar{r}_1 - a\bar{r}_3 = 0,$$

which is possible only if $\bar{r}_1 = \bar{r}_3 = 0$, because $a^2 + b^2 \neq 0$. This implies $\bar{r}_2 = \pm 1$. Therefore $Ur = (0, \pm 1, 0)^T$ and

$$2S^T S + \overline{R}S - S^T \overline{R} = \begin{pmatrix} 0 & 0 & 0 \\ 0 & 2(a^2 + \alpha^2 + \beta^2 \mp b & 2\alpha\beta \\ 0 & 2\alpha\beta & 2(a^2 + \beta^2 + b^2 \mp b) \end{pmatrix}.$$

This proves condition (2). The validity of conditions (1) and (3) is established in the same way by immediate calculations.

Sufficiency This part of the proof follows from the equivalence of conditions (1)–(3) of Theorem 5.2.4 and conditions (i)–(iii) and the correlation $URU^T = \overline{R}$, where

$$\overline{R} = \begin{pmatrix} 0 & 0 & 0 \\ 0 & 0 & \pm 1 \\ 0 & \mp 1 & 0 \end{pmatrix},$$

which is implied by the fact that $Ur = (0, \pm 1, 0)^T$. This completes the proof of Theorem 5.2.5.

5.3 Stability of Uncertain Discrete-Time Neural Networks

The aim of this Section is to develop a method of exponential stability analysis of neural systems with nonperturbed and perturbed equilibrium based on the hierarchical Liapunov function under one-level decomposition.

The efficiency of the proposed algorithm follows from the comparison of numerical results for the system investigated on the basis of scalar and vector approaches.

5.3.1 Stability of uncertain discrete-time systems
Discrete-time uncertain systems are satisfactory models for investigation of real phenomena in populational dynamics, macroeconomics, for simulation of chemical reactions, and also for analysis of discrete Markov processes, finite and probabilistic automatous and others.

We consider a discrete-time system with uncertainties and perturbations of the form

$$(5.3.1) \qquad x(\tau + 1) = (A + \Delta A)x(\tau) + g(x(\tau)),$$

where $\tau \in \mathcal{T} = \{t_0 + k, \ k = 0, 1, 2, \ldots\}$, $t_0 \in R$, $x \in R^n$, $x_e \equiv 0$ is an equilibrium of (5.3.1), $g \colon U \to R^n$ is a continuous vector function, $U \subseteq R^n$ is an open subset containing x_e, $A \in R^{n \times n}$ is a constant matrix and $\Delta A \in R^{n \times n}$ is an uncertain matrix. The only knowledge we have regarding the matrix ΔA is that it lies in the known compact set $S \subset R^{n \times n}$.

Assume that the system (5.3.1) is decomposed into two interconnected subsystems

$$(5.3.2) \quad \begin{aligned} x_1(\tau+1) &= (A_1 + \Delta A_1)\,x_1(\tau) + (B_1 + \Delta B_1)\,x_2(\tau) + g_1(x(\tau)), \\ x_2(\tau+1) &= (A_2 + \Delta A_2)\,x_2(\tau) + (B_2 + \Delta B_2)\,x_1(\tau) + g_2(x(\tau)). \end{aligned}$$

Here $x_i \in R^{n_i}$, A_i and B_i are submatrices of the known matrix

$$(5.3.3) \qquad A = \begin{pmatrix} A_1 & B_1 \\ B_2 & A_2 \end{pmatrix},$$

ΔA_i and ΔB_i are submatrices of the uncertain matrix

$$(5.3.4) \qquad \Delta A = \begin{pmatrix} \Delta A_1 & \Delta B_1 \\ \Delta B_2 & \Delta A_2 \end{pmatrix},$$

with $A_i, \Delta A_i \in R^{n_i \times n_i}$, $B_i, \Delta B_i \in R^{n_i \times n_j}$, $i,j = 1,2$, $i \neq j$, $g = (g_1^{\mathrm{T}}, g_2^{\mathrm{T}})^{\mathrm{T}}$, $g_i \colon U \to R^{n_i}$ are continuous vector functions.

5. APPLICATIONS

Assumption 5.3.1 We assume that:

(1) the nominal subsystems

(5.3.5) $$x_i(\tau + 1) = A_i x_i(\tau), \quad i = 1, 2,$$

are asymptotically stable, i.e. there exist unique symmetric and positive definite matrices $P_i \in R^{n_i \times n_i}$, which satisfy the Liapunov matrix equations

(5.3.6) $$A_i^T P_i A_i - P_i = -G_i, \quad i = 1, 2,$$

where $G_i \in R^{n_i \times n_i}$ are arbitrary symmetric and positive definite matrices;

(2) there exists a constant $\gamma \in (0; 1)$ such that

(5.3.7) $$\|B_1\| \|B_2\| < \gamma^2 \mu_1 \mu_2,$$

where $\mu_i = \left(\sigma_M^{1/2}(P_i - I_{n_i})\sigma_M^{1/2}(P_i) + \sigma_M(P_i)\right)^{-1}$, P_i are solutions of the Liapunov matrix equations (5.3.6) for the matrices $G_i = I_{n_i}$, I_{n_i} are $n_i \times n_i$ identity matrices, $i = 1, 2$;

(3) $\lim\limits_{\|x\| \to 0} \|g(x)\|/\|x\| = 0$

Here $\|B_i\| = \sup_{\|x_i\| \le 1} \|B_i x_i\|$, $\|x_i\| = (x_i^T x_i)^{1/2}$ are the Euclidean norms of vectors x_i, $\sigma_M(P_i)$ are the maximum eigenvalues of P_i.

Let P_i be determined as solutions of the Liapunov matrix equations (4.4.6) for $G_i = I_{n_i}$. We define the constants

(5.3.8) $$\begin{aligned} \alpha_i &= \sigma_M^{1/2}(P_i)\mu_i = \left(\sigma_M^{1/2}(P_i - I_{n_i}) + \sigma_M^{1/2}(P_i)\right)^{-1}, \quad i = 1, 2, \\ a &= \sigma_M^{1/2}(P_1)\sigma_M^{1/2}(P_2), \quad b = \sigma_M^{1/2}(P_1)\sigma_M^{1/2}(P_2)(\|B_1\| + \|B_2\|), \\ c &= \gamma^2 \alpha_1 \alpha_2 - \sigma_M^{1/2}(P_1)\sigma_M^{1/2}(P_2)\|B_1\| \|B_2\|, \\ \epsilon &= \frac{1}{2a}((b^2 + 4ac)^{1/2} - b). \end{aligned}$$

Theorem 5.3.1 *We assume that for the uncertain system (5.3.1) the decomposition (5.3.2)–(5.3.4) takes place and all conditions of Assumption 5.3.1 are satisfied. If the inequalities*

(5.3.9) $$\|\Delta A_i\| \le (1 - \gamma)\mu_i \quad \text{and} \quad \|B_i\| < \epsilon, \quad i = 1, 2.$$

are true, then the equilibrium $x_e = 0$ of (5.3.1) is exponentially stable.

Proof For nominal subsystems (5.3.5) we construct the norm-like functions

(5.3.10) $$v_i(x_i) = (x_i^T P_i x_i)^{1/2}, \quad i = 1, 2.$$

and the function

(5.3.11) $$v(x) = d_1 v_1(x_1) + d_2 v_2(x_2),$$

where d_1, d_2 are some positive constants.

For the first differences $\Delta v_i(x_i)$ of the functions (5.3.10) along the solutions of the system (5.3.1) we obtain the estimates

$$\begin{aligned}\Delta v_i(x_i)\big|_{\hat{S}_i} &= v_i(A_i x_i) - v_i(x_i) + v_i((A_i + \Delta A_i)x_i) - v_i(A_i x_i) \\ &+ v_i((A_i + \Delta A_i)x_i + (B_i + \Delta B_i)x_j + g_i(x)) - v_i((A_i + \Delta A_i)x_i) \\ &\leq (x_i^T A_i^T P_i A_i x_i)^{1/2} - (x_i^T P_i x_i)^{1/2} + \sigma_M^{1/2}(P_i)\|\Delta A_i x_i\| \\ &+ \sigma_M^{1/2}(P_i)\|(B_i + \Delta B_i)x_j + g_i(x)\| \leq \\ &\leq \frac{x_i^T A_i^T P_i A_i x_i - x_i^T P_i x_i}{(x_i^T A_i^T P_i A_i x_i)^{1/2} + (x_i^T P_i x_i)^{1/2}} + \sigma_M^{1/2}(P_i)\|\Delta A_i\|\,\|x_i\| \\ &+ \sigma_M^{1/2}(P_i)(\|B_i\| + \|\Delta B_i\|)\|x_j\| + \sigma_M^{1/2}(P_i)\|g_i(x)\| \\ &\leq -(\alpha_i - \sigma_M^{1/2}(P_i)\|\Delta A_i\|)\|x_i\| + \sigma_M^{1/2}(P_i)(\|B_i\| + \|\Delta B_i\|)\|x_j\| \\ &+ \sigma_M^{1/2}(P_i)\|g_i(x)\|, \quad i, j = 1, 2, \quad i \neq j.\end{aligned}$$

Here we have used the known inequality $(\xi^T P \xi)^{1/2} - (\nu^T P \nu)^{1/2} \leq \sigma_M^{1/2}(P)\|\xi - \nu\|$, if $P \in R^{n \times n}$, $P^T = P > 0$, $\xi, \nu \in R^n$. Then the following estimate holds

(5.3.12)
$$\begin{aligned}\Delta v(x)\big|_S &= d_1 \Delta v_1(x_1)\big|_{\hat{S}_1} + d_2 \Delta v_2(x_2)\big|_{\hat{S}_2} \leq \\ &\leq d_1\big[-(\alpha_1 - \sigma_M^{1/2}(P_1)\|\Delta A_1\|)\|x_1\| + \sigma_M^{1/2}(P_1)(\|B_1\| + \|\Delta B_1\|)\|x_2\| \\ &+ \sigma_M^{1/2}(P_1)\|g_1(x)\|\big] + d_2\big[-(\alpha_2 - \sigma_M^{1/2}(P_2)\|\Delta A_2\|)\|x_2\| \\ &+ \sigma_M^{1/2}(P_2)(\|B_2\| + \|\Delta B_2\|)\|x_1\| + \sigma_M^{1/2}(P_2)\|g_2(x)\|\big] \\ &= -\big[(\alpha_1 - \sigma_M^{1/2}(P_1)\|\Delta A_1\|)d_1 - \sigma_M^{1/2}(P_2)(\|B_2\| + \|\Delta B_2\|)d_2\big]\|x_1\| \\ &- \big[(\alpha_2 - \sigma_M^{1/2}(P_2)\|\Delta A_2\|)d_2 - \sigma_M^{1/2}(P_1)(\|B_1\| + \|\Delta B_1\|)d_1\big]\|x_2\| \\ &+ d_1 \sigma_M^{1/2}(P_1)\|g_1(x)\| + d_2 \sigma_M^{1/2}(P_2)\|g_2(x)\| = \tilde{d}^T W z + \tilde{g}(x),\end{aligned}$$

5. APPLICATIONS

where $\tilde{d} = (d_1, d_2)^{\mathrm{T}}$, $z = (\|x_1\|, \|x_2\|)^{\mathrm{T}}$, $W \in R^{2 \times 2}$ is a matrix with the elements

$$w_{ij} = \begin{cases} \alpha_i - \sigma_M^{1/2}(P_i)\|\Delta A_i\|, & \text{if } i = j, \\ -\sigma_M^{1/2}(P_i)(\|B_i\| + \|\Delta B_i\|), & \text{if } i \neq j, \end{cases}$$

the function $\tilde{g}\colon R^n \to R_+$ is such that $\lim_{\|x\|\to 0} \|\tilde{g}(x)\|/\|x\| = 0$.

The matrix W is positive definite. Indeed, it follows from (5.3.9) that $w_{ii} > 0$. Moreover

$$\begin{aligned}
w_{11}w_{22} - w_{12}w_{21} &= \left[\alpha_1 - \sigma_M^{1/2}(P_1)\|\Delta A_1\|\right]\left[\alpha_2 - \sigma_M^{1/2}(P_2)\|\Delta A_2\|\right] \\
&\quad - \sigma_M^{1/2}(P_1)\sigma_M^{1/2}(P_2)(\|B_1\| + \|\Delta B_1\|)(\|B_2\| + \|\Delta B_2\|) \\
&> [\alpha_1 - \sigma_M^{1/2}(P_1)(1-\gamma)\mu_1][\alpha_2 - \sigma_M^{1/2}(P_2)(1-\gamma)\mu_2] \\
&\quad - \sigma_M^{1/2}(P_1)\sigma_M^{1/2}(P_2)(\|B_1\| + \epsilon)(\|B_2\| + \epsilon) \\
&= \gamma^2 \alpha_1 \alpha_2 - \sigma_M^{1/2}(P_1)\sigma_M^{1/2}(P_2)(\|B_1\| + \epsilon)(\|B_2\| + \epsilon) \\
&= -\sigma_M^{1/2}(P_1)\sigma_M^{1/2}(P_2)\epsilon^2 - \sigma_M^{1/2}(P_1)\sigma_M^{1/2}(P_2)(\|B_1\| + \|B_2\|)\epsilon \\
&\quad + \gamma^2\alpha_1\alpha_2 - \sigma_M^{1/2}(P_1)\sigma_M^{1/2}(P_2)\|B_1\|\,\|B_2\| = -a\epsilon^2 - b\epsilon + c.
\end{aligned}$$

From the condition (2) of Assumption 5.3.1, we get

$$\begin{aligned}
c &= \gamma^2 \alpha_1 \alpha_2 - \sigma_M^{1/2}(P_1)\sigma_M^{1/2}(P_2)\|B_1\|\,\|B_2\| \\
&= \sigma_M^{1/2}(P_1)\sigma_M^{1/2}(P_2)\left[\gamma^2 \mu_1\mu_2 - \|B_1\|\,\|B_2\|\right] > 0,
\end{aligned}$$

therefore by (5.3.8) $-a\epsilon^2 - b\epsilon + c = 0$ and $w_{11}w_{22} - w_{12}w_{21} > 0$.

The matrix W is positive definite and has negative off-diagonal elements, i.e. it is M-matrix. Then there exist positive constants d_1 and d_2 such that the vector $\tilde{d}^{\mathrm{T}} W$ has positive elements.

Using the trivial inequalities

$$\|x_i\| \geq v_i(x_i)/\sigma_M^{1/2}(P_i), \quad i = 1, 2,$$

for the first difference of the function $v(x)$ along the solutions of the system

(5.3.1) we get

$$\Delta v(x)\big|_S \leq -(f_1 - g_2)\frac{1}{d_1\sigma_M^{1/2}(P_1)}d_1v_1(x_1) - (f_2 - g_1)\frac{1}{d_2\sigma_M^{1/2}(P_2)}d_2v_2(x_2)$$

$$+ \tilde{g}(x) = -\left[\mu_1 - \|\Delta A_1\| - \frac{d_2\sigma_M^{1/2}(P_2)}{d_1\sigma_M^{1/2}(P_1)}(\|B_2\| + \|\Delta B_2\|)\right]d_1v_1(x_1)$$

$$- \left[\mu_2 - \|\Delta A_2\| - \frac{d_1\sigma_M^{1/2}(P_1)}{d_2\sigma_M^{1/2}(P_2)}(\|B_1\| + \|\Delta B_1\|)\right]d_2v_2(x_2) + \tilde{g}(x)$$

(5.3.13)
$$\leq -\omega(d_1v_1(x_1) + d_2v_2(x_2)) + \tilde{g}(x) = -\omega v(x) + \tilde{g}(x),$$

where

$$f_i = (\alpha_i - \sigma_M^{1/2}(P_i)\|\Delta A_i\|)d_i, \quad g_i = \sigma_M^{1/2}(P_i)(\|B_i\| + \|\Delta B_i\|)d_i, \quad i = 1, 2;$$

$$\omega = \min_{\substack{i,j=1,2 \\ i \neq j}} \{\mu_i - \|\Delta A_i\| - \frac{d_j\sigma_M^{1/2}(P_j)}{d_i\sigma_M^{1/2}(P_i)}(\|B_j\| + \|\Delta B_j\|)\}.$$

The choice of the constants d_1 and d_2 implies $\omega > 0$. Let us assume that $\omega \geq 1$, then

(5.3.14)
$$\mu_1 - \|\Delta A_1\| - \frac{d_2\sigma_M^{1/2}(P_2)}{d_1\sigma_M^{1/2}(P_1)}(\|B_2\| + \|\Delta B_2\|) \geq 1,$$

$$\mu_2 - \|\Delta A_2\| - \frac{d_1\sigma_M^{1/2}(P_1)}{d_2\sigma_M^{1/2}(P_2)}(\|B_1\| + \|\Delta B_1\|) \geq 1.$$

If $\|\Delta A_1\| = \|\Delta A_2\| = \|\Delta B_1\| = \|\Delta B_2\| = \|B_1\| = \|B_2\| = 0$, the system (5.3.1) is written in the form

$$x_1(\tau + 1) = A_1x_1(\tau) + g_1(x(\tau)),$$
$$x_2(\tau + 1) = A_2x_2(\tau) + g_2(x(\tau)),$$

It is known that the equilibrium $x = 0$ of this system is exponentially stable.

Let at least one of the numbers $\|\Delta A_i\|, \|\Delta B_i\|, \|B_i\|$ nor equal zero, for example, $\|\Delta A_1\|$. Then the first inequality (5.3.14) gives

$$\mu_1 \geq 1 + \|\Delta A_1\| + \frac{d_2\sigma_M^{1/2}(P_2)}{d_1\sigma_M^{1/2}(P_1)}(\|B_2\| + \|\Delta B_2\|) > 1,$$

but
$$\mu_1 = \frac{1}{\sigma_M^{1/2}(P_1 - I_{n_1})\sigma_M^{1/2}(P_1) + \sigma_M(P_1)} \leq 1,$$

since $\sigma_M(P_1) \geq 1$. We get the contradiction, from which it follows $0 < \omega < 1$.

Using the condition (3) of Assumption 5.3.1 for the function $\tilde{g}(x)$ we get the estimate

$$\tilde{g}(x) = d_1\sigma_M^{1/2}(P_1)\|g_1(x_1)\| + d_2\sigma_M^{1/2}(P_2)\|g_2(x_2)\|$$
$$\leq (d_1\sigma_M^{1/2}(P_1) + d_2\sigma_M^{1/2}(P_2))\|g(x)\| \leq a(d_1\sigma_M^{1/2}(P_1) + d_2\sigma_M^{1/2}(P_2))\|x\|$$
$$\leq a(d_1\sigma_M^{1/2}(P_1) + d_2\sigma_M^{1/2}(P_2))(\|x_1\| + \|x_2\|)$$
$$\leq a(d_1\sigma_M^{1/2}(P_1) + d_2\sigma_M^{1/2}(P_2))\left(\frac{v_1(x_1)}{\sigma_m^{1/2}(P_1)} + \frac{v_2(x_2)}{\sigma_m^{1/2}(P_2)}\right)$$
$$\leq a(d_1\sigma_M^{1/2}(P_1) + d_2\sigma_M^{1/2}(P_2))\max\left\{\frac{1}{d_1\sigma_m^{1/2}(P_1)}, \frac{1}{d_2\sigma_m^{1/2}(P_2)}\right\}v(x),$$

where $\sigma_m^{1/2}(P_i)$ are minimum eigenvalues of the matrices P_i, a is a small positive number such that for the constant

$$\tilde{\omega} = \omega - a(d_1\sigma_M^{1/2}(P_1) + d_2\sigma_M^{1/2}(P_2))\max\left\{\frac{1}{d_1\sigma_m^{1/2}(P_1)}, \frac{1}{d_2\sigma_m^{1/2}(P_2)}\right\}$$

the inequality $0 < \tilde{\omega} < 1$ holds.

Using (5.3.13) we get the estimate

$$\Delta v(x)\big|_S \leq -\tilde{\omega}v(x)$$

for all x belonging to a sufficiently small neighborhood of the origin $\tilde{U} \subseteq U$, from where exponential stability of the equilibrium $x = 0$ of (5.3.1) follows by Theorem 2.3.3.

5.3.2 Exponential stability of a discrete-time neural system

We consider discrete-time neural networks described by the equations

(5.3.15)
$$x_1(\tau + 1) = G_1 x_1(\tau) + C_1 s_1(T_{11} x_1(\tau) + T_{12} x_2(\tau) + I_1),$$
$$x_2(\tau + 1) = G_2 x_2(\tau) + C_2 s_2(T_{21} x_1(\tau) + T_{22} x_2(\tau) + I_2),$$

where $\tau \in \mathcal{T} = \{t_0 + k, k = 0, 1, 2, \dots\}$, $t_0 \in R$, $x_i \in R^{n_i}$, $x_i = (x_{i1}, x_{i2}, \dots x_{in_i})^T$, x_{ij} represents the state of the ij-th neuron, $x_{ij} \in R$,

$s_i\colon R^{n_i} \to R^{n_i}$, $s_i(x_i) = (s_{i1}(x_{i1}), s_{i2}(x_{i2}), \ldots, s_{in_i}(x_{in_i}))^{\mathrm T}$, $s_{ij}\colon R \to (-1;1)$, $T_{ij} \in R^{n_i\times n_j}$, $G_i = \mathrm{diag}\,\{g_{i1}, g_{i2}, \ldots g_{in_i}\}$, $g_{ij} \in [-1;1]$, $C_i = \mathrm{diag}\,\{c_{i1}, c_{i2}, \ldots, c_{in_i}\}$, $c_{ij} \neq 0$ for all $i = 1, 2$, $j = 1, 2, \ldots, n_i$. The functions s_{ij} are twice continuously differentiable functions, they are monotonically increasing and odd.

Together with the system (5.3.15) we consider the uncertain system (5.3.16)
$$\begin{aligned}x_1(\tau+1) &= (G_1 + \Delta G_1)x_1(\tau) + (C_1 + \Delta C_1)s_1\big((T_{11} + \Delta T_{11})x_1(\tau) \\ &\quad + (T_{12} + \Delta T_{12})x_2(\tau) + (I_1 + \Delta I_1)\big), \\ x_2(\tau+1) &= (G_2 + \Delta G_2)x_2(\tau) + (C_2 + \Delta C_2)s_2\big((T_{21} + \Delta T_{21})x_1(\tau) \\ &\quad + (T_{22} + \Delta T_{22})x_2(\tau) + (I_2 + \Delta I_2)\big).\end{aligned}$$

Here $\Delta G_i, \Delta C_i \in R^{n_i\times n_i}$, $\Delta T_{ij} \in R^{n_i\times n_j}$, $\Delta I_i \in R^{n_i}$ are uncertain matrices and a vector.

Let $x_e = (x_{1e}^{\mathrm T}, x_{2e}^{\mathrm T})^{\mathrm T}$ denote the equilibrium of (5.3.15),

$$\begin{aligned}s_i'(x_i) &= \mathrm{diag}\,\{s_{i1}'(x_{i1}), s_{i2}'(x_{i2}), \ldots, s_{in_i}'(x_{in_i})\}, \\ s_i''(x_i) &= \mathrm{diag}\,\{s_{i1}''(x_{i1}), s_{i2}''(x_{i2}), \ldots, s_{in_i}''(x_{in_i})\}, \\ L_{i1} &= \sup_{x_i\in R^{n_i}} \|s_i'(x_i)\|, \quad L_{i2} = \sup_{x_i\in R^{n_i}} \|s_i''(x_i)\|.\end{aligned}$$

All indicated above assumptions concerning the matrices G_i, C_i, T_{ij}, the vectors I_i and the functions s_i are similar to the assumptions under which a scalar Liapunov function is applied to the neural systems of (5.3.15) type in the paper by Feng and Michel [1]. Below we need assumptions connected with decomposition of neural system.

Let us introduce the matrices

(5.3.17)
$$\begin{aligned}A_i &= G_i + C_i s_i'(T_{i1}x_{1e} + T_{i2}x_{2e} + I_i)T_{ii}, \\ B_i &= C_i s_i'(T_{i1}x_{1e} + T_{i2}x_{2e} + I_i)T_{ij}, \quad i,j = 1,2, \quad i \neq j.\end{aligned}$$

Assumption 5.3.2 Assume that:

(1) for the matrices (5.3.17) all conditions of Assumption 5.3.1 are satisfied;

(2) x_e is an equilibrium of both (5.3.15) and (5.3.16).

5. APPLICATIONS

We set
(5.3.18)
$$\beta_i = 1 + (\|C_i\| + \|T_{ii}\|)L_{i1} + (1 + \|x_{1e}\| + \|x_{2e}\|)\|C_i\|\,\|T_{ii}\|L_{i2},$$
$$\delta_i = (\|C_i\| + \|T_{ij}\|)L_{i1} + (1 + \|x_{1e}\| + \|x_{2e}\|)\|C_i\|\,\|T_{ij}\|L_{i2},$$
$$K_i = \min\{((\beta^2 + 4(1-\gamma)\mu_i L_{i1})^{1/2} - \beta_i)/2L_{i1},$$
$$((\delta^2 + 4\epsilon L_{i1})^{1/2} - \delta_i)/2L_{i1}\},$$

where $i, j = 1, 2$, $i \neq j$, the constants μ_i and ϵ are computed by (5.3.7), (5.3.8) for the matrices (5.3.17).

Theorem 5.3.2 *Let for the system (5.3.16) all conditions of Assumption 5.3.2 be satisfied. If the inequalities*

(5.3.19) $\max\{\|\Delta G_i\|, \|\Delta C_i\|, \|\Delta T_{i1}\|, \|\Delta T_{i2}\|, \|\Delta I_i\|\} < K_i, \quad i = 1, 2,$

are true, then the equilibrium x_e of (5.3.16) is exponentially stable.

Proof We denote
(5.3.20)
$$f_i(x) = G_i x_i + C_i s_i(T_{i1}x_1 + T_{i2}x_2 + I_i),$$
$$h_i(x) = \Delta G_i x_i + (C_i + \Delta C_i)s_i((T_{i1} + \Delta T_{i1})x_1 + (T_{i2} + \Delta T_{i2})x_2$$
$$+ (I_i + \Delta I_i)) - C_i s_i(T_{i1}x_1 + T_{i2}x_2 + I_i)$$

and write the system (5.3.15) in the form

(5.3.21) $\qquad x_i(\tau + 1) = f_i(x(\tau)), \quad i = 1, 2,$

and the system (5.3.16) in the form

(5.3.22) $\qquad x_i(\tau + 1) = f_i(x(\tau)) + h_i(x(\tau)), \quad i = 1, 2.$

For the functions f_i we have
(5.3.23)
$$\frac{\partial f_i(x_e)}{\partial x_i} = G_i + C_i s'_i(T_{i1}x_{1e} + T_{i2}x_{2e} + I_i)T_{ii} = A_i,$$
$$\frac{\partial f_i(x_e)}{\partial x_j} = C_i s'_i(T_{i1}x_{1e} + T_{i2}x_{2e} + I_i)T_{ij} = B_i, \quad i, j = 1, 2, \quad i \neq j,$$

where

$$\frac{\partial f_i}{\partial x_i} = \begin{pmatrix} \frac{\partial f_{i1}}{\partial x_{i1}} & \frac{\partial f_{i1}}{\partial x_{i2}} & \cdots & \frac{\partial f_{i1}}{\partial x_{in_i}} \\ \frac{\partial f_{i2}}{\partial x_{i1}} & \frac{\partial f_{i2}}{\partial x_{i2}} & \cdots & \frac{\partial f_{i2}}{\partial x_{in_i}} \\ \cdots & \cdots & \cdots & \cdots \\ \frac{\partial f_{in_i}}{\partial x_{i1}} & \frac{\partial f_{in_i}}{\partial x_{i2}} & \cdots & \frac{\partial f_{in_i}}{\partial x_{in_i}} \end{pmatrix},$$

$$\frac{\partial f_i}{\partial x_j} = \begin{pmatrix} \frac{\partial f_{i1}}{\partial x_{j1}} & \frac{\partial f_{i1}}{\partial x_{j2}} & \cdots & \frac{\partial f_{i1}}{\partial x_{jn_j}} \\ \frac{\partial f_{i2}}{\partial x_{i1}} & \frac{\partial f_{i2}}{\partial x_{j2}} & \cdots & \frac{\partial f_{i2}}{\partial x_{jn_j}} \\ \cdots & \cdots & \cdots & \cdots \\ \frac{\partial f_{in_i}}{\partial x_{j1}} & \frac{\partial f_{in_i}}{\partial x_{j2}} & \cdots & \frac{\partial f_{in_i}}{\partial x_{jn_j}} \end{pmatrix}.$$

As the functions f_i and h_i are twice continuously differentiable functions in the neighborhood of the equilibrium x_e, the equations (5.3.21) can be written in the equivalent form

$$\begin{aligned}(5.3.24)\\ x_i(\tau+1) - x_e &= f_i(x(\tau)) + h_i(x(\tau)) - f_i(x_e) - h_i(x_e) \\ &= \frac{\partial f_i(x_e)}{\partial x_i}(x_i(\tau) - x_{ie}) + \frac{\partial f_i(x_e)}{\partial x_j}(x_j(\tau) - x_{je}) \\ &+ \frac{\partial h_i(x_e)}{\partial x_i}(x_i(\tau) - x_{ie}) + \frac{\partial h_i(x_e)}{\partial x_j}(x_j(\tau) - x_{je}) + g_i(x(\tau) - x_e),\end{aligned}$$

where $g_i(x(\tau) - x_e)$ are the higher-order terms with respect to $(x(\tau) - x_e)$. Let

$$\Delta A_i = \frac{\partial h_i(x_e)}{\partial x_i}, \quad \Delta B_i = \frac{\partial h_i(x_e)}{\partial x_j}, \quad y(\tau) = x(\tau) - x_e,$$

then the equations (5.3.24) are written in the form

$$(5.3.25) \quad y_i(\tau+1) = (A_i + \Delta A_i) y_i(\tau) + (B_i + \Delta B_i) y_j(\tau) + g_i(y(\tau))$$

and the state $y = 0$ is the equilibrium of the system (5.3.25).
  Letting

$$\begin{aligned} t_i &= T_{i1} x_{1e} + T_{i2} x_{2e} + I_i, \\ z_i &= (T_{i1} + \Delta T_{i1}) x_{1e} + (T_{i2} + \Delta T_{i2}) x_{2e} + (I_i + \Delta I_i), \quad i = 1, 2,\end{aligned}$$

we find

(5.3.26)
$$\begin{aligned}\Delta A_i &= \Delta G_i + (C_i + \Delta C_i)s'_i(z_i)(T_{ii} + \Delta T_{ii}) - C_i s'_i(t_i)T_{ii} \\ &= \Delta G_i + C_i s'_i(z_i)\Delta T_{ii} + \Delta C_i s'_i(z_i)(T_{ii} + \Delta T_{ii}) \\ &\quad + C_i(s'_i(z_i) - s'_i(t_i))T_{ii} = \Delta G_i + C_i s'_i(z_i)\Delta T_{ii} \\ &\quad + \Delta C_i s'_i(z_i)(T_{ii} + \Delta T_{ii}) + C_i Q_i(z_i, t_i)\Lambda_i(z_i - t_i)T_{ii}.\end{aligned}$$

Here we have used the formula

$$f(a) - f(b) = (a - b)\int_0^1 f'(a + \xi(b-a))\,d\xi$$

for the functions $f = s_{ij}$,

$$Q_i(z_i, t_i) = \operatorname{diag}\left\{\int_0^1 s''_{i1}(z_i + \xi(t_i - z_i))\,d\xi,\ \int_0^1 s''_{i2}(z_i + \xi(t_i - z_i))\,d\xi, \ldots,\right.$$

$$\left.\int_0^1 s''_{in_i}(z_i + \xi(t_i - z_i))\,d\xi\right\},$$

$$\Lambda_i(z_i - t_i) = \operatorname{diag}\left\{z_{i1} - t_{i1},\ z_{i2} - t_{i2},\ \ldots,\ z_{in_i} - t_{in_i}\right\},\quad i = 1, 2.$$

It is easy to see that

(5.3.27)
$$\|Q_i(z_i, t_i)\| = \sup_{j=1,2,\ldots,n_i}\left|\int_0^1 s''_{ij}(z_i + \xi(t_i - z_i))\,d\xi\right| \leq L_{i2},$$

$$\|\Lambda_i(z_i - t_i)\| \leq \|z_i - t_i\| \leq (1 + \|x_{1e}\| + \|x_{2e}\|)K_i.$$

Using (5.3.18), (5.3.19) and (5.3.27), we get
(5.3.28)
$$\begin{aligned}\|\Delta A_i\| &< K_i + \|C_i\|L_{i1}K_i + L_{i1}(\|T_{ii}\| + K_i)K_i \\ &\quad + \|C_i\|L_{i2}\|z_i - t_i\|\,\|T_{ii}\| \leq L_{i1}K_i^2 + (1 + (\|C_i\| + \|T_{ii}\|)L_{i1} \\ &\quad + \|C_i\|\,\|T_{ii}\|L_{i2}(1 + \|x_{1e}\| + \|x_{2e}\|))K_i \\ &= L_{i1}K_i^2 + \beta_i K_i \leq (1 - \gamma)\mu_i.\end{aligned}$$

Similarly for $i \neq j$

$$\begin{aligned}\Delta B_i &= (C_i + \Delta C_i)s'_i(z_i)(T_{ij} + \Delta T_{ij}) - C_i s'_i(t_i)T_{ij} \\ &= C_i s'_i(z_i)\Delta T_{ij} + \Delta C_i s'_i(z_i)(T_{ij} + \Delta T_{ij}) + C_i(s'_i(z_i)s'_i(t_i))T_{ij} \\ &= C_i s'_i(z_i)\Delta T_{ij} + \Delta C_i s'_i(z_i)(T_{ij} + \Delta T_{ij}) + C_i Q_i(z_i, t_i)\Lambda_i(z_i - t_i)T_{ij}\end{aligned}$$

and

(5.3.29)
$$\|\Delta B_i\| < L_{i1}K_i^2 + ((\|C_i\| + \|T_{ij}\|)L_{i1}$$
$$+ \|C_i\|T_{ij}L_{i2}(1 + \|x_{1e}\| + \|x_{2e}\|))K_i = L_{i1}K_i^2 + \delta_i K_i \le \epsilon.$$

It follows from (5.3.28), (5.3.29) and Assumption 5.3.2 that, for the system (5.3.25) all conditions of Theorem 5.3.2 are satisfied. Hence the equilibrium $y = 0$ of the system (5.3.25) is exponentially stable, and it is equivalent to exponential stability of the equilibrium x_e of the system (5.3.16). The theorem is proved.

5.3.3 Neural system with perturbed equilibrium

Assumption 5.3.3 We assume that:

(1) for the matrices (5.3.17) all conditions of Assumption 5.3.1 are satisfied;

(2) x_e is an equilibrium of (5.3.15), $\overline{x}_e$ is an equilibrium of (5.3.16), $x_e \ne \overline{x}_e$.

We denote

(5.3.30)
$$K_i = \min\left\{\left((\beta_i^2 + 2(1-\gamma)\mu_i L_{i1})^{1/2} - \beta_i\right)/2L_{i1},\right.$$
$$\left.\left((\delta_i^2 + 2\epsilon L_{i1})^{1/2} - \delta_i\right)/2L_{i1}\right\},$$
$$r_{ij} = L_{i2}(\|C_i\| + K_i)(\|T_{ij}\| + K_i)(\|T_{i1}\| + \|T_{i2}\| + 2K_i),$$
$$\Delta < \min\left\{\frac{(1-\gamma)\mu_1}{2r_{11}}, \frac{(1-\gamma)\mu_2}{2r_{22}}, \frac{\epsilon}{2r_{12}}, \frac{\epsilon}{2r_{21}}\right\},$$

where $i, j = 1, 2$, the constants μ_1, μ_2 and ϵ are computed by (5.3.22) for the matrices (5.3.17).

Theorem 5.3.3 *Let for the system (5.3.2) all conditions of Assumption 5.3.3 be satisfied. If the inequalities*

(5.3.31)
$$\max\left\{\|\Delta G_i\|, \|\Delta C_i\|, \|\Delta T_{i1}\|, \|\Delta T_{i2}\|, \|\Delta I_i\|\right\} < K_i,$$
$$\|x_{ie} - \overline{x}_{ie}\| < \Delta, \quad i = 1, 2,$$

are true, then the equilibrium $\overline{x}_e$ of (5.3.16) is exponentially stable.

Proof As in the proof of the previous theorem, using the functions (5.3.5), we write the systems (5.3.15) and (5.3.16) in the form (5.3.21)

5. APPLICATIONS

and (5.3.22) respectively and get the identities (5.3.23). We rewrite the equation (5.3.22) in the equivalent form

$$\begin{aligned}
x_i(\tau+1) - \overline{x}_e &= f_i(x(\tau)) + h_i(x(\tau)) - f_i(\overline{x}_e) - h_i(\overline{x}_e) \\
&= \left(\frac{\partial f_i(\overline{x}_e)}{\partial x_i} + \frac{\partial h_i(\overline{x}_e)}{\partial x_i}\right)(x_i(\tau) - \overline{x}_{ie}) \\
&\quad + \left(\frac{\partial f_i(\overline{x}_e)}{\partial x_j} + \frac{\partial h_i(\overline{x}_e)}{\partial x_j}\right)(x_j(\tau) - \overline{x}_{je}) + g_i(x(\tau) - \overline{x}_e) \\
&= \left(\frac{\partial f_i(x_e)}{\partial x_i} + \frac{\partial f_i(\overline{x}_e)}{\partial x_i} + \frac{\partial h_i(\overline{x}_e)}{\partial x_i} - \frac{\partial f_i(x_e)}{\partial x_i}\right)(x_i(\tau) - \overline{x}_{ie}) \\
&\quad + \left(\frac{\partial f_i(x_e)}{\partial x_j} + \frac{\partial f_i(\overline{x}_e)}{\partial x_j} + \frac{\partial h_i(\overline{x}_e)}{\partial x_j} - \frac{\partial f_i(x_e)}{\partial x_j}\right)(x_j(\tau) - \overline{x}_{je}) \\
&\quad + g_i(x(\tau) - x_e),
\end{aligned}$$

$i, j = 1, 2$, $i \neq j$, $g_i(x(\tau) - \overline{x}_e)$ are the higher-order terms with respect to $(x(\tau) - \overline{x}_e)$.

Let us denote

$$\Delta A_i = \frac{\partial f_i(\overline{x}_e)}{\partial x_i} + \frac{\partial h_i(\overline{x}_e)}{\partial x_i} - \frac{\partial f_i(x_e)}{\partial x_i},$$

$$\Delta B_i = \frac{\partial f_i(\overline{x}_e)}{\partial x_j} + \frac{\partial h_i(\overline{x}_e)}{\partial x_j} - \frac{\partial f_i(x_e)}{\partial x_j},$$

$$y(\tau) = x(\tau) - \overline{x}_e, \quad i, j = 1, 2, \quad i \neq j.$$

The equations (5.3.22) are written in the form (5.3.25) and the state $y = 0$ is an equilibrium of the system (5.3.25).

Similarly to the proof of the previous theorem, by (5.3.30) and (5.3.31), we have:

(5.3.32)
$$\left\|\frac{\partial h_i(x_e)}{\partial x_i}\right\| < L_{i1} K_i^2 + \beta_i K_i \leq \frac{1}{2}(1-\gamma)\mu_i,$$

$$\left\|\frac{\partial h_i(x_e)}{\partial x_j}\right\| < L_{i1} K_i^2 + \delta_i K_i \leq \frac{1}{2}\epsilon, \quad i, j = 1, 2, \quad i \neq j.$$

Let

$$\bar{t}_i = T_{i1}\overline{x}_{1e} + T_{i2}\overline{x}_{2e} + I_i,$$
$$\bar{z}_i = (T_{i1} + \Delta T_{i1})\overline{x}_{1e} + (T_{i2} + \Delta T_{i2})\overline{x}_{2e} + (I_i + \Delta I_i), \quad i = 1, 2.$$

Then

$$\Delta A_i = \frac{\partial f_i(\bar{x}_e)}{\partial x_i} + \frac{\partial h_i(\bar{x}_e)}{\partial x_i} - \frac{\partial f_i(x_e)}{\partial x_i} - \frac{\partial h_i(x_e)}{\partial x_i} + \frac{\partial h_i(x_e)}{\partial x_i}$$

(5.3.33)
$$= (C_i + \Delta C_i)(s_i'(\bar{z}_i) - s_i'(z_i))(T_{ii} + \Delta T_{ii}) + \frac{\partial h_i(x_e)}{\partial x_i}$$

$$= (C_i + \Delta C_i) Q_i(\bar{z}_i, z_i) \Lambda_i(\bar{z}_i - z_i)(T_{ii} + \Delta T_{ii}) + \frac{\partial h_i(x_e)}{\partial x_i}.$$

As above

(5.3.34)
$$\|Q_2^{(i)}(\bar{z}_i, z_i)\| \le L_{i2},$$
$$\|\Lambda_i(\bar{z}_i - z_i)\| < \Delta(\|T_{i1}\| + \|T_{i2}\| + 2K_i).$$

It follows from (5.3.32)–(5.3.34) that

(5.3.35)
$$\|\Delta A_i\| < L_{i2}(\|C_i\| + K_i)(\|T_{i1}\| + \|T_{i2}\| + 2K_i)(\|T_{ii}\| + K_i)\Delta + \\ + (1-\gamma)\mu_i/2 = r_{ii}\Delta + (1-\gamma)\mu_i/2 < (1-\gamma)\mu_i.$$

Similarly, for $i \ne j$

$$\Delta B_i = \frac{\partial f_i(\bar{x}_e)}{\partial x_j} + \frac{\partial h_i(\bar{x}_e)}{\partial x_j} - \frac{\partial f_i(x_e)}{\partial x_j} - \frac{\partial h_i(x_e)}{\partial x_j} + \frac{\partial h_i(x_e)}{\partial x_j}$$

$$= (C_i + \Delta C_i)(s_i'(\bar{z}_i) - s_i'(z_i))(T_{ij} + \Delta T_{ij}) + \frac{\partial h_i(x_e)}{\partial x_j}$$

$$= (C_i + \Delta C_i) Q_i(\bar{z}_i, z_i) \Lambda_i(\bar{z}_i - z_i)(T_{ij} + \Delta T_{ij}) + \frac{\partial h_i(x_e)}{\partial x_j}$$

and

(5.3.36)
$$\|\Delta B_i\| < L_{i2}(\|C_i\| + K_i)(\|T_{i1}\| + \|T_{i2}\| + 2K_i)(\|T_{ij}\| + K_i)\Delta \\ + \epsilon/2 = r_{ij}\Delta + \epsilon/2 < \epsilon.$$

It follows from (5.3.35), (5.3.36) and Assumption 5.3.3 that for the system (5.3.25) all conditions of Theorem 5.3.1 are satisfied. Hence the equilibrium $y = 0$ of the system (5.3.25) is exponentially stable, and it is equivalent to exponential stability of the equilibrium $\bar{x}_e$ of the system (5.3.16). The theorem is proved.

5.3.4 Numerical results

5. APPLICATIONS

Example 5.3.1 Assume that the system has the form

(5.3.37)
$$x_1(\tau+1) = \frac{4}{\pi}\arctan\left(x_1(\tau) + 0.005\, x_2(\tau) - 0.005\right),$$
$$x_2(\tau+1) = -0.75\, x_2(\tau) + \frac{7}{\pi}\arctan\left(\frac{1}{350}x_1(\tau) + x_2(\tau) - \frac{1}{350}\right),$$

where $x_1, x_2 \in R$.

(a) Scalar approach We will find robust bound for the system (5.3.37). In this case

$$x_e = (1,1)^T, \quad s(x) = \left(\frac{2}{\pi}\arctan x_1, \frac{2}{\pi}\arctan x_2\right)^T,$$
$$G = \mathrm{diag}\,\{0;\ -0.75\}, \quad C = \mathrm{diag}\,\{2;\ 3.5\},$$
$$A = \begin{pmatrix} \frac{2}{\pi} & \frac{1}{100\pi} \\ \frac{1}{100\pi} & \frac{14-3\pi}{4\pi} \end{pmatrix}, \quad T = \begin{pmatrix} 1 & 0.005 \\ \frac{1}{350} & 1 \end{pmatrix},$$
$$P = \begin{pmatrix} 1.68153 & 0.00617 \\ 0.00617 & 1.15285 \end{pmatrix}.$$

The constants have the following values: $R_0 = 1$, $L_1 = L_2 = 2/\pi$, $|A|_2 = 0.73339$, $|P|_2 = 2.0388$, $|G|_\infty = 0.75$, $|C|_\infty = 3.5$, $|T|_\infty = 1.005$, $\beta = 8.34659$, $\sigma_1 = 0.28068$, $\sigma = 0.2806$, $K_0 = 0.03353$. Here

$$|A|_2 = \left(a_{11}^2 + a_{12}^2 + a_{21}^2 + a_{22}^2\right), \quad |T|_\infty = \max_{i=1,2}\{|t_{i1}| + |t_{i2}|\}.$$

(b) Vector approach Now we have:

$$x_{1e} = x_{2e} = 1, \quad S_1(x_1) = \frac{2}{\pi}\arctan x_1, \quad S_2(x_2) = \frac{2}{\pi}\arctan x_2,$$
$$L_{11} = L_{12} = L_{21} = L_{22} = 2/\pi,$$
$$T_{11} = T_{22} = 1, \quad T_{12} = 0.005, \quad T_{21} = 1/350,$$
$$A_1 = \frac{2}{\pi}, \quad A_2 = \frac{14-3\pi}{14}, \quad B_1 = \frac{1}{100\pi}, \quad B_2 = \frac{1}{100\pi},$$
$$\gamma = 0.2, \quad P_1 = 1.68147, \quad P_2 = 1.15281, \quad \mu_1 = 0.36338,$$
$$\mu_2 = 0.63591, \quad \epsilon = 0.09295, \quad \beta_1 = 6.72957, \quad \beta_2 = 10.54929,$$
$$\delta_1 = 1.29552, \quad \delta_2 = 2.24908, \quad K_1 = 0.04302, \quad K_2 = 0.04085.$$

In order to compare obtained results we consider the system
(5.3.38)
$$x_1(\tau+1) = \frac{4}{\pi}\arctan\left(1.04\,x_1(\tau) + 0.005\,x_2(\tau) - 0.045\right),$$
$$x_2(\tau+1) = -0.73\,x_2(\tau) + \frac{7.08}{\pi}\arctan\left(\frac{4}{175}x_1(\tau) + 1.02\,x_2(\tau) - \frac{3}{70}\right).$$

For this system the state $x_e = (1,1)^{\mathrm{T}}$ is an equilibrium,

$$\Delta G = \mathrm{diag}\,\{0; 0.02\}, \quad \Delta C = \mathrm{diag}\,\{0; 0.04\},$$
$$\Delta I = \mathrm{diag}\,(-0.04; -0.04)^{\mathrm{T}}, \quad \Delta T = \begin{pmatrix} 0.04 & 0 \\ 0.02 & 0.02 \end{pmatrix}.$$

As $|\Delta C|_\infty = |\Delta T|_\infty = |\Delta I|_\infty = 0.04 > K_0$, we can not conclude that the equilibrium $x_e = (1,1)^{\mathrm{T}}$ of the system (5.3.38) is exponentially stable. But

$$\max\{\,\|\Delta G_1\|, \|\Delta C_1\|, \|\Delta T_{11}\|, \|\Delta T_{12}\|, \|\Delta I_1\|\,\} = 0.04 < K_1,$$
$$\max\{\,\|\Delta G_2\|, \|\Delta C_2\|, \|\Delta T_{21}\|, \|\Delta T_{22}\|, \|\Delta I_2\|\,\} = 0.04 < K_2,$$

and it follows from Theorem 5.3.2 that the equilibrium $x_e = (1;1)^{\mathrm{T}}$ of the system (5.3.38) is exponentially stable.

Example 5.3.2 Let us consider again the system (5.3.37) with the equilibrium $x_e = (1,1)^{\mathrm{T}}$ and the perturbed system
(5.3.39)
$$x_1(\tau+1) = \frac{4.04}{\pi}\arctan\left(1.02\,x_1(\tau) - 0.0052\,x_2(\tau) - 0.025\right),$$
$$x_2(\tau+1) = -0.75\,x_2(\tau) + \frac{7}{\pi}\arctan\left(\frac{1}{350}x_1(\tau) + 1.02\,x_2(\tau) - \frac{801}{35000}\right)$$

with the equilibrium $\bar{x}_e = (1.01; 1)^{\mathrm{T}}$.

(a) In the framework of scalar approach, by Šiljak [1], we have

$$K_1 = 0.01678, \quad \bar{\epsilon}_1 = 0.09428,$$
$$\Delta G = \mathrm{diag}\,\{0; 0\}, \quad \Delta C = \mathrm{diag}\,\{0.02; 0\},$$
$$\Delta I = \left(-0.02; -\frac{701}{35000}\right)^{\mathrm{T}}, \quad \Delta T = \begin{pmatrix} 0.02 & -0.0102 \\ 0 & 0.02 \end{pmatrix}.$$

As $|\Delta T|_\infty = 0.0302 > K_1$, we can not conclude about exponential stability of the equilibrium $\bar{x}_e = (1.01,\,1)^{\mathrm{T}}$ of the system (5.3.39).

(b) In the framework of the vector approach the constants computed by (5.3.30) are:

$$K_1 = 0.02155, \quad K_2 = 0.02054, \quad \Delta = 0.10548.$$

Then

$$\max\{\,\|\Delta G_1\|,\ \|\Delta C_1\|,\ \|\Delta T_{11}\|,\ \|\Delta T_{12}\|,\ \|\Delta I_1\|\,\} = 0.02 < K_1,$$
$$\max\{\,\|\Delta G_2\|,\ \|\Delta C_2\|,\ \|\Delta T_{21}\|,\ \|\Delta T_{22}\|,\ \|\Delta I_2\|\,\} = 0.0200286 < K_2,$$
$$\|\overline{x}_{1e} - x_{1e}\| = 0.01 < \Delta, \qquad \|\overline{x}_{2e} - x_{2e}\| = 0 < \Delta$$

and, by Theorem 5.3.3, we can conclude that the equilibrium $\overline{x}_e = (1.01,\,1)^{\mathrm{T}}$ of the system (5.3.39) is exponentially stable.

5.4 Stability in the Extended Dynamical Systems

Recently it was established that a class of linear (quasilinear) systems exists whose first approximation does not allow application of the method of vector Liapunov functions even after their extension by the method of overlapping decomposition (see Ikeda and Siljak [1]). We recall that necessary condition of applicability of the vector Liapunov functions to a dynamical system is the exponential decreasing of the norm of solutions for linear (nonlinear) subsystems of the system under consideration. In the cases when this condition is not satisfied for the initial system, the problem of stability can be solved either by extension of the initial system via the method of generalized decomposition or by application of matrix-valued functions.

In this section we present the general approach of matrix-valued functions application for stability analysis of dynamical system in the context of the method of generalized decomposition.

5.4.1 Preliminary results
Let the system of perturbed motion equations

(5.4.1) $$\frac{dx}{dt} = f(t, x), \quad x(t_0) = x_0,$$

be given, where $x \in R^n$ and $f \in C(R_+ \times R^n,\ R^n)$, which possesses under initial conditions $(t_0, x_0) \in \mathrm{int}\,(R_+ \times R^n)$ a unique solution $x(t) = x(t; t_0, x_0)$ determined for all $t \geq t_0$, $t_0 \geq 0$.

Together with system (5.4.1) we consider a nonlinear continuously differentiable transformation

$$(5.4.2) \qquad y = \Phi(t, x), \quad \Phi \in C^{(1,1)}(R_+ \times R^n, R^m)$$

for which a reverse one exists and is determined by the formula

$$(5.4.3) \qquad x = \Pi(t, y), \quad \Pi \in C^{(1,1)}(R_+ \times R^m, R^n).$$

Using (5.4.2), and (5.4.3) we reduce system (5.4.1) to the form

$$(5.4.4) \qquad \frac{dy}{dt} = \tilde{f}(t, y), \quad y(t_0) = \Phi(t_0, x_0),$$

where $y \in R^m$, $\tilde{f}: R_+ \times R^m \to R^m$ and

$$(5.4.5) \qquad \tilde{f}(t, y) = \frac{\partial \Phi}{\partial t}(t, \Pi(t, y)) + \frac{\partial \Phi}{\partial x}(t, \Pi(t, y)) f(t, \Pi(t, y)).$$

In view of assumptions on system (5.4.1) and transformations (5.4.2) and (5.4.3) it is clear that $\tilde{f} \in C(R_+ \times R^m, R^m)$.

By the appropriate choice of transformation (5.4.2) one can obtain either an extension of system (5.4.1) or its contraction. We set out the following definition.

Definition 5.4.1 The system (5.4.4) is an *extension* of system (5.4.1), if for $m > n$ there exists the transformation (5.4.2) together with the reverse transformation (5.4.3) and for any initial conditions $(t_0, x_0) \in \text{int}\,(R_+ \times R^n)$ for the system (5.4.1) the condition $y_0 = \Phi(t_0, x_0)$ implies

$$(5.4.6) \qquad x(t; t_0, x_0) = \Pi(t, y(t; t_0, y_0))$$

for all $t \geq t_0$.

Definition 5.4.2 The system (5.4.4) is a *contraction* of the system (5.4.1), if for $m < n$ there exist the transformations (5.4.2) and (5.4.3) and relation (5.4.6) holds true under the conditions mentioned in Definition 5.4.1.

Note that for $m = 2n$, transformation (5.4.2) yields maximal nondegenerate extension of system (5.4.1) and for $m = 1$, transformation (5.4.2) makes maximal contraction of system (5.4.1). Recall that the Liapunov function $v = v(t,x)$, $v \in C^{(1,1)}(R_+ \times R^n, R_+)$, $v(t,0) = 0$, makes maximal contraction of system (5.4.1) up to the scalar equation.

Partial case of transformation (5.4.2) is the linear transformation

$$(5.4.7) \qquad y = Tx,$$

where T is $m \times n$ constant matrix with full rank columns. The extension of system (5.4.1) by means of transformation (5.4.7) was proposed by Ikeda and Šiljak [1] in context with vector Liapunov function.

In this case vector function (5.4.5) is of the form

$$(5.4.8) \qquad \tilde{f}(t,y) = Tf(t,T^I y) + \tilde{m}(t,y),$$

where $\tilde{m}\colon R_+ \times R^m \to R^m$ is a correcting function.

Theorem 5.4.1 *System (5.4.4) with vector function (5.4.8) is an extension of system (5.4.1), if there exists transformation (5.4.7) with generalized reverse transformation T^I and one of the conditions below is satisfied*

(a) $\tilde{m}(t,Tx) = 0$ *for all* $(t,x) \in R_+ \times R^n$;
(b) $T^I \tilde{m}(t,y) = 0$ *for all* $(t,y) \in R_+ \times R^m$.

For the *proof* of this Theorem see Ikeda and Šiljak [1].

Theorem 5.4.2 (cf. Ikeda and Šiljak [1]) *Assume that*

(1) *conditions of Theorem 5.4.1 are satisfied;*
(2) *systems (5.4.1) and (5.4.4) with vector function (5.4.8) have a unique equilibrium state $x = 0$ and $y = 0$ respectively, i.e. $t \in R_+$;*

Then certain type of stability (asymptotic stability) of the equilibrium state $y = 0$ of system (5.4.4) implies the corresponding type of stability (asymptotic stability) of the equilibrium state $x = 0$ of system (5.4.1) for both $m > n$, and $m < n$.

The *proof* of Theorem 5.4.2 is based on the use of correlation (5.4.7) and the fact that for the extended (contracted) system (5.4.4) the assumption on stability (asymptotic stability) of its equilibrium state implies the existence of a corresponding Liapunov function (see Yoshizawa [1], etc.).

5.4.2 Application of Liapunov's matrix function to extended systems
The methods of stability analysis of large-scale dynamical systems via one-level decomposition of the system and vector Liapunov functions

were summarized in a series of monographs. In this section an approach of construction of a matrix-valued Liapunov function for an extended system is proposed.

We consider an extended system (5.4.4) with a finite number of degrees of freedom whose motion is described by the equations

$$(5.4.9) \qquad \frac{dy_i}{dt} = f_i(y_i) + \sum_{\substack{j=1 \\ j \neq i}}^{m} g_{ij}(t, y_i, y_j) + G_i(t, y), \quad i = 1, 2, \ldots, s.$$

where $y_i \in R^{m_i}$, $y_j \in R^{m_j}$, $f_i \in C(R^{m_i}, R^{m_i})$, $g_{ij} \in C(R_+ \times R^{m_i} \times R^{m_j}, R^{m_j})$, $G_i \in C(R_+ \times R^m, R^{m_i})$, and besides $f_i(0) = y_{ij}(t, 0, 0) = G_i(t, 0)$ for all $i, j = 1, 2, \ldots, s$, $t \in R_+$. System (5.4.1) is a partial case of system (5.4.4) whose independent subsystems are autonomous.

In order to extend the method of matrix Liapunov functions to systems (5.4.9) it is necessary to estimate the variation of matrix-valued function elements and their total derivatives along solutions of the corresponding systems.

The approach presented in Section 1.3 together with Theorem 5.4.2 allows investigation of class of systems (5.4.1) which after extension are reduced to (5.4.9) form.

5.4.3 Stability of linear large scale systems We consider the dynamical system

$$(5.4.10) \qquad \frac{dx}{dt} = Ax, \quad x(t_0) = x_0,$$

where $x(t) \in R^n$, $t \in R_+$, A is $n \times n$-constant matrix. Vector x is divided into three subvectors x_i, $i = 1, 2, 3$, so that $x = (x_1^T, x_2^T, x_3^T)^T \in R^n$ and $x_i \in R^{n_i}$, $n = n_1 + n_2 + n_3$.

Matrix A of system (5.4.10) is presented in the block form

$$A = \begin{pmatrix} A_{11} & A_{12} & A_{13} \\ A_{21} & A_{22} & A_{23} \\ A_{31} & A_{32} & A_{33} \end{pmatrix},$$

where submatrices $A_{ij} \in R^{n_i \times n_j}$ for all $i, j = 1, 2, 3$. Further three components of vector x are transformed into two components of vector y according to the rule: $y_1 = (x_1^T, x_2^T)^T$, and $y_2 = (x_2^T, x_3^T)^T$. Besides $y = (y_1^T, y_2^T)^T \in R^{\tilde{n}}$, where $\tilde{n} = n_1 + 2n_2 + n_3$.

5. APPLICATIONS

By means of linear nondegenerate transform

(5.4.11) $$y = Tx,$$

where T is $n \times n$-matrix of the form

$$T = \begin{pmatrix} I_1 & 0 & 0 \\ 0 & \frac{1}{2}I_2 & 0 \\ 0 & \frac{1}{2}I_2 & 0 \\ 0 & 0 & I_3 \end{pmatrix},$$

I_1, I_2, and I_3 are identity matrices whose dimensions correspond to the dimensions of subvectors x_1, x_2, and x_3 of vector x. We reduce system (5.4.10) to the form

(5.4.12)
$$\frac{dy_1}{dt} = \widetilde{A}_{11} y_1 + \widetilde{A}_{12} y_2,$$
$$\frac{dy_2}{dt} = \widetilde{A}_{21} y_1 + \widetilde{A}_{22} y_2.$$

Here

$$\widetilde{A}_{22} = \begin{pmatrix} A_{11} & A_{12} \\ A_{21} & A_{22} \end{pmatrix}, \quad \widetilde{A}_{12} = \begin{pmatrix} 0 & A_{13} \\ 0 & A_{23} \end{pmatrix},$$

$$\widetilde{A}_{22} = \begin{pmatrix} A_{22} & A_{23} \\ A_{32} & A_{33} \end{pmatrix}, \quad \widetilde{A}_{21} = \begin{pmatrix} A_{21} & 0 \\ A_{31} & 0 \end{pmatrix}.$$

Designate $\widetilde{A} = (\widetilde{A}_{ij})$, $i, j = 1, 2$, and note that

$$\widetilde{A} = T A T^{\mathrm{T}} + M,$$

where

$$T = \begin{pmatrix} I_1 & 0 & 0 \\ 0 & \frac{1}{2}I_2 & 0 \\ 0 & \frac{1}{2}I_2 & 0 \\ 0 & 0 & I_3 \end{pmatrix}, \quad T^I = \begin{pmatrix} I_1 & 0 & 0 & 0 \\ 0 & I_2 & I_2 & 0 \\ 0 & 0 & 0 & I_3 \end{pmatrix},$$

and

$$M = \begin{pmatrix} 0 & \frac{1}{2}A_{12} & -\frac{1}{2}A_{12} & 0 \\ 0 & \frac{1}{2}A_{22} & -\frac{1}{2}A_{22} & 0 \\ 0 & \frac{1}{2}A_{22} & -\frac{1}{2}A_{22} & 0 \\ 0 & \frac{1}{2}A_{32} & -\frac{1}{2}A_{32} & 0 \end{pmatrix}.$$

Definition 5.4.3 System (5.4.12) is an *extension* of system (5.4.10) if there exists a linear transformation of maximal rank (5.4.11) such that for $y_0 = T x_0$
$$x(t, x_0) = T^I y(t, y_0), \quad t \geq t_0,$$
where $y(t, y_0)$ is a solution of system (5.4.12).

Remark 5.4.2 It is noted by Ikeda and Šiljak [1], and Šiljak [1] that the extension procedure of system (5.4.10) or the general nonlinear system can result in the applicability of the vector Liapunov function while its immediate application to the initial system is difficult or impossible. Below we cite an example showing that this is not a common case, i.e. there exist systems of (5.4.10) type to which it is impossible to apply the vector Liapunov function even after their extension in the sense of Definition 5.4.1 to the form (5.4.12).

We select out of the extended system (5.4.12) two independent subsystems

$$\frac{dz_1}{dt} = \widetilde{A}_{11} z_1, \quad z_1(t_0) = z_{10},$$

$$\frac{dz_2}{dt} = \widetilde{A}_{22} z_2, \quad z_2(t_0) = z_{20},$$

and assume that for the given sign-definite matrices G_{11} and G_{22} the algebraic Liapunov equations

(5.4.13) $$\widetilde{A}_{11}^{\mathrm{T}} P_{11} + P_{11} \widetilde{A}_{11} = G_{11},$$

(5.4.14) $$\widetilde{A}_{22}^{\mathrm{T}} P_{22} + P_{22} \widetilde{A}_{22} = G_{22}$$

have the solutions in the form of sign-definite symmetric matrices P_{11} and P_{22} respectively.

Let matrices P_{11} and P_{22} be determined by equations (5.4.13) and (5.4.14). Assume that there exist constants $\eta_1, \eta_2 > 0$ such that the algebraic equation

(5.4.15) $$\widetilde{A}_{11}^{\mathrm{T}} P_{12} + P_{12} \widetilde{A}_{22} = -\frac{\eta_1}{\eta_2} P_{11} \widetilde{A}_{12} - \frac{\eta_2}{\eta_1} \widetilde{A}_{21}^{\mathrm{T}} P_{22}$$

has bounded matrix P_{12} as its solution.

Proposition 5.4.1 If and only if matrices $\widetilde{A}_{11}$ and $-\widetilde{A}_{22}$ do not have common eigenvalues, then the algebraic equation (5.4.15) has a unique solution in the form of bounded matrix P_{12}.

5. APPLICATIONS

Proposition 5.4.1 follows from Theorem 85.1 of Lankaster [1].

Equations (5.4.13)–(5.4.14) form the basis of the proposed new approach of Liapunov matrix-valued function construction.

Now we construct a two-index system of functions

$$U(y) = [v_{ij}(y_1, y_2)], \quad i, j = 1, 2$$

with the elements

(5.4.16) $\quad v_{11}(y_1) = y_1^T P_{11} y_1, \quad v_{22}(y_2) = y_2^T P_{22} y_2,$

(5.4.17) $\quad v_{12}(y_1, y_2) = v_{21}(y_1, y_2) = y_1^T P_{12} y_2.$

Here matrices P_{11}, P_{22}, and P_{12} are determined by the equations (5.4.22)–(5.4.24).

For quadratic forms (5.4.16) and bilinear form (5.4.17) the estimates

$$v_{11}(y_1) \geq \lambda_m(P_{11}) \|y_1\|^2,$$
$$v_{22}(y_2) \geq \lambda_m(P_{22}) \|y_2\|^2,$$
$$v_{12}(y_1, y_2) \geq -\lambda_m^{1/2}(P_{12} P_{12}^T) \|y_1\| \|y_2\|,$$

are true, where $\lambda_m(P_{11})$ and $\lambda_m(P_{22})$ are minimal eigenvalues of matrices P_{11} and P_{22}, and $\lambda_M^{1/2}(\cdot)$ is the norm of matrix $P_{12} P_{12}^T$, coordinated with the vector norm.

It is easy to show that the function

(5.4.18) $\quad v(y, \eta) = \eta^T U(y) \eta, \quad \eta \in R_+^2, \quad \eta > 0,$

satisfies for all $y \in R^{\tilde{n}}$ the estimate

(5.4.19) $\quad v(y, \eta) \geq u^T H^T C H u,$

where $u = (\|y_1\|, \|y_2\|)^T$, $H = \mathrm{diag}\,(\eta_1, \eta_2)$, and

$$C = \begin{pmatrix} \lambda_m(P_{11}) & -\lambda_M^{1/2}(P_{12} P_{12}^T) \\ -\lambda_M^{1/2}(P_{12} P_{12}^T) & \lambda_m(P_{22}) \end{pmatrix}.$$

The variation of the total derivative of function (5.4.18) along solutions of system (5.4.12)

$$\begin{aligned} Dv(y,\eta) &= y_1^T (\tilde{A}_{11}^T P_{11} + P_{11} \tilde{A}_{11}) y_1 \eta_1^2 \\ &+ 2 y_1^T \big[(\tilde{A}_{11}^T P_{12} + P_{12} \tilde{A}_{22}) \eta_1 \eta_2 + \eta_1^2 P_{11} \tilde{A}_{11} + \eta_2^2 \tilde{A}_{22} P_{22} \big] y_2 \\ &+ y_2^T (\tilde{A}_{22}^T P_{22} + P_{22} \tilde{A}_{22}) y_2 + \eta_1 \eta_2 y_1^T (P_{12} \tilde{A}_{21} + \tilde{A}_{21}^T P_{12}^T) y_1 \\ &+ \eta_1 \eta_2 y_2^T (\tilde{A}_{12}^T P_{12} + P_{12}^T \tilde{A}_{12}) y_2 \end{aligned}$$

is estimated in view of equation (5.4.15). Denote by

(5.4.20)
$$\lambda_M = \lambda_M(\widetilde{A}_{11}^T P_{11} + P_{11}\widetilde{A}_{11}); \quad \beta_M = \beta_M(\widetilde{A}_{22}^T P_{22} + P_{22}\widetilde{A}_{22});$$
$$\varkappa_M = \varkappa_M(P_{12}\widetilde{A}_{21} + \widetilde{A}_{21}^T P_{12}^T); \quad \chi_M = \chi_M(\widetilde{A}_{12}^T P_{12} + P_{12}^T \widetilde{A}_{12})$$

the maximal eigenvalues of the corresponding matrices. In view of designations (5.4.32) for all $(y_1, y_2) \in N_1 \times N_2$

$$Dv(y, \eta)\big|_{(5.4.14)} \leq \lambda_M \eta_1^2 \|y_1\|^2 + \varkappa_M \eta_1 \eta_2 \|y_1\|^2$$
$$+ \beta_M \eta_2^2 \|y_2\|^2 + \chi_M \eta_1 \eta_2 \|y_2\|.$$

Hence it follows that for all $(y_1, y_2) \in N_1 \times N_2$ the inequality

(5.4.21)
$$Dv(y, \eta)\big|_{(5.4.14)} \leq u^T S u$$

holds true, where the matrix $S = [\sigma_{ij}]$, $i, j = 1, 2$, has the elements

$$\sigma_{11} = \lambda_M \eta_1^2 + \varkappa_M \eta_1 \eta_2, \quad \sigma_{22} = \beta_M \eta_2^2 + \chi_M \eta_1 \eta_2, \quad \sigma_{12} = \sigma_{21} = 0.$$

Estimates (5.4.19) of function (5.4.18) and its total derivative (5.4.21) enable us to establish a new stability test for system (5.4.10) as follows.

Theorem 5.4.3 *Assume that the equations (5.4.10) are such that the following conditions are satisfied:*

(1) *system (5.4.12) is the extension of system (5.4.10) in the sense of Definition 5.4.1;*
(2) *there exist solutions to algebraic equations (5.4.13) – (5.4.15);*
(3) *in estimate (5.4.19) matrix C is positive definite;*
(4) *in estimate (5.4.21) matrix S is negative semi-definite (negative definite).*

Then the equilibrium state $x = 0$ of system (5.4.10) is uniformly stable in the whole (uniformly asymptotically stable in the whole).

The proof of Theorem 5.4.3 is similar to the proof of Theorem 4.1 by Martynyuk and Slyn'ko [1].

Theorem 5.4.4 *Assume that conditions (1) – (3) of Theorem 5.4.3 are satisfied and there exists a vector $\eta = (\eta_1, \eta_2) > 0$ such that instead of condition (4) of Theorem 5.4.3 the following inequalities are fulfilled*

(5.4.22) $\eta_1^2(\widetilde{A}_{11}^T P_{11} + P_{11}\widetilde{A}_{11}) + \eta_1 \eta_2 (P_{12}\widetilde{A}_{21} + \widetilde{A}_{21}^T P_{12}^T) < 0,$

(5.4.23) $\eta_2^2(\widetilde{A}_{22}^T P_{22} + P_{22}\widetilde{A}_{22}) + \eta_1 \eta_2 (\widetilde{A}_{12}^T P_{12} + P_{12}^T \widetilde{A}_{12}) < 0.$

5. APPLICATIONS   303

Then the equilibrium state $x = 0$ of system (5.4.10) is uniformly asymptotically stable in the whole.

Proof Together with conditions (1)–(3) of Theorem 5.4.3 inequalities (5.4.22), and (5.4.23) ensure satisfaction of all conditions of Theorem 2.11 by Ikeda and Šiljak [1] and Theorem 25.2 by Hahn [1]. Hence the assertion of Theorem 5.4.4 is proved.

5.4.4 Numerical example The proposed technique of Liapunov matrix-valued function construction for the extended system is illustrated by the example of the third order system

$$(5.4.24) \qquad \frac{dx}{dt} = \begin{pmatrix} -3 & 4 & 6 \\ 4 & -5 & 4 \\ -10 & -4 & -3 \end{pmatrix} x,$$

where $x = (x_1, x_2, x_3)^{\mathrm{T}}$. Diagonal blocks of the matrix are taken as the matrices of coefficients of independent subsystems of the extended system, i.e. system (5.4.24) is extended to

$$(5.4.25) \qquad \frac{dy}{dt} = \begin{pmatrix} -3 & 4 & 0 & 6 \\ 4 & -5 & 0 & 4 \\ 4 & 0 & -5 & 4 \\ -10 & 0 & -4 & -3 \end{pmatrix} y,$$

where $y = (y_1, y_2)^{\mathrm{T}}$, $y_1 = (x_1, x_2)^{\mathrm{T}}$, $y_2 = (x_2, x_3)^{\mathrm{T}}$ are the state vectors of two second order subsystems

$$\frac{dy_1}{dt} = \begin{pmatrix} -3 & 4 \\ 4 & -5 \end{pmatrix} y_1 + \begin{pmatrix} 0 & 6 \\ 0 & 4 \end{pmatrix} y_2,$$

$$\frac{dy_2}{dt} = \begin{pmatrix} -5 & 4 \\ -4 & -3 \end{pmatrix} y_2 + \begin{pmatrix} 4 & 0 \\ -10 & 0 \end{pmatrix} y_1.$$

According to the adopted notation we have

$$\tilde{A}_{11} = \begin{pmatrix} -3 & 4 \\ 4 & -5 \end{pmatrix}, \quad \tilde{A}_{22} = \begin{pmatrix} -5 & 4 \\ -4 & -3 \end{pmatrix},$$

$$\tilde{A}_{12} = \begin{pmatrix} 0 & 6 \\ 0 & 4 \end{pmatrix}, \quad \tilde{A}_{21} = \begin{pmatrix} 4 & 0 \\ -10 & 0 \end{pmatrix}.$$

Assume that $P_{11} = P_{22} = E$, $\eta = (1,1)^T$ and take

$$v_{11}(y_1) = y_1^T y_1, \quad v_{22}(y_2) = y_2^T y_2,$$
$$v_{12}(y_1, y_2) = v_{21}(y_1, y_2) = y_1^T P_{12} y_2.$$

as the elements of matrix-valued function $U(y)$. Here matrix P_{12} is determined by the equation

$$\begin{pmatrix} -3 & 4 \\ 4 & -5 \end{pmatrix} P_{12} + P_{12} \begin{pmatrix} -5 & 4 \\ -4 & -3 \end{pmatrix} = \begin{pmatrix} -4 & 4 \\ 0 & -4 \end{pmatrix}$$

corresponding to equation (5.4.15). It is easy to verify that $P_{12} = \dfrac{1}{2} E$, where E is 2×2 identity matrix. Since $v_{11} = \|y_1\|^2$, $v_{22} = \|y_2\|^2$, $v_{12} \geq -\dfrac{1}{2}\|y_1\|\|y_2\|$, the matrix C in estimate (5.4.19) reads

$$C = \begin{pmatrix} 1 & -1/2 \\ -1/2 & 1 \end{pmatrix}$$

and is positive definite. It is easy to see that conditions (1)–(3) of Theorem 5.4.4 are satisfied and conditions (5.4.22), and (5.4.23) of this theorem have the form

$$\begin{pmatrix} -6 & 8 \\ 8 & -10 \end{pmatrix} + \frac{1}{2} \begin{pmatrix} 8 & -10 \\ -10 & 0 \end{pmatrix} = \begin{pmatrix} -2 & 3 \\ 3 & -10 \end{pmatrix} = S_1,$$

$$\begin{pmatrix} -10 & 0 \\ 0 & -6 \end{pmatrix} + \frac{1}{2} \begin{pmatrix} 0 & 6 \\ 6 & 8 \end{pmatrix} = \begin{pmatrix} -10 & 3 \\ 3 & -2 \end{pmatrix} = S_2.$$

One can easily check that the matrices S_1 and S_2 are negative definite. Consequently, the equilibrium state $y = 0$ of system (5.4.25) is uniformly asymptotically stable in the whole. Since all conditions of Theorem 5.4.4, are fulfilled, the equilibrium state $x = 0$ of system (5.4.24) possesses the same type of stability.

Example (5.4.24) with the extension (5.4.25) is the one to which vector Liapunov function is not applicable. This can be verified easily by the method proposed by Ikeda and Šiljak [1] for the proof of vector Liapunov function nonapplicability to the non-extended system in one case. The method proposed by us for Liapunov matrix-valued function construction in conjunction with the extension method of the state space of dynamical

5. APPLICATIONS

systems enlarges the area of the direct Liapunov's method application in linear and nonlinear dynamics of systems.

5.5 Interval Stability of Mechanical Systems

In this section we present a new approach to the problem of interval stability analysis of dynamical systems. The essence of this approach lies in conjunction with the concept of maximal extension of the interval system with direct Liapunov method. Efficiency of the proposed approach is illustrated by the example of a general type mechanical system with finite degrees of freedom.

5.5.1 Auxiliary results
We consider the linear differential system

$$(5.5.1) \qquad \frac{dx}{dt} = Ax, \quad x(t_0) = x_0,$$

where $x \in R^n$, $t \in R_+ = [0, \infty)$, and A is a constant $n \times n$ matrix.

By means of the linear nondegenerate transform

$$(5.5.2) \qquad y = Tx,$$

where T is an $m \times n$ constant matrix with full column rank we represent the system (5.5.1) as

$$(5.5.3) \qquad \frac{dy}{dt} = \tilde{A}y, \quad y(t_0) = y_0,$$

where $y \in R^m$, $t \in R_+$, $\tilde{A}$ is an $m \times m$ constant matrix, $m = 2n$.

Dynamical system (5.5.3) is a maximum extension of system (5.5.1), if and only if the linear transformation (5.5.2), for which $y_0 = Tx_0$, provides a relation between the solutions of (5.5.1) and (5.5.2) as

$$(5.5.4) \qquad x(t; t_0, x_0) = T^I y(t; t_0, y_0) \quad \text{for all} \quad t \geq t_0,$$

where T^I is the generalized inverse transformation.

Note that when $n \leq m$ the meaning of (5.5.4) is the same as that given in Definition 2.4 of the paper by Ikeda and Šiljak [1].

From this paper it follows that system (5.5.3) is the maximum extension of system (5.5.1) if and only if the matrix $M = \tilde{A} - TAT^I$ satisfies one of the relations $MT = 0$ or $T^I M = 0$.

5.5.2 Interval stability of linear system

We consider the dynamic system

$$(5.5.5) \qquad \frac{dx}{dt} = (A_0 + pA_1)x, \quad x(t_0) = x_0,$$

where $x \in R^n$, A_0 and A_1 are constant $n \times n$-matrices, and $p \in S \subseteq R$ is a scalar parameter of uncertainty, and S is a compact set. By the linear transformation (5.5.2), where

$$T = (I, I)^{\mathrm{T}}$$

we transform system (5.5.5) to the form

$$(5.5.6) \qquad \begin{aligned} \frac{dy_1}{dt} &= A_0 y_1 + pA_1 y_2, & y_1(t_0) &= y_{10}, \\ \frac{dy_2}{dt} &= (p - \overline{p})A_1 y_1 + (A_1 + \overline{p}A_1)y_2, & y_2(t_0) &= y_{20}, \end{aligned}$$

where $y_1 \in R^n$, $y_2 \in R^n$, and $\overline{p} \in S$ is some fixed value of the parameter p, $\overline{p} > 0$.

It is easy to verify that system (5.5.6) is the maximum extension of system (5.5.5), since the condition $MT = 0$, where $T = (I, I)^{\mathrm{T}}$, is satisfied.

Assume that the matrices A_0 and $A_0 + \overline{p}A_1$ in (5.5.6) are stable. i.e., there exist symmetric positive definite matrices B_1 and B_2, that satisfy the Liapunov matrix equations

$$(5.5.7) \qquad A_0^{\mathrm{T}} B_1 + B_1 A_0 = -2Q_1,$$

$$(5.5.8) \qquad (A_0 + \overline{p}A_1)^{\mathrm{T}} B_2 + B_2(A_0 + \overline{p}A_1) = -2Q_2$$

for some symmetric positive definite matrices Q_1 and Q_2.

Definition 5.5.1 System (5.5.5) is *intervally stable*, if the equilibrium state $x = 0$ is asymptotically Liapunov stable for each $p \in [0, \overline{p}]$.

The Theorem below includes the interval-stability conditions for system (5.5.5).

Theorem 5.5.1 *If system (5.5.5) is such that*

(1) *the matrices A_0 and $A_0 + \overline{p}A_1$ are stable for some $\overline{p} \in S$;*
(2) *value $\overline{p}$ of $p \in S$ satisfies the inequality*

$$(5.5.9) \qquad \overline{p}^2 < \frac{4\lambda_m(Q_1)\lambda_m(Q_2)}{\|B_1 A_1\| \|B_2 A_1\|},$$

where $\lambda_m(\cdot)$ are minimum eigenvalues of the matrices Q_1 and Q_2 in equations (5.5.7) and (5.5.8) respectively.

5. APPLICATIONS 307

Then system (5.5.5) is intervally stable.

Proof Let us construct a vector Liapunov function for extended system (5.5.6)
$$v(y) = (v_1(y_1), v_(y_2))^{\mathrm{T}}$$
with the components $v_1(y_1) = y_1^{\mathrm{T}} B_1 y_1$ and $v_2(y_2) = y_2^{\mathrm{T}} B_2 y_2$.

Consider the total derivatives of the components of the vector Liapunov function along the solutions of system (5.5.6)

$$\left.\frac{dv_1}{dt}\right|_{(5.5.6)} = 2y_1^{\mathrm{T}} B_1(A_0 + pA_1 y_2) \leq -2\lambda_m(Q_1)\|y_1\|^2 + 2|p|\|B_1 A_1\|,$$

$$\left.\frac{dv_2}{dt}\right|_{(5.5.6)} = 2y_2^{\mathrm{T}} B_2((A_0 + pA_1)y_2 + (p-\bar{p})A_1 y_1)$$
$$\leq -2\lambda_m(Q_2)\|y_2\|^2 + 2|p-\bar{p}|\|B_2 A_1\|\,\|y_1\|\,\|y_2\|.$$

It is well known that for system (5.5.6) to be asymptotically stable, it is sufficient that the matrix

$$S = \begin{pmatrix} -\lambda_m(Q_1) & |p|\|B_1 A_1\| \\ |p-\bar{p}|\|B_2 A_1\| & -\lambda_m(Q_2) \end{pmatrix}$$

be an M-matrix, i.e.

(5.5.10) $\qquad \lambda_m(Q_1)\lambda_m(Q_2) - |p|\|\bar{p} - p\|\|B_1 A_1\|\,\|B_2 A_1\| > 0.$

We note that for $p \in [0, \bar{p}]$ the inequality $|p|\|\bar{p} - p| \leq \frac{1}{4}\bar{p}^2$ holds, therefore taking condition (5.5.2) of Theorem 5.5.1 into account, we arrive at inequality (5.5.7). Thus, the theorem is proved, because the stability of system (5.5.5) follows from the stability of system (5.5.6) according to Theorem 5.4.2.

5.5.3 Interval stability conditions for a mechanical system Consider the finite-degree-of-freedom system (see Cao and Shu [1])

(5.5.11) $\qquad \left(M + p\widetilde{M}\right)\dfrac{d^2 x}{dt^2} + \left(D + p\widetilde{D}\right)\dfrac{dx}{dt} + \left(K + p\widetilde{K}\right)x = 0,$

where $x \in R^n$, M, D, K, $\widetilde{M}$, $\widetilde{D}$, and $\widetilde{K}$ are constant $n \times n$ -matrices. Assume that $\det M \neq 0$ and $|p| < \|M^{-1}\widetilde{M}\|^{-1}$. Rearrange system (5.5.11) into the form

(5.5.12) $\qquad \dfrac{dy}{dt} = (A_0 + pA_1 + \cdots + p^k A_k + \ldots)y,$

where $y = \left(x, \dfrac{dx}{dt}\right)^{\mathrm{T}}$, $y \in R^{2n}$ and A_k are constant $2n \times 2n$-matrices,

$$A_0 = \begin{pmatrix} 0 & I \\ -M^{-1}K & -M^{-1}D \end{pmatrix},$$

$$A_k = (-1)^k F \begin{pmatrix} (M^{-1}\widetilde{M})^{k-1} & 0 \\ 0 & (M^{-1}\widetilde{M})^{k-1} \end{pmatrix} G, \quad k \geq 1,$$

where

$$F = \begin{pmatrix} 0 & 0 \\ 0 & M^{-1} \end{pmatrix}, \quad \text{and} \quad G = \begin{pmatrix} 0 & 0 \\ \widetilde{K} - M^{-1}\widetilde{M}K & \widetilde{D} - M^{-1}\widetilde{M}D \end{pmatrix}.$$

Assume that the matrices A_0 and $\overline{A}$ are stable, where

$$\overline{A} = \begin{pmatrix} 0 & I \\ -\overline{M}^{-1}\overline{K} & -\overline{M}^{-1}\overline{D} \end{pmatrix},$$

$$\overline{M} = M + \overline{p}\widetilde{M}, \quad \overline{K} = K + \overline{p}\widetilde{K}, \quad \overline{D} = D + \overline{p}\widetilde{D}.$$

Then, for positive definite matrices Q_1 and Q_2, the Liapunov equations

$$A_0^{\mathrm{T}} P_1 + P_1 A_0 = -2Q_1,$$
$$\overline{A}^{\mathrm{T}} P_2 + P_2 \overline{A} = -2Q_2,$$

have solutions in the form of symmetric, positive definite matrices P_1 and P_2.

Theorem 5.5.2 *If the interval system (5.5.11) is such that*

(1) *the matrices A_0 and $\overline{A}$ are stable;*

(2) *there exists a value of parameter $\overline{p}$, $\overline{p} \in S$ such that*

$$\overline{p}^2 \leq \dfrac{4\lambda_m(Q_1)\lambda_m(Q_2)(1 - \overline{p}\|M^{-1}\widetilde{M}\|)^3}{\|P_1 F\| \|P_2 F\| \|G\|^2}.$$

Then system (5.5.11) is intervally stable.

Proof System (5.5.11) by transformation (5.5.2) is rearranged to the form

(5.5.13)
$$\dfrac{dz_1}{dt} = A_0 z_1 + \sum_{k=1}^{\infty} p^k A_k z_2,$$

$$\dfrac{dz_2}{dt} = \overline{A} z_2 + \sum_{k=1}^{\infty} (p^k - \overline{p}^k) A_k z_1.$$

5. APPLICATIONS

Let us construct a vector Liapunov function for system (5.5.13) $v(z) = (v_1(z_1), v_2(z_2))^T$ where $v_1(z_1) = z_1^T P_1 z_1$ and $v_2(z_2) = z_2^T P_2 z_2$ and estimate the total derivatives of the components of vector Liapunov function along the solutions of system (5.5.13)

$$\frac{dv_1}{dt}\bigg|_{(5.5.13)} = 2z_1^T P_1 \left(A_0 z_1 + \sum_{k=1}^{\infty} p^k A_k z_2 \right) \le -2\lambda_m(Q_1)\|z_1\|^2$$

$$+ 2|p|\|P_1 F\|\|G\| \sum_{k=1}^{\infty} \|M^{-1}\widetilde{M}\|^{k-1} \|z_1\| \|z_2\|$$

$$= -2\lambda_m(Q_2)\|z_1\|^2 + \frac{2|p|\|P_1 F\|\|G\|}{1 - \overline{p}\|M^{-1}\widetilde{M}\|} \|z_1\| \|z_2\|,$$

$$\frac{dv_2}{dt}\bigg|_{(5.5.13)} = 2z_1^T P_1 \left(\overline{A} z_2 + \sum_{k=1}^{\infty} (p^k - \overline{p}^k) A_k z_1 \right) \le -2\lambda_m(Q_2)\|z_2\|^2$$

$$+ 2\|P_2 F\|\|G\| \sum_{k=1}^{\infty} \|M^{-1}\widetilde{M}\|^{k-1} |p^k - \overline{p}^k| \|z_1\| \|z_2\|$$

$$\le -2\lambda_m(Q_2)\|z_2\|^2$$

$$+ 2\|P_2 F\|\|G\| |p - \overline{p}| \sum_{k=1}^{\infty} k \overline{p}^{k-1} \|M^{-1}\widetilde{M}\|^{k-1} \|z_1\| \|z_2\|$$

$$= -2\lambda_m(Q_2)\|z_2\|^2 + 2\|P_2 F\|\|G\| \frac{|p - \overline{p}|}{(1 - \overline{p}\|M^{-1}\widetilde{M}\|)^2} \|z_1\| \|z_2\|.$$

The system (5.5.13) is stable if the matrix

$$S = \begin{pmatrix} \lambda_m(Q_1) & \dfrac{\|P_1 F\|\|G\||p|}{1 - \overline{p}\|M^{-1}\widetilde{M}\|} \\ \dfrac{\|P_2 F\|\|G\||p - \overline{p}|}{(1 - \overline{p}\|M^{-1}\widetilde{M}\|)^2} & \lambda_m(Q_2) \end{pmatrix}$$

is an M-matrix, i.e.

$$(5.5.14) \qquad \det S = \lambda_m(Q_1)\lambda_m(Q_2) - \frac{\overline{p}^2 \|P_1 F\| \|P_2 F\| \|G\|^2}{4(1 - \overline{p}\|M^{-1}\widetilde{M}\|)^3} > 0.$$

The further proof of Theorem 5.5.2 is similar to that of Theorem 5.5.1.

The theory of stability of interval systems has been intensively developed in the last decades. The concept of maximal extension of the initial interval system with subsequent application of multicomponent Liapunov function is a new and nonstudied area. The estimates of the values $\bar{p}$ of parameter p, obtained from inequalities (5.5.10) and (5.5.14) have wider limits than those mentioned in the literature (see, for example, Cao and Shu [1], Chung-Li Jiang [1], etc.).

5.6 Analysis of Population Growth Models of Kolmogorov Type

For unperturbed and perturbed Kolmogorov models of population dynamics, new sufficient conditions of stability and boundedness with respect to two criteria (measures) are provided. An application to a generalized Lotka-Volterra equation is given.

5.6.1 Generalized Kolmogorov model
We consider the system of ordinary differential equations
(5.6.1)
$$\frac{dx_i}{dt} = \beta_i(x_i) F_i(t, x_1, \ldots, x_n, \mu), \quad x_i(t_0) = x_{i0} \geq 0, \quad i = 1, 2, \ldots, n.$$

Here β_i are infinitely differentiable functions on $R_+ = [0, \infty)$, $\beta_i(0) = 0$, $\beta_i'(x_i) > 0$ for $x_i > 0$, $\beta_i^j(x_i) \geq 0$ $(j = 2, 3, \ldots)$, and $F_i \in C(R_+ \times R_+^n \times M^k, R)$, where $M^k = [0, 1] \times \ldots \times [0, 1]$ $(i = 1, 2, \ldots, n)$. System (5.6.1) represents a multiplicatively and additively perturbed Kolmogorov equation serving for modeling dynamics of population. Alongwith (5.6.1), we consider the unperturbed Kolmogorov system,

(5.6.2) $\quad \dfrac{dx_i}{dt} = K_i x_i f_i(t, x_1, \ldots, x_n), \quad x_i(t_0) = x_{i0} \geq 0, \quad i = 1, 2, \ldots, n.$

Here $K_i > 0$ and $f_i(t, x_1, \ldots, x_n) = F_i(t, x_1, \ldots, x_n, 0)$ $(i = 1, 2, \ldots, n)$. In what follows, x stands for a vector with coordinates $x_1, \ldots, x_n$.

5.6.2 Boundedness with respect to two measures
A qualitative analysis of systems (5.6.1) and (5.6.2) will involve two measures of boundedness, ρ and ρ_0. These are elements from

$$\mathcal{M} = \left\{ \rho \in C(R_+ \times R_+^n, R_+) : \inf_{(t,x) \in R_+ \times R_+^n} \rho(t, x) = 0 \right\}.$$

5. APPLICATIONS

Definition 5.6.1 The solutions of system (5.6.2) are said to be

(i) (ρ_0, ρ)-*bounded* if there exists a nonnegative function $\delta(t_0, \alpha)$ on R_+^2 continuous in t_0, satisfying $\delta(t_0, \alpha) > 0$ for $\alpha > 0$, and such that for every $\alpha > 0$ and every $(t_0, x_0) \in R_+ \times R_+^n$, solution $x(t)$ of (5.6.2) satisfies

$$\rho(t, x(t)) < \delta(t_0, \alpha) \quad \text{for all} \quad t \geq t_0$$

whenever $\rho_0(t_0, x_0) < \alpha$;

(ii) *uniformly* (ρ_0, ρ)-*bounded* if $\delta(t_0, \alpha)$ in (i) does not depend on t_0.

Definition 5.6.2 The solutions of system (5.6.1) are said to be

(i) (ρ_0, ρ, μ)-*bounded* if there exist nonnegative functions $\delta(t_0, \alpha)$ and $\mu^*(t_0, \alpha)$ on R_+^2 such that $\delta(t_0, \alpha)$ is continuous in t_0, $\delta(t_0, \alpha) > 0$, $\mu^*(t_0, \alpha) > 0$ for $\alpha > 0$, and for every $\varepsilon > 0$ and every $(t_0, x_0) \in R_+ \times R_+^n$, solution $x(t, \mu)$ of (5.6.1) satisfies

$$\rho(t, x(t, \mu)) < \delta(t_0, \alpha) \quad \text{for all} \quad t \geq t_0$$

whenever $\rho_0(t_0, x_0) < \alpha$ and $\mu_i < \mu^*(t_0, \alpha)$, $i = 1, 2, \ldots, k$;

(ii) *uniformly* (ρ_0, ρ)-*bounded* if $\delta(t_0, \alpha)$ in (i) does not depend on t_0.

5.6.2.1 Unperturbed Kolmogorov system We shall provide a (ρ_0, ρ)-boundedness criterion for system (5.6.2) using the direct Liapunov method. We present the following definitions.

Definition 5.6.3 Let $\rho, \rho_0 \in \mathcal{M}$. We say that

(i) ρ is *continuous with respect to* ρ_0 if there exist $\Delta > 0$ and $\varphi \in C(R_+^2, R)$ such that $s \to \varphi(t, s)$ belongs to class KR for each $t \in R_+$, and $\rho(t, x) < \varphi(t, \rho_0(t, x))$ holds whenever $\rho_0(t, x) < \Delta$;

(ii) ρ is *uniformly continuous with respect to* ρ_0 if $\varphi(t, s)$ in (i) does not depend on t.

Consider the matrix function

$$U(t, x) = [u_{ij}(t, x)], \quad i, j = 1, 2, \ldots, m,$$

where $u_{ij} \in C(R_+ \times R_+^n, R)$. Take $y \in R^m$ and introduce the Liapunov function

(5.6.3) $$v(t, x, y) = y^T U(t, x) y.$$

Define the derivative of V along the vector field (5.6.2) at point (t,x) by

$$D^+v(t,x,y)|_{(5.6.2)} = \limsup_{\theta \to 0+}[v(t+\theta,\, x+\theta\,\mathrm{diag}\,(Kx)f(t,x),y)-v(t,x,y)]\theta^{-1}.$$

Here $\mathrm{diag}\,(Kx)$ is a diagonal $n \times n$ matrix with $K_1x_1, \ldots, K_nx_n$ on the diagonal. The next Theorem formulates a sufficient condition for (ρ_0, ρ)-boundedness in terms of the Liapunov function $v(t,x,y)$.

Theorem 5.6.1 *Suppose that*

(1) $\rho, \rho_0 \in \mathcal{M}$ *and ρ is continuous with respect to ρ_0;*
(2) $v(t,x,y)$ *is locally Lipschitz in x, $a \in KR$, $w \in C(R_+^2, R_+)$, and*

$$a(\rho(t,x)) \leq v(t,x,y) \leq w(t,\rho_0(t,x))$$

for all $(t,x) \in R_+ \times R_+^n$;
(3) *for all $(t,x) \in R_+ \times R_+^n$*

$$D^+v(t,x,y)|_{(5.6.2)} \leq 0.$$

Then the solutions of system (5.6.2) are (ρ_0, ρ)-bounded.

See Freedman and Martynyuk [3] for the proof of Theorem 5.6.1.

Let us give a criterion for the uniform (ρ_0, ρ)-boundedness. In what follows we use notations

$$S(\rho_0, h) = \{(t,x) \in R_+ \times R_+^n : \rho_0(r,x) < h\},$$
$$S(\rho, h) = \{(t,x) \in R_+ \times R_+^n : \rho(r,x) < h\};$$

$S^c(\rho_0, h)$ stands for the complement of $S(\rho_0, h)$.

Theorem 5.6.2 *Suppose that*

(1) $\rho, \rho_0 \in \mathcal{M}$ *and ρ is continuous with respect to ρ_0;*
(2) $v(t,x,y)$ *is locally Lipschitz in x, $a \in KR$, $w \in C(R_+, R_+)$, and*

$$a(\rho(t,x)) \leq v(t,x,y) \leq w(t,\rho_0(t,x))$$

for all $(t,x) \in S^c(\rho_0, h)$ with some $h > 0$;
(3) *for all $(t,x) \in S^c(\rho_0, h)$*

$$D^+v(t,x,y)|_{(5.6.2)} \leq 0.$$

5. APPLICATIONS

Then the solutions of system (5.6.2) are uniformly (ρ_0, ρ)-bounded.

Proof For every $\alpha > 0$ choose $\gamma > 0$ so that

(5.6.4) $\qquad a(\gamma) > \max\{w(\max\{\alpha, h\}),\, a^{-1}(\varphi(h))\}.$

Here φ is a function introduced in Definition 5.6.4(ii). Since by Definition 5.6.3 $\lim_{r \to \infty} a(r) = \infty$, there is γ satisfying (5.6.4). Let $t_0 \in R_+$ and $\rho_0(t_0, x_0) < \alpha$. Assume that for some solution $x(t)$ of system (5.6.2) there is $t^* \geq t_0$ such that

$$\rho(t^*, x(t^*)) \geq \gamma.$$

By (5.6.4) $\gamma \geq \varphi(h)$, hence by the definition of φ, we have $\rho_0(t^*, x(t^*)) > h$. Then by the continuity of $x(t)$ there exist t_1 and t_2, $t_0 \leq t_1 < t_2 \leq t^*$, such that

$$\rho_0(t_1, x(t_1)) = \max\{\alpha, h\},$$

$$(t, x(t)) \in S(\rho, \gamma) \cap S^c(\rho_0, \max\{\alpha, h\}) \quad \text{for all} \quad t \in [t_1, t_2].$$

By condition (2)

(5.6.5) $\qquad v(t_1, x(t_1), y) \leq w(\rho_0(t_1, x(t_1))) = w(\max\{\alpha, h\})$

and

(5.6.6) $\qquad a(\gamma) = a(\rho(t_2, x(t_2))) \leq V(t_2, x(t_2), y).$

By condition (3)

(5.6.7) $\qquad v(t_2, x(t_2), y) \leq v(t_1, x(t_1), y).$

Estimates (5.6.5)–(5.6.7) imply that

$$a(\gamma) \leq w(\max\{\alpha, h\})$$

which contradicts (5.6.4). Therefore $\rho(t, x(t)) < \gamma$ for all $t \geq t_0$ whenever $\rho_0(t_0, x_0) \leq \alpha$, i.e., the solutions of system (5.6.2) are uniformly (ρ_0, ρ)-bounded.

5.6.2.2 Perturbed Kolmogorov system For the perturbed system (5.6.2) we shall utilize the $m \times m$ matrix functions
(5.6.8)
$$U_1(t, x) = [u_{1ij}(t, x)], \quad U_2(t, x, \mu) = [u_{2ij}(t, x, \mu)], \quad i, j = 1, 2, \ldots, m,$$

where $u_{1ij} \in C(R_+ \times R_+^n, R)$, $u_{2ij} \in C(R_+ \times R_+^n \times M^k, R)$. Take $\eta \in R^m$ and set

$$v_1(t, x, \eta) = \eta^T U_1(t, x)\eta, \quad v_2(t, x, \mu, \eta) = \eta^T U_2(t, x, \mu)\eta.$$

Define the derivatives of v_1 and v_2 along the vector field (5.6.1) at point (t, x) by

$$D^+ v_1(t, x, \eta)|_{(5.6.1)}$$
$$= \limsup_{\theta \to 0+} [v_1(t + \theta, x + \theta \operatorname{diag}(\beta(x))F(t, x, \mu), \eta) - v_1(t, x, \eta)]\theta^{-1},$$
$$D^+ v_2(t, x, \mu, \eta)|_{(5.6.1)}$$
$$= \limsup_{\theta \to 0+} [v_2(t + \theta, x + \theta \operatorname{diag}(\beta(x))F(t, x, \mu), \eta) - v_1(t, x, \mu, \eta)]\theta^{-1}.$$

Here $\operatorname{diag}(\beta(x))$ is a diagonal $n \times n$ matrix with $\beta_1(x_1), \ldots, \beta_n(x_n)$ on the diagonal.

Theorem 5.6.3 *Suppose that*

(1) *$\rho, \rho_0 \in \mathcal{M}$ and ρ is continuous with respect to ρ_0;*
(2) *$v_1(t, x, \eta)$ is locally Lipschitz in x*

$$a(\rho(t, x)) \leq v_1(t, x, \eta) \leq w(t, \rho_0(t, x))$$

for all $(t, x) \in R_+ \times R_+^n$, where $a \in KR$, $w \in C(R_+^2, R_+)$, and

$$D^+ v_1(t, x, \eta)|_{(5.6.2)} \leq g_1(t, v_1(t, x, \eta))$$

for all $(t, x) \in R_+ \times R_+^n$, where $g_1 \in C(R_+^2, R)$ and $g_1(t, u)$ is nondecreasing in u for all $t > t_0'$;
(3) *for all $(t, x) \in S^c(\rho_0, h)$ with some $h > 0$ and all $\mu \in M^k$*

$$|v_2(t, x, \mu, \eta)| < c(\mu),$$

where $c \in C(M^k, R_+)$, $c(0) = 0$, and

$$D^+ v_1(t, x, \eta)|_{(5.6.1)} + D^+ v_2(t, x, \mu, \eta)|_{(5.6.1)}$$
$$\leq g_2(t, v_1(t, x, \eta) + v_2(t, x, \mu, \eta), \mu)$$

where $g_2 \in C(R_+^2 \times M^k, R)$ and $g_2(t, u, \mu)$ is nondecreasing in u for all $t > t_0'$;

(4) *every solution of the equation*

$$\frac{du}{dt} = g_1(t,u), \quad u(t_0) = u_0 \geq 0$$

is bounded;

(5) *the solutions of the equation*

$$\frac{dw}{dt} = g_2(t,w,\mu), \quad w(t_0) = w_0 \geq 0$$

are bounded uniformly in $\mu \in M^k$.

Then the solutions of system (5.6.1) are (ρ, ρ_0, μ)-bounded.

For the proof we refer to Freedman and Martynyuk [3].

5.6.3 Stability with respect to two measures

5.6.3.1 Unperturbed Kolmogorov system

Definition 5.6.5 System (5.6.2) is called (ρ_0, ρ)-*stable if there exists a nonnegative function* $\delta(t_0, \varepsilon)$ *on* R_+^2 *continuous in* t_0, *satisfying* $\delta(t_0, \varepsilon) > 0$ *for* $\varepsilon > 0$, *and such that for every* $\varepsilon > 0$ *and every* $(t_0, x_0) \in R_+ \times R_+^n$, *solution* $x(t)$ *of (5.6.2) satisfies*

$$\rho(t, x(t)) < \varepsilon \quad \text{for all} \quad t \geq t_0$$

whenever $\rho_0(t_0, x_0) < \delta(t_0, \varepsilon)$.

Again, we use the Liapunov function $v(t, x, y)$.

Theorem 5.6.4 *Suppose that*

(1) $\rho, \rho_0 \in \mathcal{M}$ *and* ρ *is continuous with respect to* ρ_0;
(2) $v(t, x, y)$ *is locally Lipschitz in* x, $a \in KR$, $w \in C(R^2, R_+)$, $w(t, 0) = 0$, *and*

$$a(\rho(t,x)) \leq v(t,x,y) \leq w(t, \rho_0(t,x))$$

for all $(t, x) \in R_+ \times R_+^n$;
(3) *for all* $(t, x) \in S(\rho, h)$ *with some* $h > 0$

$$D^+ v(t,x,y)|_{(5.6.2)} \leq 0.$$

Then the solutions of system (5.6.2) are (ρ_0, ρ)-stable.

This theorem follows from Theorem 5.6.4 by Freedman and Martynyuk [1, 2].

5.6.3.2 Perturbed Kolmogorov system

Definition 5.6.6 System (5.6.1) is called (ρ_0, ρ, μ)-stable if there exist a nonnegative function $\delta(t_0, \varepsilon)$ on R_+^2 continuous in t_0 and satisfying $\delta(t_0, \varepsilon) > 0$ for $\varepsilon > 0$, and a function $\mu^*(t_0, \varepsilon) \colon R_+^2 \to M^k \setminus \{0\}$ such that for every $\varepsilon > 0$, every $(t_0, x_0) \in R_+ \times R_+^n$, and every $\mu \in M^k$, the solution $x(t, \mu)$ of (5.6.1) satisfies

$$\rho(t, x(t, \mu)) < \varepsilon \quad \text{for all} \quad t \geq t_0$$

whenever $\rho_0(t_0, x_0) < \delta(t_0, \varepsilon)$ and $\mu_i < \mu_i^*(t_0, \varepsilon)$, $i = 1, 2, \ldots, k$.

Below, we refer to the Liapunov functions v_1 and v_2.

Theorem 5.6.5 *Suppose that*
(1) ρ, $\rho_0 \in \mathcal{M}$ *and ρ is continuous with respect to ρ_0;*
(2) $v_1(t, x, \eta)$ *is locally Lipschitz in x and for all $(t, x) \in R_+ \times R_+^n$*

$$a(\rho(t, x)) \leq v_1(t, x, \eta) \leq w(t, \rho_0(t, x)),$$

where $a \in K$, $w \in C(R_+^2, R_+)$, $w(t, 0) = 0$, and for all $(t, x) \in S(\rho, h)$ with some $h > 0$

$$D^+ v_1(t, x, \eta)|_{(5.6.2)} \leq g_1(t, v_1(t, x, \eta)),$$

where $g_1 \in C(R_+^2, R)$, $g_1(t, u)$ is nondecreasing in u and $g_1(t, 0) = 0$ for $t > t_0'$;
(3) *for all $(t, x) \in S(\rho, h) \cup S(\rho_0, h)$ and all $\mu \in M^k$*

$$|v_2(t, x, \mu, \eta)| < c(\mu),$$

where $c \in C(M^k, R_+)$, $c(0) = 0$, and

$$D^+ v_1(t, x, \eta)|_{(5.6.1)} + D^+ v_2(t, x, \mu, \eta)|_{(5.6.1)}$$
$$\leq g_2(t, v_1(t, x, \eta) + v_2(t, x, \mu, \eta), \mu)$$

where $g_2 \in C(R_+^2 \times M^k, R)$, $g_2(t, u, \mu)$ is nondecreasing in u and $g_2(t, 0, \mu) = 0$ for all $t > t_0'$;
(4) *the zero solution of the equation*

$$\frac{du}{dt} = g_1(t, u), \quad u(t_0) = u_0 \geq 0,$$

is stable;
(5) *the zero solutions of the equations*

$$\frac{dw}{dt} = g_2(t, w, \mu), \quad w(t_0) = w_0 \geq 0,$$

are stable uniformly in $\mu \in M^k$.

Then system (5.6.1) is ρ_0, ρ, μ-stable.

The proof is similar to that of Theorem 5.6.3.

5.6.4 Application to a generalized Lotka–Volterra system
Consider a generalized Lotka–Volterra equation,

$$(5.6.9) \qquad \frac{dx}{dt} = \operatorname{diag}(x)[b + A(x)x], \quad x(t_0) = x_0, \quad i = 1, 2,$$

as an example of system (5.6.2). Here $x \in R_+^2$, $A(x) = [a_{ij}(x)]$, $i, j = 1, 2$, $a_{ij} \in C(R_+^2, R)$, $b \in R^2$. Let $x^* \in R_+^2$ be a nonzero equilibrium state of system (5.6.9). Assume it to be unique. Introduce Liapunov variables

$$y_1 = x_1 - x_1^*, \quad y_2 = x_2 - x_2^*.$$

Define the matrix function

$$U(y) = [u_{ij}(y)], \quad i, j = 1, 2,$$

by setting

$$u_{11}(y) = \alpha y_1^2, \quad u_{12}(y) = u_{21}(y) = -\gamma y_1 y_2, \quad u_{22}(y) = \beta y_2^2$$

with $\alpha, \beta > 0$. Let $\eta \in R_+^2$ and

$$v(y, \eta) = \eta^\mathrm{T} U(y) \eta.$$

The following estimates hold

$$(5.6.10) \qquad v(y, \eta) \geq u^\mathrm{T}(y) H^\mathrm{T} P H u(y),$$

where $u^\mathrm{T}(y) = (|y_1|, |y_2|)$, $H = \operatorname{diag}(\eta)$, and

$$P = \begin{pmatrix} \alpha & -\gamma \\ -\gamma & \beta \end{pmatrix}.$$

Further,

$$(5.6.11) \qquad D^+ v(y, \eta)|_{(5.6.9)} \leq u^\mathrm{T}(y)[C(y) + G(y)] u(y),$$

where

$$C(y) = \begin{pmatrix} C_{11}(y) & C_{12}(y) \\ C_{21}(y) & C_{22}(y) \end{pmatrix}, \quad G(y) = \begin{pmatrix} \sigma_{11}(y) & \sigma_{12}(y) \\ \sigma_{21}(y) & \sigma_{22}(y) \end{pmatrix},$$

$$C_{11}(y) = 2\eta_1(\alpha\eta_1 a_{11}(y+x^*)x_1^* + \gamma\eta_2 a_{21}(y+x^*)x_2^*),$$

$$C_{22}(y) = 2\eta_2(\beta\eta_2 a_{22}(y+x^*)x_2^* + \gamma\eta_1 a_{12}(y+x^*)x_1^*),$$

$$C_{12}(y) = C_{21}(y) = \alpha\eta_1^2 |a_{12}(y+x^*)|x_1^* + \beta\eta_2^2 |a_{21}(y+x^*)|x_2^*$$
$$+ \eta_1\eta_2 |\gamma(a_{11}(y+x^*)x_1^* + a_{21}(y+x^*)x_2^*)|,$$

and

$$\sigma_{11}(y) = 2\alpha\eta_1^2 |a_{11}(y+x^*)|\,|y_1|,$$
$$\sigma_{22}(y) = 2\beta\eta_2^2 |a_{22}(y+x^*)|\,|y_2|,$$

$$\sigma_{12}(y) = \sigma_{21}(y)$$
$$= [\alpha\eta_1^2 |a_{12}(y+x^*)| + \eta_1\eta_2|\gamma|\,|a_{21}(y+x^*) + a_{11}(y+x^*)|]\,|y_1|$$
$$+ [\beta\eta_2^2 |a_{21}(y+x^*)| + \eta_1\eta_2|\gamma|\,|a_{22}(y+x^*) + a_{12}(y+x^*)|]\,|y_2|.$$

We apply Theorem 5.6.1 with

(5.6.12) $$\rho(t,y) = \rho_0(t,y) = \|y\|.$$

Note that

$$\lambda_m u^{\mathrm{T}}(y)u(y) \leq v(y,\eta) \leq \lambda_M u^{\mathrm{T}}(y)u(y),$$

where λ_m and λ_M are the minimum and maximum eigenvalues of matrix $H^{\mathrm{T}}PH$ respectively.

Theorem 5.6.6 *Assume that*

(1) *ρ and ρ_0 are defined by (5.6.12);*
(2) *$0 < \lambda_m < \lambda_M$,*
(3) *there exist 2×2 matrices $\overline{C}$ and $\overline{G}$ such that $\overline{C} + \overline{G}$ is negative semi-definite, and estimates*

$$w^{\mathrm{T}}C(y)w \leq w^{\mathrm{T}}\overline{C}w, \quad w^{\mathrm{T}}G(y)w \leq w^{\mathrm{T}}\overline{G}w$$

hold for all $y, w \in R_+^2$.

Then the solutions of system (5.6.9) are (ρ_0, ρ)-bounded.

ProofP One can easily see that, in view of estimates (5.6.10) and (5.6.11) and assumptions (1)–(3), all conditions of Theorem 5.6.1 are satisfied.

The next statement follows from Theorem 5.6.4.

Theorem 5.6.7 *Assume that*

(1) *conditions (1) and (2) of Theorem 5.6.6 are satisfied;*
(2) *there exist 2×2 matrices C^* and G^* such that $C^* + G^*$ is negative semi-definite, and estimates*

$$w^{\mathrm{T}} C(y) w \leq w^{\mathrm{T}} C^* w, \quad w^{\mathrm{T}} G(y) w \leq w^{\mathrm{T}} G^* w$$

hold for all $y, w \in S(\rho, h)$ with some $h > 0$.
Then system (5.6.12) is (ρ_0, ρ)-stable.

5.7 Notes and References

Section 5.1 Many applications of classical version of the direct Liapunov method and its generalizations can be found in published monographs (see References for Chapters 1–4 of this book, etc.) and numerous papers. The approaches developed in recent years allow application of the method of matrix-valued Liapunov functions to various problems of mechanics, automatic control, stochastic oscillating systems, singularly perturbed systems and systems of mathematical biology equations.

Section 5.2 The problems on orientation of a rigid body rotating around a fixed point are scrutinized by Zubov [1] and others. In this section we use the results by Martynyuk and Slyn'ko [2] and apply matrix-valued function to investigation of the general problem. Some general results by Appell [1], Lur'e [1] and those of Chapter 1 of this book are applied.

Section 5.3 One of the topics actively worked out in the recent years is the dynamics of neural systems which describe discrete-time equations (see Feng and Michel [1] and references therein). The results presented in this Section are due to Lukyanova and Martynyuk [1, 2]. The proof we give is different from the original.

Section 5.4 The approach to stability analysis based on the idea of state space extension of dynamical system was discussed in several papers (see Ikeda and Šiljak [1], Šiljak [1], Leela [1], Martynyuk [1, 2], etc.).

This section encorporates some results by Martynyuk [1, 7] and Martynyuk and Slyn'ko [1].

Section 5.5 This section provides a new approach to the theory of interval stability based on the idea of maximal extension of dynamical system with subsequent application of the multicomponent Liapunov functions method. In a paper by Martynyuk and Slyn'ko [3], the proposed approach is realized in terms of vector Liapunov functions. The obtained result is compared with some estimates of robust stability boundary of mechanical system cited by Cao and Shu [1]. We note that a traditional approach to this problem is the application of linear matrix inequalities (see Bliman [1], Šiljak and Stipanović [1] and references therein).

Section 5.6 This section is based on the results by Martynyuk [6], and Freedman and Martynyuk [1–3], who first investigated the problems of population dynamics in terms of two measures and matrix-valued Liapunov functions. There are a number of good books on mathematical biology. A voluminous monograph by Murray [1] includes useful chapters on enzyme kinetics (as well as the Belousov reaction and the spruce budworm model). Other books worthy to note are those by Freedman [1], Segel [1] and Rubinow [1].

References

Appell, P.E
[1] *Traite de mecanique rationelle.* V. I and II. Paris: Gauthier-Villars, 1909.

Bliman, P.-A.
[1] A convex approach to robust stability for linear systems with uncertain scalar parameters. *SIAM J. Control Optim.* bf 42(6) (2004) 2016–2042.

Cao, D.Q. and Shu, Z.Z.
[1] Robust stability bounds for multi-degree-of-freedom linear systems with structured perturbations. *Dynamics and Stability of Systems* **9**(1) (1994) 79–87.

Chung-Li Jiang
[1] Other sufficient conditions for the stability of interval matrices. *Int. J. of Control* **47**(1) (1988) 181–186.

Feng, Zhaoshu and Michel, A.N.
[1] Robustness analysis of a class of discrete-time systems with applications to neural networks. *Nonlinear Dynamics and Systems Theory* **3**(1) (2003) 75–86.

Freedman, H.I.
[1] *Deterministic Mathematical Models in Population Ecology.* Edmonton: HIFR Consulting, 1987.

Freedman, H.I. and Martynyuk, A.A.
[1] On stability with respect to two measures in Kolmogorov's model of dynamics of populations. *Dokl. Akad. Nauk Russia* **329** (1993) 423–425. [Russian]
[2] Stability analysis with respect to two measures for population growth models of Kolmogorov type. *Nonlin. Analysis* **25** (1995) 1221–1230.

[3] Boundedness criteria for solutions of perturbed Kolmogorov population models. *Canadian Appl. Math. Quarterly* **3** (1995) 203–217.

Gantmacher, F.R.
[1] *Theory of Matrices.* Second Edition. Moscow: Nauka, 1966. [Russian]

Lankaster, P.
[1] *Theory of Matrices.* Moscow: Nauka, 1978. [Russian]

Hahn, W.
[1] *Stability of Motion.* Berlin: Springer-Verlag, 1967.

Ikeda, M. and Šiljak, D.D.
[1] Generalized decomposition of dynamic systems and vector Lyapunov functions. *IEEE Trans. Autom. Control* **AC-26** (1981) 1118–1125.

Leela, S.
[1] Large-scale systems, cone-valued Lyapunov functions and quasi-solution. In: *Trends in Theory and Practice of Nonlinear Differential Equations.* New York: Marcel Dekker, 1984, P.323–330.

Liapunov, A.M.
[1] *General Problem of Motion Stability.* Moscow: Gostekhizdat, 1950. [Russian]

Lukyanova, T.A. and Martynyuk, A.A.
[1] Robust stability: three approaches for discrete-time systems. *Nonlinear Dynamics and Systems Theory* **2**(1) (2002) 45–55.
[2] Hierarchical Lyapunov functions for stability analysis of discrete-time systems with applications to the neural networks. *Nonlinear Dynamics and Systems Theory* **4**(1) (2004) 31–49.

Lur'e, A.I.
[1] *Analytical Mechanics.* Moscow: Fizmatgiz, 1961. [Russian]

Martynyuk, A.A.
[1] Extension of the state space of dynamical systems and the problem of stability. In: *Colloquia Mathematica Societaties Janos Bolyai 47. Differential Equations: Qualitative Theory*, Szeged (Hungary), 1984, P.711–749.
[2] A contraction principle. In: *Advances in Nonlinear Dynamics* (Eds.: S. Sivasundaram and A.A. Martynyuk). Amsterdam: Gordon and Breach Science Publishers, 1997, P.99–105.
[3] *Stability by Liapunov's Matrix Functions Method with Applications.* New York: Marcel Dekker, 1998.
[4] *Qualitative Methods in Nonlinear Dynamics. Novel Approaches to Liapunov's Matrix Functions.* New York: Marcel Dekker, 2002.
[5] Qualitative analysis of nonlinear systems by the method of matrix Lyapunov functions. *Rocky Mountain J. of Math.* **25** (1995) 397–415.
[6] Qualitative analysis with respect to two measures for population growth models of Kolmogorov type. In *Dynamics amd Control* (Eds.: G. Leitmann, F.E. Udwadia and A.V. Kryazhimskii). Amsterdam: Gordon and Breach Science Publishers, 1999, P.109–117.
[7] Direct Liapunov's matrix functions method and overlapping decomposition of large scale systems. *Dynamics of Continuous, Discrete and Impulsive Systems, Series A: Mathematical Analysis* **11** (2004) 205–217.

Martynyuk, A.A. and Slyn'ko, V.I.
[1] Matrix-valued Lyapunov function for extended systems. *Int. Appl. Mech.* **37** (2001) 1083–1088.
[2] Stabilization of motion of a rigid body in the given direction, (to appear).
[3] Choosing the parameters of a mechanical system with interval stability. *Int. Appl. Mech.* **39**(9) (2003) 1089–1092.

Murray, J.D.
[1] *Mathematical Biology.* Berlin: Springer-Verlag, 1989.

Rubinow, S.I.
[1] *Introduction to Mathematical Biology.* New York: John Wiley, 1975.

Rumiantsev, V.V. and Oziraner, A.S.
[1] *Stability and Stabilization of Motion with Respect to a Part of Variables.* Moscow: Nauka, 1987. [Russian]

Segel, L.A.
[1] *Modeling Dynamic Phenomena in Molecular and Cellular Biology.* Cambridge: Cambridge University Press, 1986.

Šiljak, D.D.
[1] *Decentralized Control of Complex Systems.* Boston: Academic Press, 1991.

Šiljak, D.D. and Stipanović, D.M.
[1] Robust stabilization of nonlinear systems: The LMI approach. *Mathematical Problems in Engineering* **6** (2000) 461–493.

Yoshizawa, T.
[1] *Stability Theory by Liapunov's Second Method.* Tokyo: The Mathematical Society of Japan, 1966.

Zubov, V.I.
[1] *Lectures on Control Theory.* Moscow: Nauka, 1975. [Russian]